"十四五"时期国家重点出版物出版专项规划项目
材料研究与应用丛书

Numerical Analysis of Vibration and Enhanced Heat Transfer Performance of Elastic Tube Bundle Heat Exchangers

Jiadong Ji
Baojun Shi
Haishun Deng

INTRODUCTION

To solve the problems existing in the research and application of elastic tube bundle (ETB) heat exchangers, vibration and heat transfer of ETBs in heat exchangers under shell-side fluid induction are systematically studied in this book. There are nine chapters, including introduction, numerical calculation method, vibration of ETBs induced by uniform shell-side fluid, vibration and heat transfer analysis of ETBs induced by actual shell-side fluid, design of a shell-side distributed pulsating flow generator, effect of pulsating flow generator on vibration and heat transfer of ETBs, effect of baffles on vibration and heat transfer of TETBs, research on vibration-enhanced heat transfer of IETB heat exchanger, and effect of baffle structure on the performance of IETB heat exchanger.

This book can be used as a reference for researchers and engineering technicians engaged in energy recycling, enhanced heat transfer, vibration control and other fields. This book can also be used as a reference for teaching courses such as "Advanced Engineering Fluid Mechanics" "Modern Design Methods of Fluid Machinery" for graduate students of related majors in colleges and universities, and "Thermal Engineering Foundation" "Heat Transfer" for senior undergraduate students.

图书在版编目(CIP)数据

换热器内弹性管束振动和传热性能的数值研究= Numerical Analysis of Vibration and Enhanced Heat Transfer Performance of Elastic Tube Bundle Heat Exchangers:英文/季家东,史宝军,邓海顺著. 哈尔滨:哈尔滨工业大学出版社,2025. 1. —(材料研究与应用丛书). —ISBN 978-7-5767-1760-0

Ⅰ. TK172

中国国家版本馆 CIP 数据核字第 20246HN324 号

Jointly published with Springer Nature Singapore Pte Ltd.

The printed edition is not for sale outside of China. Customers outside of China please order the print book from Springer Nature Singapore Pte Ltd.

策划编辑　甄淼淼　许雅莹
责任编辑　周一瞳　宋晓翠　甄淼淼
封面设计　刘　乐
出版发行　哈尔滨工业大学出版社
社　　址　哈尔滨市南岗区复华四道街 10 号　邮编 150006
传　　真　0451-86414749
网　　址　http://hitpress. hit. edu. cn
印　　刷　哈尔滨博奇印刷有限公司
开　　本　720 mm×1 000 mm　1/16　印张 16. 75　字数 363 千字
版　　次　2025 年 1 月第 1 版　2025 年 1 月第 1 次印刷
书　　号　ISBN 978-7-5767-1760-0
定　　价　98. 00 元

PREFACE

Heat exchangers are equipments to realize heat exchange, and are widely used in geothermal development, waste heat recovery, battery heat dissipation, wastewater treatment, solar power generation, nuclear energy development, etc. The failure of heat transfer elements caused by flow-induced vibration is one of the most prominent problems faced in the design and application of heat exchangers. Flow-induced vibration is an unfavorable factor affecting the service life of heat exchangers, and also a positive factor for enhancing heat transfer. Elastic tube bundle (ETB) heat exchangers replace traditional rigid heat transfer elements with elastic heat transfer elements, utilizing fluid induced internal ETBs vibration to achieve composite enhanced heat transfer, opening up new directions and ideas for the application of passive enhancement technology in heat exchangers.

However, after extensive investigation by the authors of this book and their research team, it has been found that there are still some shortcomings in the current research on flow-induced vibration, vibration enhanced heat transfer and vibration/heat transfer control of ETB heat exchangers, mainly including the followings.

(1) Due to the influence of ETB structure, ETB heat exchangers have a small heat transfer area per unit volume, resulting in poor overall heat transfer performance and thus affecting its market competition.

(2) Flow-induced vibration in ETB heat exchangers is a complex fluid solid interaction problem involving multi field fluid. At present, there is a lack of systematic research on numerical analysis of flow-induced vibration of ETBs under complex flow conditions.

(3) In practical engineering applications, there is a phenomenon of uneven vibration for ETBs in ETB heat exchangers. In this way, ETBs with severe vibration are prone to vibration damage, while ETBs with mild vibration have poor heat transfer efficiency, which further affects the service life and heat transfer efficiency of ETB heat exchangers. At present, there is no good solution to this problem.

In response to the above shortcomings, starts from the vibration response of a single and multi row of ETBs induced by uniform shell-side fluid, the vibration responses of ETBs in heat exchanger under actual shell-side fluid induction are investigated. Then, by selecting appropriate vortex element and improving the branch domain channel structure, a distributed pulsating flow generator is designed to conduct numerical analysis and experimental research on the vibration responses of ETBs induced by the coupling of shell-side fluid and distributed pulsating fluid. In addition, based on the ETB structural improvement and the baffle installation, the vibration and heat transfer performances of the ETBs in heat exchangers are compared and studied. And also, the influences of the baffle structure parameters (height and curvature) on the vibration-enhanced heat transfer performance of an improved ETB heat exchanger are systematically studied under different conditions. The research work in this book has important theoretical and engineering significance for improving the heat transfer performance of ETB heat exchangers and effectively stimulating and controlling the vibration of ETBs.

The content of the book is a summary of the authors and their research team members' research work over the past decade, and is a reference book in the field of enhanced heat transfer. Among them, Chapter 1 was complated by associate professor Jiadong Ji (Anhui University of Science and Technology) and professor Baojun Shi (Hebei University of Technology), Chapter 2 was complated by associate professor Jiadong Ji and professor Haishun Deng (Anhui University of Science and Technology), Chapters 3 to 7 were complated by associate professor Jiadong Ji, Chapter 8 was complated by associate professor Jiadong Ji and professor Baojun Shi, and Chapter 9 was complated by associate professor Jiadong Ji and professor Haishun Deng. The total number of words writtend by associate professor Jiadong Ji, professor Baojun Shi and professor Haishun Deng are 300 thousand words, 40 thousand words and 20 thousand words. This book is drafted by associate professor Jiadong Ji.

Special thanks to professor Peiqi Ge (Shandong University) for his selfless guidance and assistance in the research project. During this book, professors Gang Shen (Anhui University of Science and Technology), Qinghua Chen (Anhui University of Science and Technology) and Ping Liu (Anhui University of Science and Technology) gave warm support and help. The postgraduate students Yuling Pan, Xuwang Ni, Jingwei Zhang, Feiyang Li, Jinhui Zhao, Pengcheng Li, Chengzhi Duan, Fan Ye, Xiyi Sun and Nana Li participated in the draft writing. The authors express their gratitude to them.

This book was supported by the National Natural Science Foundation of China (Grant No. 52175070), the Natural Science Research Project of Anhui Education Committee (Grant No. 2024AH040068, KJ2018A0080), the Anhui Provincial Natural

Science Foundation (Grant No. 1908085ME160), the Medical Special Cultivation Project of Anhui University of Science and Technology (Grant No. YZ2023H2B011), and the Excellent Research and Innovation Team of Anhui Province (Grant No. 2022AH010052). We would like to express our sincere gratitude.

Despite caution, due to the limitations of the author's level, there are inevitably shortcomings in the book. Readers are kindly requested to provide criticism and correction, please contact jjd1006@163.com.

Jiadong Ji
Baojun Shi
Haishun Deng
November, 2024

Nomenclature

A	amplitude/heat transfer area
a	acceleration
C	damping matrix
C_d	drag coefficient
C_l	lift coefficient
C_p	specific heat at constant pressure
C_μ	turbulence model constant
$C_{\varepsilon 1}$	turbulence model constant
$C_{\varepsilon 2}$	turbulence model constant
D	diameter/mark of flow direction
d	diameter/width/hypotenuse length
E	elastic modulus
F	force/face/ Fanning friction factor
f	frequency
g	displacement
H	pitch/height/row spacing
h	heat transfer coefficient
h^s	node displacement
J	Colburn factor
K	stiffness matrix
k	turbulent kinetic energy equation
L	length
l	transverse dimension
M	mass matrix
m	number of nodes
n	normal vector/number
Nu	Nusselt number
Pr	Prandtl number
p	pressure
Q	convection heat exchange
q	heat flow density
R	radius/maximum radius
Re	Reynolds number
r	minimum radius
S	vibration displacement/distance
St	Strauhal number
T	temperature
t	time
u	flow velocity
W	width
w	longitudinal dimension
x, y, z	axes

Greek letters

α	installation angle/included angle	λ	thermal conductivity
β	transition angle	μ	dynamic viscosity
γ	direction angle	υ	kinematic viscosity coefficient/ Poisson's ratio
δ	wall thickness	ρ	density
ε	turbulent kinetic energy dissipation equation	σ	diffusion Prandtl number
ζ	root mean square	τ	stress
η	elliptic ratio/height factor	φ	taper/position angle
θ	swing angle/baffle curvature		

Subscripts

a	average/ actuarial calculation	out	outlet
b	baffle	pul	pulsating flow
c	constant	r	rough calculation
f	fluid	s	structural
in	inlet	t	turbulent fluid
l	laminar fluid	v	vibration/velocity
n	node	w	wall

Abbreviations

ETB	elastic tube bundle	IETB	improved ETB heat exchanger
FFT	fast Fourier transform	NB	no baffle
FSI	fluid solid interaction	PEC	performance evaluation criteria
GGI	general grid interface	RMS	root mean square
HB	have baffle	TETB	traditional ETB heat exchanger

CONTENTS

Chapter 1 Introduction

Heat exchangers[1-3], as shown in Fig. 1. 1, are equipment to realize heat exchange, and are widely used in geothermal development, waste heat recovery, battery heat dissipation, wastewater treatment, solar power generation, nuclear energy development, biological refrigerant storage, etc[4-5].

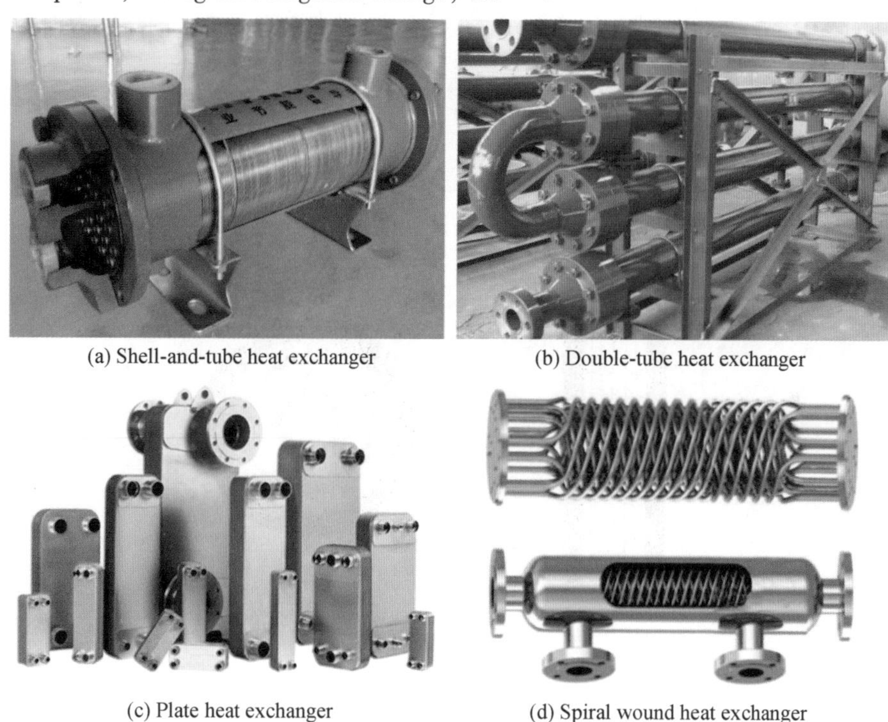

(a) Shell-and-tube heat exchanger

(b) Double-tube heat exchanger

(c) Plate heat exchanger

(d) Spiral wound heat exchanger

Fig. 1. 1 Physical images of four types of common heat exchangers

Taking mine ventilation as an example, after the fresh air from the ground is fed into the intake shaft, it supplies fresh air to the working face and also absorbs heat dissipation from the surrounding rock, mechanical equipment, coal oxidation, personnel and other aspects. When the air is discharged from the return shaft, the temperature of the mine return air is much higher than that of the intake air. In addition, the mine return air has a large amount of low-temperature heat energy, which is not utilized and directly discharged into the atmosphere. It will cause great waste of thermal energy[6]. Using heat exchanger to recover waste heat resources and use them for mine fresh air heating, building heating and other occasions can effectively reduce energy consumption and carbon emissions, so it has outstanding social and economic benefits[7]. On the other hand, the abundant underground space created during thc mining process of the deposits provides the basic conditions for the large-scale storage of renew energy such as solar energy. A novel method of realizing seasonal storage of renewable energy, such as solar energy, is the contruction of backfill heat exchangers with a heat storage/release function by inserting ground heat exchangers into backfill bodies in mines, as shown in Fig. 1. 2. Therefore, heat exchanger equipment and related technologies play an important role in industrial production.

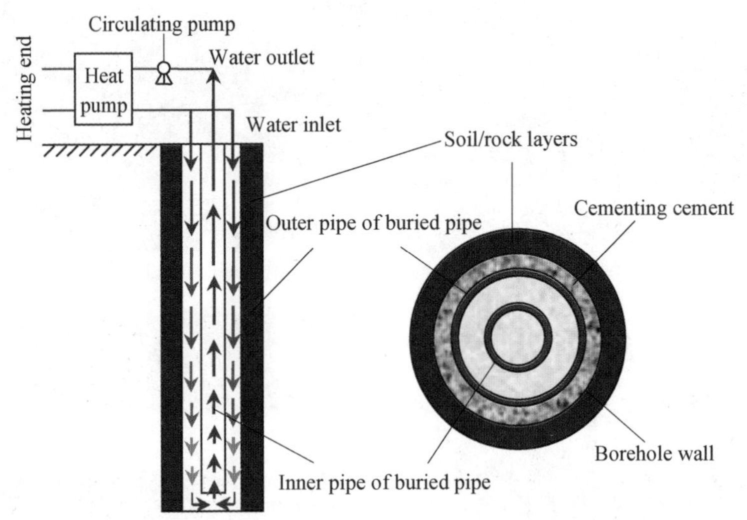

Fig. 1.2 Deep shaft heat exchange system

For modern chemical industry, the cost of heat exchanger equipment accounts for about 30% of the total cost. In refining equipment, the heat exchanger equipment accounts for about 40% of the total cost. In seawater desalination, the process equipment is almost all the heat exchanger equipment[2]. On the other hand, the heat exchanger energy consumption accounts for about 13% – 15% of the total industrial energy consumption[8]. It can be said that the heat transfer performance and service life of the heat exchanger is directly related to the effective use of energy and the normal operation of equipment, and has a direct impact on the economic benefits of high energy consumption industry[9]. Therefore, it is of great significance to develop efficient and energy-saving heat transfer technology and equipment to save energy and reduce energy consumption.

The rapid development of heat exchangers and related technology in recent decades have made a lot of fruitful progress[10-13]. At the same time, some urgent problems such as flow-induced vibration and fatigue failure of heat transfer elements have also emerged, restricting the further development of heat exchangers and related technologies[1].

It is well known that flow-induced vibration will cause fatigue damage of heat transfer elements in heat exchangers, and then affect the service life of heat exchangers[11-13]. The failure of heat transfer element caused by flow-induced vibration is one of the most prominent problems in modern heat exchanger design and practical application. There are many examples of vibration failure of heat exchangers[17]: in the early period, the tubular exchanger manufacturers association (TEMA) investigated 42 heat exchangers and found that 24 of them had vibration failure. The heat transfer research institute (HTRI) conducted a survey of 66 heat exchangers and found that 58 had suffered vibration failure. In recent years, the reports of vibration damage of heat exchangers have been numerous.

The failure of heat transfer element caused by flow-induced vibration can be alleviated or reduced to a certain extent by increasing the stiffness of heat transfer element. However, it is impossible to completely eliminate vibration in the heat exchangers, and the method of increasing the stiffness of the heat transfer elements is not very effective.

On the basis of reasonable use of vibration and taking into account heat transfer performance and fatigue life of components, heat transfer element of elastic tube bundle (ETB)[1, 4, 9, 15-16] can enhance heat transfer by using flow-induced vibration of ETBs (Fig. 1.3). The core idea of the design of the ETB is to replace the rigid heat transfer element with the elastic heat transfer element. The vibration of the ETBs induced by the fluid flow around is used to strengthen the heat transfer, and the vibration of the ETBs is controlled to ensure that it has a certain service life[1].

Accordingly, the research on heat transfer enhancement, flow-induced vibration, vibration excitation and control has become the focus of academic circles[18-20]. The following is literature review on vibration enhanced heat transfer and vibration control, and then the main research content of this book is introduced.

Fig. 1.3 Manufacturing process and finished products of ETB heat exchangers

1.1 Research Status of Vibration Enhanced Heat Transfer

Enhanced heat transfer technology is an advanced technology to improve the heat transfer performance of heat transferelements. The main objectives are to reduce equipment costs, protect high-temperature components and realize rational utilization of energy by improving heat transfer efficiency[21-23].

There are many ways to enhance heat transfer. Especially in recent years, with the rapid development of heat transfer technology, a large number of new, efficient and practical technologies and methods have emerged[24-25]. Based on the research content of this book, the convection enhanced heat transfer technology is mainly discussed.

Based on the classification standard proposed by Bergles[26], heat transfer enhancement by convection can be divided into active enhancement technology, passive enhancement technology and composite enhancement technology. The introduction is as follows.

(1) Active enhancement technology refers to the technology that requires the consumption of external high-quality energy such as mechanical force and electromagnetic force to achieve enhanced heat transfer. It mainly includes mechanical agitation[27-28], heat transfer wall vibration or fluid pulsation[29-31], electromagnetic field[32-34], suction or jet impact[35-36], etc.

(2) Passive enhancement technology refers to the technology of heat transfer enhancement without consuming additional high-quality energy or power. It mainly

includes surface treatment[37-40], surface roughness[41-43], surface extension[44-46], swirl element[47-50], tension structure[51], additive[52], etc.

(3) Composite enhancement technology refers to the combination of two or more kinds of strengthening technology. It mainly includes composite strengthening technology of helical groove tube and bond[53-54], composite strengthening technology of helical groove tube and inlet cyclone[55], composite heat transfer enhancement of inner fin tube and bond[56], etc.

1.1.1 Active Enhancement Technology

It is well known that the heat transfer surface strengthens the disturbance of the surrounding fluid through vibration and destroys the boundary layer of the wall, thus achieving enhanced heat transfer. Early research on enhanced heat transfer by vibration mainly focused on enhanced heat transfer experiments based on forced vibration. The vibration forms mainly include mechanical vibration or motor driven eccentric device and ultrasonic excitation vibration. With the rapid development of computer technology and numerical analysis methods, people begin to use various numerical simulation methods to study the heat transfer characteristics of heat transfer components under forced vibration conditions.

As shown in Fig. 1.4, mechanical vibration device or eccentric device is simple in structure and easy to adjust the amplitude and frequency, and the heat transfer performance of heat transfer components under different conditions can be studied in depth. Among them, Deaver et al.[57], Lemlich et al.[58], Penny et al.[59], Hsieh et al.[60], Dawood et al.[61], Saxena et al.[62], Leung et al.[63], Katinas et al.[64-65], Takahashi et al.[66], Karanth et al.[67], Cheng et al.[68], Klaczak et al.[69], Gau et al.[70], Bronfenbrener et al.[71], Fu et al.[72], Leng et al.[73], Lee et al.[74], etc., based on natural and forced convection, the heat transfer performance of vibrating tubes with different structural parameters has been studied experimentally or numerically.

Table 1.1 and Table 1.2 are the summaries of some studies. Where, d is the tube diameter; A is the vibration amplitude; f is the vibration frequency; h is the heat transfer coefficient.

It can be seen from Table 1.1 and Table 1.2 that the heat transfer performance of the vibrating tube varies significantly with the change of the fluid medium and its flow state. Under natural convection conditions, the heat transfer coefficient of vibrating tube is increased, with the highest increase of 1,228%, and the research results of

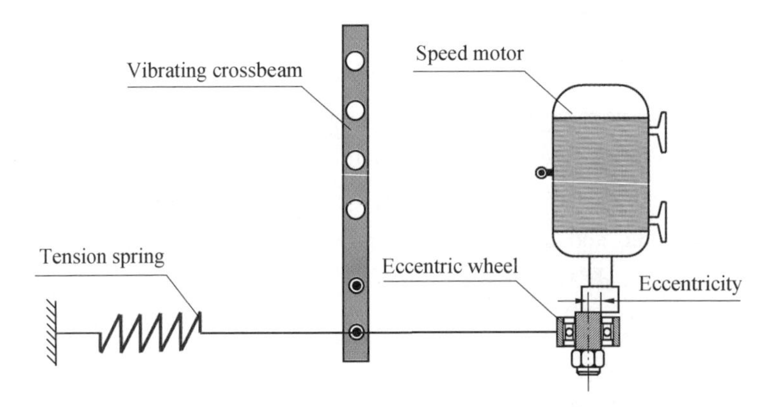

Fig. 1.4 Schematic diagram of eccentric device driven by motor

various literatures are consistent. Under forced convection conditions, the heat transfer coefficient of vibrating tube does not always increase. For example, in the gas-liquid heat transfer test conducted by Klaczak[69], the surface heat transfer coefficient decreased, with a maximum reduction of 20%. Under the condition of forced convection, the heat transfer enhancement of single-phase flow is proportional to amplitude and frequency, and the heat transfer effect decreases with the increase of *Re*. In addition, the properties of the medium also affect the effect of vibration in enhancing heat transfer. For example, in the research conducted by Lemlich[58], the heat transfer effect of water is better than that of glycerin in water.

Table 1.1 Research on heat transfer performance of vibrating tube under natural convection

Researcher	Medium	d/mm	Vibration parameter		Main conclusion
			A/mm	f/Hz	
Deaver	Water	0.18	2.54–70	4.25	h increases by 4 times
Lemlich	Air	0.64,1.00,2.06	1.40–5.87	39–122	
Lemlich	Water and glycerin in water	1.25	0.05–2.18	17–37	h increases by 12.3 times
Penny	Water	0.2	63.5	4.5	h increases by 5 times
Penny	Ethylene glycol	0.2	63.5	4.5	h increases by 2.5 times
Hsieh	Water	10	0.16–6.35	27–51	h increased by 54%
Dawood	Air	8.50, 12.7	0–17.8	0–68	h increases by 3 times

Table 1.2 Research on heat transfer performance of vibrating tube under forced convection

Researcher	Medium	Parameters (d/mm; A/mm; f/Hz)	Main conclusion
Saxena	Water	$d=22$ $A=19.6-43.8$ $f=0.4-1.2$ $Re=3,500$	The degree of heat transfer enhancement is proportional to A and f, h increases and the maximum increase is 60%
Leung	Water	$d=38$ $A=0.25$ and 0.5 $f=10$ and 30 $Re=3\times10^3-5\times10^5$	When $Re<1.5\times10^3$, h increases, but the increase is smaller When $Re>2.5\times10^3$, vibration suppresses heat transfer effect
Leng	Water	$d=20$ $A=0-2.0$ $f=6.67-24$ Re is small	The degree of heat transfer enhancement is proportional to A and f, h increases and the maximum increase is 311%
Cheng	Air	$d=16$ $A=0-10.0$ $f=0-20$ $Re=0-4,000$	The increase of Nu is proportional to A and f, and h increases by up to 30%
Klaczak	Water-vapour	$d=608$ $A=0.2-0.5$ $f=20-120$, $Re=430-2,300$	h decreased with a maximum reduction of 20%

Although mechanical vibration can enhance heat transfer significantly, its operation requires extra energy consumption and belongs to the category of active enhancement technology, which has no advantages in terms of energy saving. Studies have shown that the energy gained from improving heat transfer is only 5% of the energy consumed by external equipment[14]. Based on this, the application prospects of the mechanical vibration enhanced heat transfer are relatively bleak.

In view of the deficiency of active enhancement technology, it is necessary to explore a vibration mode which can not only avoid or reduce the consumption of extra energy, but also make effective use of the effect of vibration enhancement of heat transfer. Based on this, the technology of flow-induced vibration to enhance heat transfer without external energy consumption has gradually become a hot topic in academic research.

1.1.2 Flow-induced Vibration

It is well known that flow-induced vibration of heat transfer element in heat exchanger is a universal phenomenon. This flow-induced vibration can be expressed as various vibration phenomena caused by "fluid flow around poor fluid engineering structures"[75].

In the heat exchanger, the fluid that induces the vibration of the heat transfer element can be divided intovertical flow (flow parallel to the axis of the heat transfer tube) and horizontal flow (flow perpendicular to the axis of the heat transfer tube). Generally speaking, the vibration of heat transfer element induced by vertical flow has low intensity and little harm, which can be ignored in the research process. The intensity of horizontal flow induced vibration of heat transfer element is relatively high, and it also poses great harm to the heat transfer element, which is the main factor inducing vibration of heat transfer element.

The formation mechanism of flow-induced vibration is complex. At present, the main mechanism of horizontal flow-induced vibration in heat exchangers include vertex shedding excitation, turbulent buffeting, fluid elastic instability, acoustic resonance, etc[76]. Based on these mechanisms, researchers have carried out a lot of fruitful research, and put forward many theoretical models and empirical formulas[77-80], which have played a guiding role in the development and design of heat exchangers.

1. Planer ETB Heat Exchanger

In the design and manufacture of heat exchangers, the flow-induced vibration of heat transfer elements does not always lead to the failure of heat transfer elements. If the structure of the traditional shell-and-tube heat exchanger is changed, the flow-induced vibration can be used effectively to enhance heat transfer without causing damage to heat transfer elements.

A planer ETB (also known as traditional ETB) heat exchanger (Fig. 1.5), which is different from the design idea for the traditional shell-and-tube heat exchanger, was proposed[1, 15, 81-83] based on the idea of using heat transfer element vibration to enhance heat transfer. If not specified, the "ETB" in this book refer to the "planer ETB".

As shown in Fig. 1.5, the ETB is composed of four copper bend tubes and two stainless steel connectors (Ⅲ and Ⅳ), with the fixed ends at Ⅰ and Ⅱ. During the actual operation of the heat exchanger, the tube-side fluid enters from port Ⅰ and flows through the stainless steel connectors Ⅲ and Ⅳ multiple times before exiting from port Ⅱ. It is found that the heat exchanger achieves passively enhance heat transfer by using the small amplitude low-frequency vibration of the ETB induced by shell-side and tube-side fluid, especially under low flow rate conditions[25, 81]. At the same time, the damage of the ETB and the noise caused by violent vibration can be avoided. In addition, the ETB has a very obvious descaling effect in the vibration process.

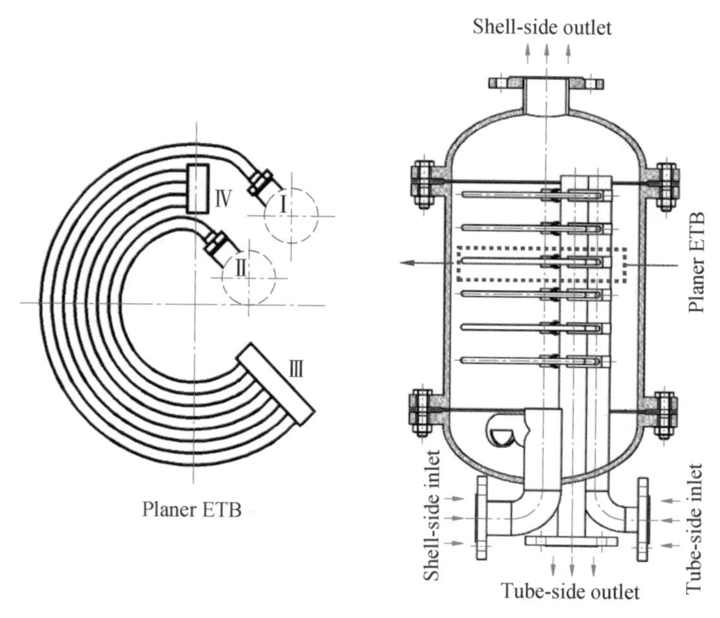

Fig. 1.5 Structure diagram of the planer ETB heat exchanger[4]

Zheng et al.[84] studied the modal frequency and mode shape of the ETB by using a component mode synthesis method. The results show that the natural vibration modes of the ETB are mainly divided into in-plane vibration (vibration in the plane of the ETB) and out-of-plane vibration (vibration perpendicular to the plane of the ETB). The vibration of the ETB is the combination of vibrations in different directions as a result of complex excitation in actual shell-side and tube-side fluid field.

Cheng et al.[85] experimentally studied the heat transfer and flow resistance characteristics of a water-water ETB heat exchanger. And also, the change rule of heat transfer coefficient of the ETB is analyzed under different working conditions. In their study, by measuring the flow resistance loss of the shell-side and tube-side fluid, a series of formula are given to calculate the flow resistance in shell-side and tube-side change.

Su et al. [86-87] experimentally studied the role of flow-induced vibration on the vibration response of a single ETB. The results show that the vibration characteristics of the ETB are strongly influenced by flow velocity, and vibration harmonics of the ETB are triggered at low fluid velocities.

Duan et al. [88-90] analyzed the effect of different structural and installation dimensions of ETBs on vibration-enhanced heat transfer. The results show that when the tube wall thickness is smaller or the distance between tubes is larger, the effect of vibration enhancing heat transfer is stronger.

Cheng et al. [91-92] experimentally studied the heat transfer characteristics of the ETB heat exchanger under steam-water heat transfer condition. The results show that the ETB can enhance heat transfer significantly at low Reynolds number compared with fixed ETB. In addition, through systematic experimental study, the correlation formula of Nusselt number at low Reynolds number was obtained as follows[92].

For the heat exchanger with pulsating flow device, we have

$$Nu = 0.911 Re^{0.633} {Pr_f}^{1/3} \left(\frac{Pr_f}{Pr_w}\right)^{0.25} \tag{1.1}$$

For the heat exchanger without pulsating flow device, we have

$$Nu = 0.711 Re^{0.645} {Pr_f}^{1/3} \left(\frac{Pr_f}{Pr_w}\right)^{0.25} \tag{1.2}$$

where Nu is the Nusselt number; Re is the Reynolds number; Pr_f is the Prandtl number using the average temperature of the fluid as the qualitative temperature; Pr_w is the Prandtl number using the wall temperature as the qualitative temperature. The subscripts "w" and "f" represent the wall and the fluid, respectively.

The applicable range of the above correlation formulas is $100<Re<500$, and the error between the correlation formula and the experimental value is ±5%.

In response to the shortcomings of low heat transfer area per unit volume and high fixed end stress in planer ETB, Jiang[93] made certain improvements on the basis of the traditional ETB and conducted experimental research on the vibration and shell-side heat transfer characteristics of the modified ETB. The results show that the modified ETB has a lower natural frequency, and the stress at the fixed end is about 1/6 of the traditional ETB with the same mass. In addition, the heat transfer area per unit volume of the modified ETB is 24.7% higher than that of the traditional ETB.

2. Space Conical Helical ETB Heat Exchanger

In view of the above problems of serious stress concentration and simple secondary flow of the traditional ETB, by taking the structural optimization of the ETB as the entry point, Yan et al.[14, 18-19, 94-97] designed a space conical helical ETB heat exchanger based on the working principle of the planer ETB heat exchanger. The structure diagram of the space conical helical ETB heat exchanger is shown in Fig. 1.6.

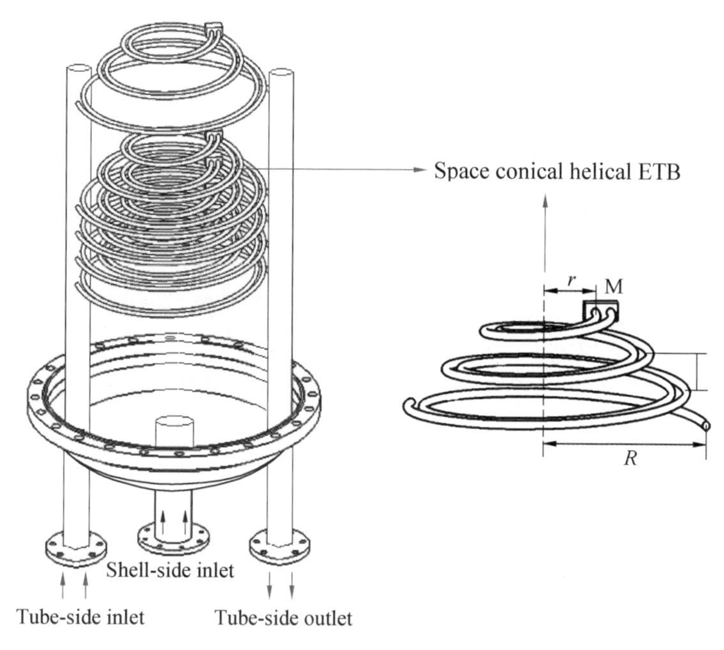

Fig. 1.6 Structure diagram of the space conical helical ETB heat exchanger[14]

The space conical helical ETB is composed of two copper helical tubes and a stainless steel connector (M). The tube-side fluid flows in from port Ⅰ, flows through the stainless steel connector M, and flows out from port Ⅱ. The installation of multi row space conical helical ETBs in the heat exchanger adopt a nested form.

Research has found that the natural vibration modes of the space conical helical ETB include longitudinal vibration and transverse vibration, with longitudinal vibration being the main mode[94]. The multi-row space conical helical ETB has a complex mode and each mode corresponds to a frequency band. In addition, the taper and thickness-diameter ratio of the space conical helical ETB have great influence on its natural frequency, while the pitch and stainless steel mass connector have little influence on its natural frequency[95-97].

Compared with the traditional ETB, the stress concentration phenomenon of the space conical helical ETB is not obvious, and the heat transfer per unit area is higher. In terms of heat transfer performance, the performance evaluation criteria (PEC) value of the space conical helical ETB is 1.4 – 1.6 times than that of the planer ETB[14].

By using linear regression method, the Nusselt number (Nu) correlation equations of the space conical helical ETB under laminar flow and turbulent flow conditions are obtained as follows[14].

For the laminar flow, we have

$$Nu = 0.54Re^{0.93}Pr^{0.33}\varphi^{0.23}\left(\frac{d}{R+r}\right)^{0.25} \tag{1.3}$$

For theturbulent flow, we have

$$Nu = 1.03Re^{0.52}Pr^{0.33}\varphi^{0.56}\left(\frac{d}{R+r}\right)^{0.26} \tag{1.4}$$

where Pr is the Prandtl number; φ is the taper; d is the section diameter; R is the maximum radius of curvature of the space conical helical ETB; r is the minimum radius of curvature of the space conical helical ETB. The applicable range of the above correlation formula is $100<Re<3,000$, $0.6<\varphi<1$ and $H=90$ mm (H is the pitch).

3. Spiral ETB Heat Exchanger

To solve the problem of uneven vibration of the traditional ETBs in heat exchanger, Ji et al.[98-103] proposed a spiral ETB heat exchanger. The structure diagram of the spiral ETB heat exchanger is shown in Fig. 1.7.

Fig. 1.7 schematically shows the layout of four spiral ETBs in the heat exchanger with four spiral deflectors. To improve the overall heat transfer performance, different numbers of spiral deflectors are installed on the heat exchanger to guide the flow direction and induce the turbulence intensity in the fluid field. Note that the inner part of the heat transfer is hollow, the four spiral deflectors are welded onto the inner and outer walls, and spiral ETBs are evenly installed on the right and left tube plates in the heat exchanger. Different numbers of spiral deflectors are alternately placed between each spiral ETB.

The cold fluid in the shell-side flows through the shell-side inlet and then flows out through the shell-side outlet; the shell-side inlet and outlet are at the bottom and top of the heat exchanger, respectively. In addition, the hot fluid in the spiral ETBs enters from the tube-side inlet, and it flows through the tube-side outlet on the right side. The spiral ETBs are vibrated with a special flow channel by the shell-side fluids during the process of heat transfer. As a result, the spiral ETB heat exchanger can realize vibration-enhanced heat transfer and achieve better overall thermal and

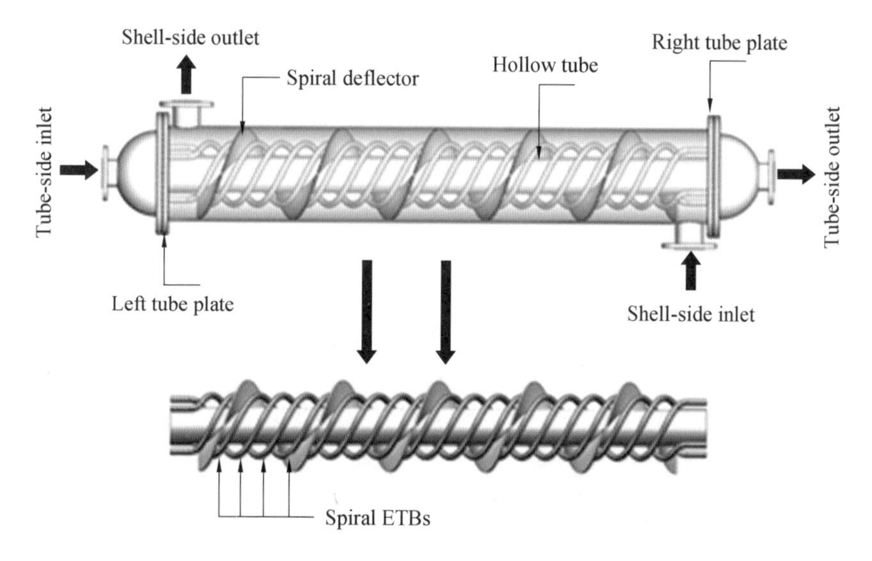

Fig. 1.7 Structure diagram of the spiral ETB heat exchanger[102]

hydraulic characteristics.

Based on the numerical results[100], the spiral ETB heat exchanger can effectively balance the vibration performance and improve the heat transfer performance. When the spiral deflector is installed on the top side of the spiral ETB heat transfer, the PEC value is the maximum and the comprehensive heat transfer performance is the best.

1.1.3 Fluid Solid Interaction

Fluid solid interaction (FSI) mechanics is a subject that studies the various behaviors of deformable solids under the action of fluids and the effects of solid configuration on the fluid domain. It is a branch of mechanics arising from the intersection of computational fluid dynamics and computational solid mechanics.

At present, widely studied FSI problems mainly include[104] vibration of liquid-filled tube, aero-elastic of aircraft wings, shaking of liquid storage containers, wind vibration of buildings, FSI of ships, FSI of underground reservoir, etc. Due to the strong nonlinear property of the FSI problems, research methods for different types of problems are different.

The engineering application of FSI vibration in tubes is very broad, and the research results can be directly applied to machinery, chemical industry, electric power, aerospace and nuclear engineering. The theoretical analysis methods mainly include dry mode method, cross iteration method, finite element method and boundary element method[105-106].

For FSI of outside tubes, the main research object is small amplitude self-oscillation induced by shell-side fluid of a cylindrical/circular tube with elastic support. The main research methods include theoretical research, experimental research and numerical simulation[107-109].

The oretical research mainly focuses on three types of semi-empirical models based on experimental results, which are wake oscillator model[110-111], single/ multi degree of freedom model[112-113] and force decomposition model[114].

The basic idea of the wake oscillator model (Fig. 1.8) is as follows: the wake is regarded as a van der Pol oscillator, and the vibration of a circular tube is simplified as a mass-spring-damping system. The lift force acting on the circular tube is described by the van der Pol equation, and the two equations are connected by coupling terms (functions of displacement, velocity or acceleration).

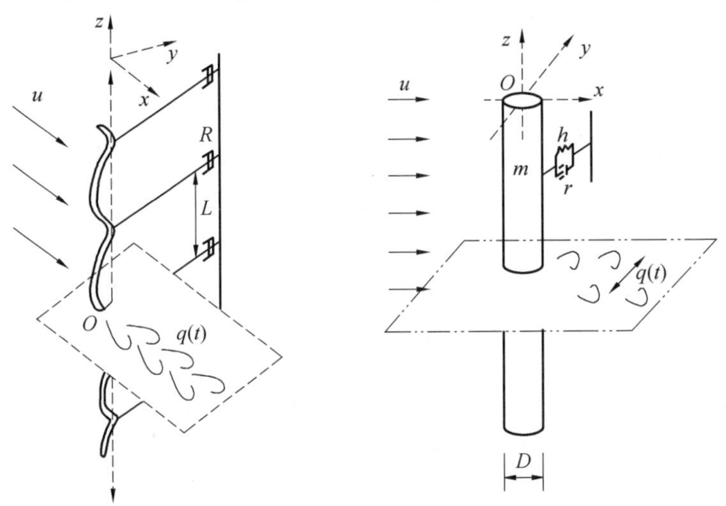

Fig. 1.8 Diagram of two wake oscillator models[110-111]

The basic idea of the single/multi degree of freedom model (Fig. 1.9) is that the vibration of the structure is described by a differential equation, and the influence of the wake on the structure is reflected by different forms of aerodynamic functions in the equation.

The basic idea of the force decomposition model (Fig. 1. 10) is to decompose the lift force acting on the tube into the inertia force and damping force of the fluid. Assuming that the displacement is a sine function, then the acceleration is also a sine function, and we can use the displacement to construct the inertial force.

In terms of experimental research, there are two types: self-excited vibration and

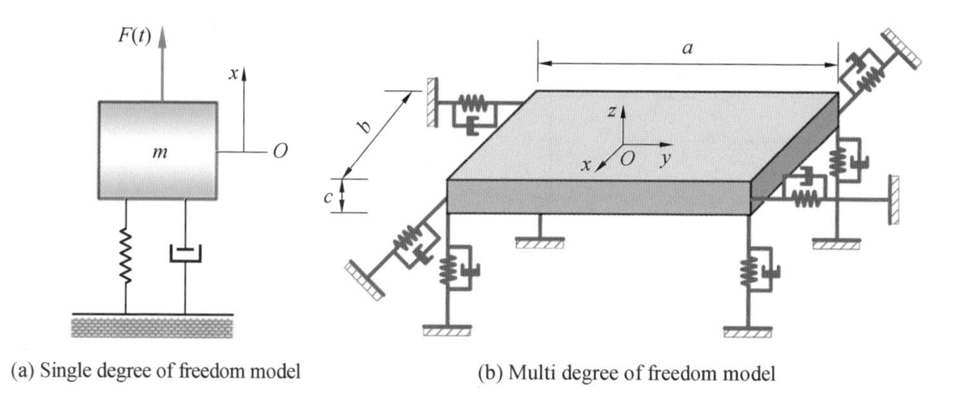

(a) Single degree of freedom model (b) Multi degree of freedom model

Fig. 1.9 Diagram of a single/multi degree of freedom model[113]

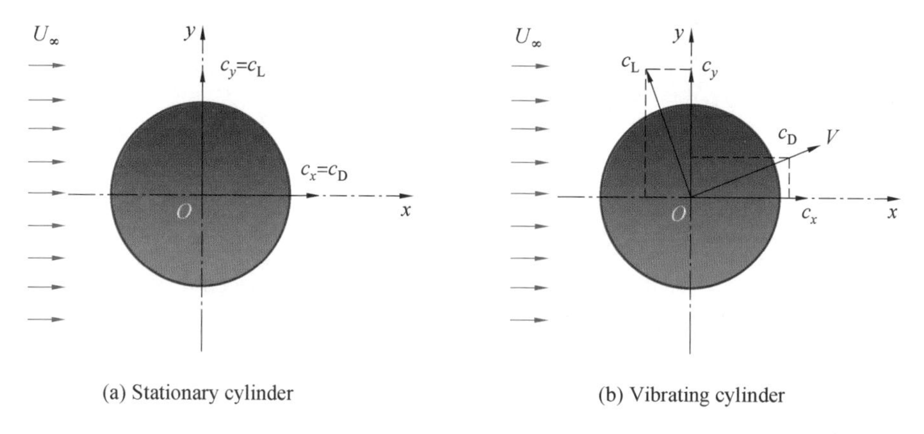

(a) Stationary cylinder (b) Vibrating cylinder

Fig. 1.10 Diagram of a force decomposition model[114]

forced vibration. For the self-excited vibration (Fig. 1.11), the cylinder is an elastic support, and its vibration is driven from the system by the vortex shedding. The limitation is that the velocity and amplitude are both small. For the forced vibration, as shown in Fig. 1.4, the cylinder is rigid support, and its vibration is driven by an external mechanical eccentric device, which can easily adjust the amplitude and frequency.

In terms of the self-excited vibration, Feng's recording and analysis of the vortex induced vibration phenomenon of a cylinder is regarded as a classic example of self-excited vibration experiments[115]. But its defect is that the mass ratio is too high, which is different from the low mass ratio problems such as the shell-side flow in heat exchanger and the Marine riser.

In recent years, researchers have conducted in-depth research on the vibration response of elastic supported tube bundle induced by fluid outside the tube. More representative studies include Ahmed et al. [116], Griffiths et al. [117], Gabbai et al. [118-119], So et al. [120], Ivanov et al. [121], Ding et al. [122].

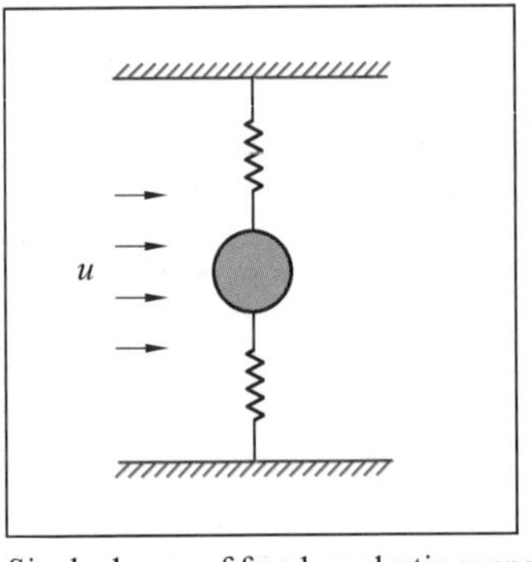

(a) Single degree of freedom elastic support cylinder diagram

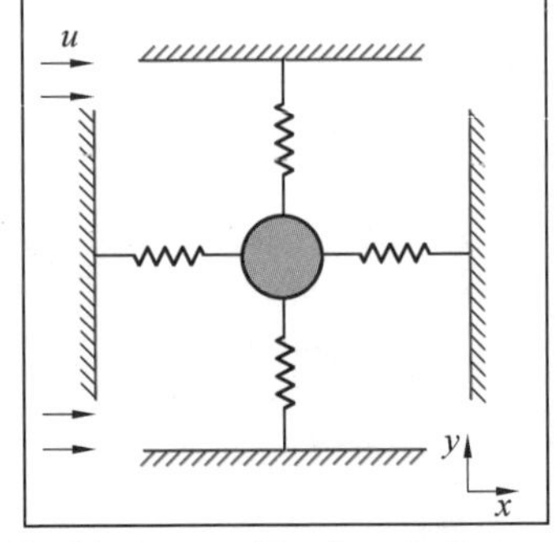

(b) Double degree of freedom elastic support cylinder diagram

Fig. 1.11 Single and double degrees of freedom elastic support cylinder diagram

1.2 Research Status of Vibration Control

The key to the design of ETB heat exchangers is the reasonable excitation and effective control of flow-induced vibration, which can then be used to conduct research on the comprehensive heat transfer performance and fatigue characteristics of the heat exchanger. Based on the structural characteristics of ETB heat exchangers and their internal heat transfer elements, the research on pulsating flow generator and baffle/deflector that can appropriately excite and control the vibration of ETBs has gradually attracted academic attention[123].

1.2.1 Studies on Flow Around Cylinder

In addition to causing the self-excited oscillation, the fluids flow through a circular cylinder will also form two rows of cross-arranged fluid vortices in its wake, namely: Karman vortex street (Fig. 1.12).

In the study of flow around a single circular cylinder, an important dimensionless parameter is Re, the expression is

$$Re = \frac{ud}{v} \tag{1.5}$$

where u is the flow velocity; d is the section diameter of the circular cylinder; v is the kinematic viscosity coefficient of fluid.

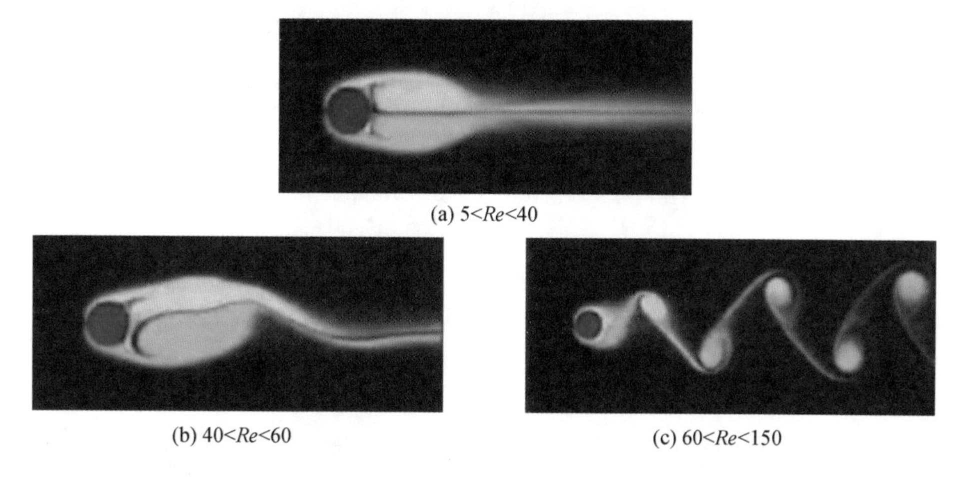

(a) $5<Re<40$

(b) $40<Re<60$

(c) $60<Re<150$

Fig. 1.12 The relationship between Karman vortex street and Re ($5<Re<150$)

According to Lenhard[124], the generation of Karman vortex street is closely related to Re. Researches have shown that when $Re<5$, no vortex is formed; when $5<Re<40$, a pair of stable "attached vortices" are formed at the wake of the circular cylinder; when $40<Re<150$, the wake of the circular cylinder forms two rows of cross-arranged fluid vortices; when $150<Re<300$, transition of layer flow to turbulent flow occurs inside the vortex; when $Re=3\times10^5$, vortex street is transformed into turbulent flow, but the boundary layer is still laminar flow, and $Re<3\times10^5$ is called subcritical region; when $3\times10^5<Re<3.5\times10^6$, the boundary layer gradually tends to the turbulent state, and there is no obvious vortex in the wake, which is called the critical state; when $Re>3.5\times10^6$, the vortex reappears and is called the supercritical region.

Another important parameter of flow around a single circular cylinder is Strauhal number (St), and the expression is

$$St = \frac{fd}{u} \tag{1.6}$$

where f is the vortex separation frequency.

Studies show that[125], when $60<Re<5,000$, relatively regular Karman vortex street can be observed, and $St\approx0.21$. When $5,000<Re<2\times10^5$, St decreased slightly. And also, when $Re\approx200$, St will appear two discontinuous breakpoints with the change of Re due to the influence of vortex shedding of two different three-dimensional structures.

The flow around a single circular cylinder is a basic problem in fluid mechanics, and the flow field characteristics of the single circular cylinder are also well understood. For in-depth analysis of flow around circular cylinders, please refer to the following literature: Uemura et al.[126], Baikov et al.[127], Rinoshika[128], Afroz et al.[129], Yan et al.[130].

The vortex element in fluid flow is not only limited to circular cylinder, but also commonly used square cylinder, triangular cylinder, elliptical cylinder, semi-elliptical cylinder (or D-shaped cylinder) and trapezoidal cylinder.

Flow around a square cylinder is a classical model, which including turbulence phenomena such as impact, separation, reattachment, encirclement and vortex. Researchers have conducted extensive and in-depth research on the flow around square cylinders in recent years. Refer to the following studies: Seol et al.[131], Song et al.[132], Basohbatnovinzad et al.[133], Dai et al.[134], Liu et al.[135], Haffner et al.[136]. Different from the flow around a circular cylinder, the separation point of the flow around a square cylinder is fixed at the corner, and the aerodynamic characteristics are basically not affected by *Re*.

Flow around atriangular cylinder is widely used in electronic components and heat exchangers[137]. There are two types of placement: sharp angle towards the incoming flow and sharp angle away from the incoming flow. According to Zeitoun et al.[138] (Fig. 1.13), when the sharp angle towards the incoming flow, the *Re* range of the stable symmetric attached vortex is $4.16 \leqslant Re \leqslant 38.03$ (R_{crit}); when the sharp angle away from the incoming flow, the *Re* range of stable symmetric attached vortex is $10 \leqslant Re \leqslant 34.7$ (R_{crit}). For both placements, the *Re* range of unstable alternating vortices is $R_{crit} \leqslant Re \leqslant 200$. In addition, the updated research references are as follows: Afroz et al.[139], Wang et al.[140-141].

In terms of the flow around a cylinder with other shapes, Faruquee et al.[142] conducted a numerical study to examine the flow field around an elliptical cylinder over a range of axis ratios from 0.3 to 1.0. The results show that a pair of steady vortices forms when axis ratio reaches a critical value of 0.34, below this value, no vortices are formed behind the elliptical cylinder. In order to study the effect of the elliptic form of the cylinder on the vortices field and the hydrodynamic forces that act on it, Daoud et al.[143] presented a numerical investigation of a two-dimensional oscillatory flow around a cylinder of different elliptic ratios (η). The results show that the longitudinal force increases with the reduction of the elliptic ratio. The transverse force appears from the elliptic ratio $\eta = 0.75$ and increases with the reduction of this ratio in the range of $0.75 \geqslant \eta \geqslant 0.4$, then decreases for $\eta < 0.4$. On the other hand, the drag coefficient is sensitive to the swirling layout while the coefficient of inertia does not seem to be much affected by the geometry of the cylinder. Palei et al.[144]

studied the flow around a semi-elliptical cylinder at low *Re* by solving two-dimensional Navier–Stokes equations. The results show that the *St* of the semi-circular cylinder is lower than that of the circular cylinder with the same *Re*.

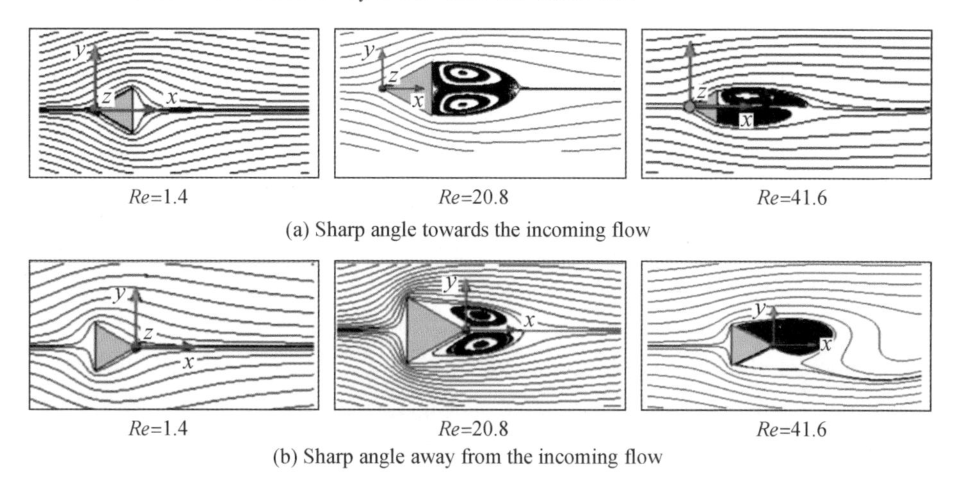

Fig. 1.13 Evolutions of the flow around a triangular cylinder[138]

In recent years, with the in-depth study of the flow around a vortex element, especially based on vortex flow meters and pulsating flow generators, the shape of vortex element has shown a diversified trend, with various shapes and sizes of vortex elements appearing. And also, the research of Karman vortex street based on vortex element with various shapes has also been widely developed[4, 123].

1.2.2 Application of Pulsating Flow Generator

In practical engineering applications, there is a phenomenon of uneven vibration in each row of ETBs in heat exchangers, with some ETBs experiencing severe vibration and some ETBs experiencing weak vibration. In this way, the ETBs with severe vibration are prone to fatigue damage, and the heat transfer performance of the ETBs with weak vibration is poor. Therefore, the key to the design of ETB heat exchangers is the reasonable excitation and effective control of flow-induced vibrations[82, 145]. Then, in response to practical problems in engineering applications, the research on pulsating flow generators that can reasonably induce and control the vibration of ETBs has gradually attracted academic attention[146-147].

Pulsating flow generator can also be divided into active pulsating flow generator and passive pulsating flow generator according to whether other auxiliary equipment is needed and external energy consumption. Active pulsating flow generators mainly include mechanical driven pulsating valve, blade type pulsating flow generation device, etc. The passive pulsating flow generators mainly include self-excited oscillating chamber, impeller driven pulsating flow generator, etc.

In terms of active pulsating flow generators, Sailor et al. [148] designed a mechanically driven pulsating valve, which uses the periodic switch of the channel to pulsate the internal flowing fluid. Li et al. [149] designed a blade type pulsation generator, which realized the periodic switch of flow passage through the relative motion of static blade and moving blade, so as to realize the pulsation excitation of fluid. Combining with the ideas of Sailor et al. and Li et al., Jiang[93] designed a motor driven blade type pulsating flow generator. The device is mainly composed of moving blades and static blades. The moving blades are driven by a motor and rotate around the central axis of the static blades. The flow passage is periodically switched through the relative motion between the moving blades and static blades, and the pulsating excitation of the flowing fluid is completed.

Due to the additional energy consumption of the active pulsating flow generators, their application prospects in engineering are greatly hindered. Therefore, research on passive pulsating flow generators has become a focus in the academic community. Gao et al. [150-151] designed a Helmholtz oscillator based on the principle of fluid self-excited oscillation. In the Helmholtz oscillator, the pressure disturbance is generated by the collision between the vortex ring and the collision wall, and then the fluid self-excited vibration is generated by the phase relationship between the disturbance waves, so as to form the pulsed jet. On the basis of the principle of the motor driven blade type pulsating flow generator, Jiang[93] designed a flow-induced impeller type pulsating flow generator. The principle of pulsating flow generated by the relative motion of the moving and static blades of the device is the same as that of the motor driven blade type pulsating flow generator, but the difference is that the driving force of the device comes from the driving impeller.

Because of its simple structure, no external energy consumption, and easy to adjust the frequency and intensity of the pulsating flow, the pulsating flow generator based on vortex element has a broad application prospect[152]. Liu et al. [153] conducted an experimental study on the vibration response of the ETB under pulsating flow excitation by setting different shapes of vortex element at the entrance of the shell pass. The results show that when the vortex element is a triangular cylinder, the vibration amplitude of the ETB increases with the increase of the water flow, and the vibration frequency changes with the law of high frequency – low frequency – high frequency. When the disturbed fluid is a trapezoidal cylinder, the vibration amplitude of the ETB increases with the increase of the water flow and gradually tends to be stable, and the vibration frequency is basically not affected by the water flow.

Based on the ideas of Liu et al., Meng[123] designed an independent pulsating flow generator (Fig. 1. 14). The independent pulsating flow generator is composed of four parts: riser tube, branch tubes, vortex elements and flange. The vortex element is a stainless steel structure triangular cylinder, which is installed on the slot in the branch tube. The independent pulsating flow generator is independent of the shell-side inlet tube, and the pulsating fluid and shell-side fluid flow into the heat exchanger through different paths.

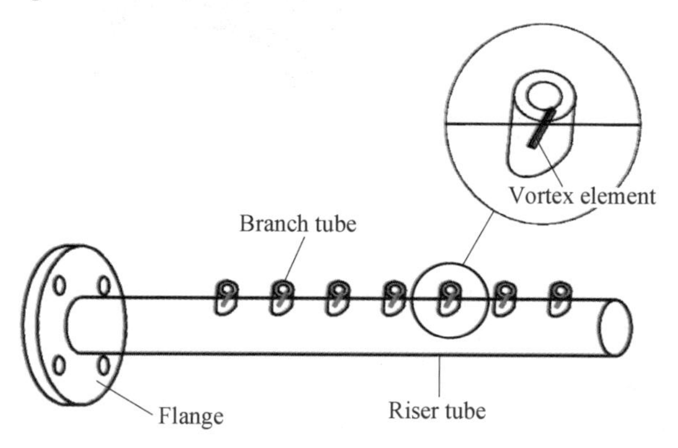

Fig. 1. 14 Schematic diagram of the independent pulsating flow generator[123]

Fig. 1. 15 shows the installation of the independent pulsating flow generator in a ETB heat exchanger. As shown in Fig. 1. 15, one end of the independent pulsating flow generator is welded to the lower head of the ETB heat exchanger, and the other end is suspended. The outlet of each branch tube corresponds to the stainless steel connector Ⅲ of the ETB (Fig. 1. 5), and the ETB vibrates with the pulsating frequency through the pulsating flow formed by each branch tube.

The analyses of the fluid domain of the independent pulsating flow generator show that the pulsating flow around the triangular cylinder, circular cylinder and square cylinder has good stability, and the pulsating flow around the triangular cylinder has high intensity. Compared with the structure of the pulsating flow generator closed at the top of the riser, the pulsating flow generator with an open hole at the top is more conducive to the formation of the pulsating flow in each branch tube[123].

Through further numerical analysis and experimental research, it is found that the independent pulsating flow generator can induce the ETB to vibrate at pulsating

Fig. 1.15 Schematic diagram of the independent pulsating flow generator

frequency to a certain extent, but the sealing structure at the top of the riser causing a flow "dead zone" in the top fluid domain. Then, the fluid into each branch tube is uneven and the flow stability is poor, and pulsating flow cannot be formed at the outlet of some branch tubes. And also, the intensity and frequency of the pulsating flow are inconsistent and the expected vibration required for heat transfer cannot be realized. In addition, controlling the flow of fluid into the pulsating flow generator by installing a valve at the riser entrance cannot solve above problems[4].

To improve the comprehensive heat transfer performance, based on a space conical helical ETB heat exchanger, in combination with an arrangement of pulsating baffles and heat transfer elements, four heat exchanger schemes were proposed by Ji et al.[154], as shown in Fig. 1.16.

The four heat exchanger schemes included two without pulsating baffle (AB-HE Ⅰ and Ⅱ), one with the same direction pulsating baffles (SD-HE) and one with opposite direction pulsating baffles (OD-HE). By using a bi-directional fluid-solid coupling calculation method, the vibration and heat transfer performances of the space conical spiral ETB were studied based on the four heat exchanger schemes.

As shown in Fig. 1.16, the high-temperature tube-side fluid inflow to the heat exchanger from the lower left tube fluid inlet, flow through the left head, the horizontal tube set up on the right tube plate, the space conical spiral ETBs, the horizontal tube set up on the left tube plate and the right head, and finally outflow from the upper right tube fluid outlet. The low-temperature shell-side fluid inflow to

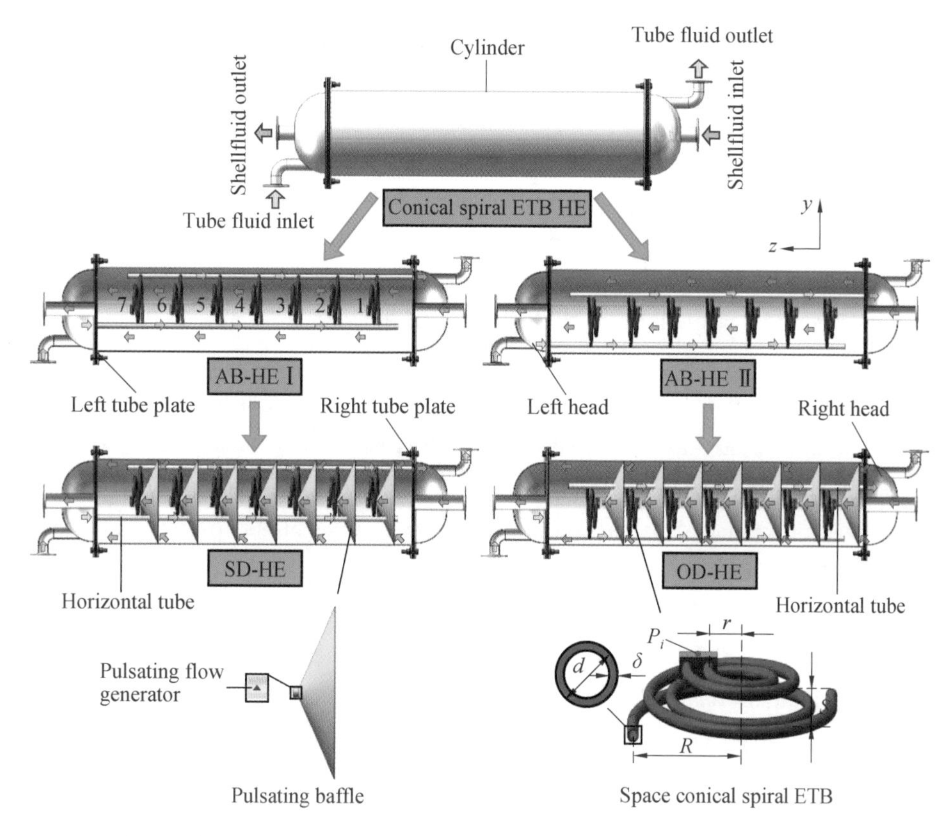

Fig. 1.16 Schematics of four space conical helical ETB heat exchanger schemes[154]

the heat exchanger from the right shell fluid inlet, impact the stainless steel connector of the space conical spiral ETB through the diversion effect of the pulsating baffle, and finally outflow from the left shell fluid outlet. The vibration of the space conical spiral ETBs induced by the pulsating flow generated by the pulsating flow generator realized the enhanced heat transfer of the composite.

The results display that the two schemes with pulsating baffles can enhance the regularity of the fluid movement and the hierarchy of the temperature field distribution. The average amplitudes of the space conical spiral ETBs in the four heat exchanger schemes are increased by 107%, 117%, 1,785% and 1,935%, respectively, when *Re* is increased in the range of 1,000–9,000. The performance evaluation criteria of space conical spiral ETBs in the four heat exchanger schemes are increased by 1.89%, 2.68%, 5.71% and 5.84%, respectively. The schemes with the pulsating baffle installed in heat exchangers can improve the heat transfer performance. Especially, the installation direction of the baffle will affect the comprehensive heat transfer performance of heat exchangers.

1.2.3 Application of Baffles/Deflectors

Due to the complex structure of the pulsating flow generator, its installation in the heat exchanger affects the further arrangement of the heat transfer elements. Therefore, the baffles or deflectors that can drain fluids have been widely adopted in engineering practice.

Chen et al. [155] experimental studied the heat transfer performance of a shell-and-tube heat exchanger based on segmental baffle, tri-flower baffle, pore plate baffle, rod baffle and segmental & pore baffle (Fig. 1.17). Based on a water-water heat transfer experiment system, the hydrodynamics and heat transfer characteristics of the five heat exchangers were compared. Based on the experimental data from Chen et al., compared with segmental baffle, the heat transfer coefficients of tri-flower baffle and pore plate baffle augment 33.8% and 17.7% at the same flow rate. The pressure drop of the tri-flower baffle, rod baffle and segmental & pore baffle decrease are 19.5%, 70.1% and 31.1% lower than that of the segmental baffle. In addition, the comprehensive performances of the tri-flower baffle, pore plate baffle and rod baffle are all superior to that of the segmental baffle.

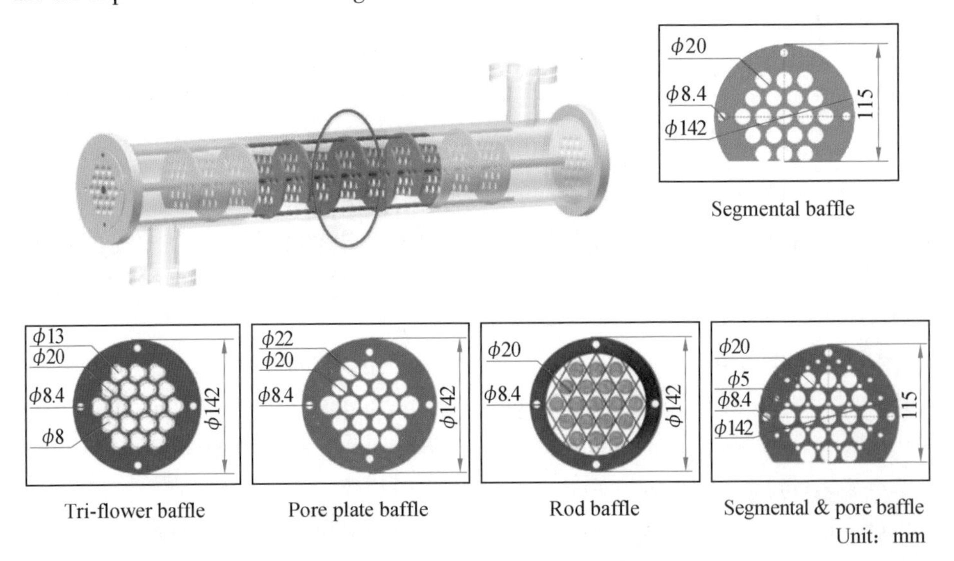

Fig. 1.17 Geometric model of five types of baffle patterns in a heat exchanger[155]

Wang et al. [156] researched the heat transfer performance of a branch baffle heat exchanger. Fig. 1.18 shows the structure of tube bundle support of the branch baffle heat exchanger. Based on their research findings, in contrast with the shell-and-tube heat exchanger with segmental baffles and shutter baffles, the pressure loss in the proposed heat exchanger with branch baffes has been dramatically improved under the same volume flow rate. In addition, the efficiency evaluation criteria of the heat exchanger with branch baffes are 28%-31%, 13.2%-14.1% higher than those with segmental baffles and shutter baffles, respectively. Further analysis in accordance with the field synergy principle illustrates that the velocity and pressure gradients of the heat exchanger with branch baffle have finer field coordination.

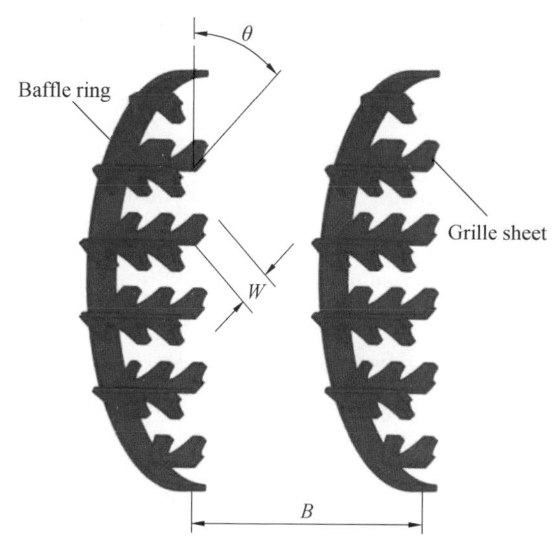

Fig. 1.18 Structure of tube bundle support of the branch baffle heat exchanger[156]

In terms of ETB heat exchangers, Sun et al. [102, 157] put forward a hollow helical deflector heat exchanger (Fig. 1.7) in order to balance the vibration performance and improve the heat transfer performance of helical heat transfer elements. And also, the effects of inlet velocity and the installed position of helical deflector on the performances of vibration and the heat transfer of the helical heat transfer elements were studied. Based on the numerical results, the vibration amplitude and heat transfer coefficient of helical heat transfer elements increase with inlet velocity. It was found that when the helical deflector is installed on the top side of the heat exchanger, the average vibration amplitude is the biggest, the average heat transfer coefficient is the smallest, and the heat transfer uniformity is the best. The PEC value of the hollow helical deflector heat exchanger is always bigger than 1.0, which shows that the hollow helical deflector heat exchanger has achieved the effect of enhancing the heat transfer. When the helical deflector is installed on the bottom, top, left and right side of the hollow helical deflector heat exchanger, the PEC value of the hollow helical deflector heat exchanger can be increased by 2.04%, 7.87%, 1.32% and 0.03% compared with that of the heat exchanger without helical deflectors, respectively.

Based on the hollow helical deflector heat exchanger, Ji et al.[103] studied the vibration enhanced heat transfer characteristics of the helical heat transfer elements under different inlet velocity and different ratio of transverse and longitudinal radius (e). Studies have indicated that for getting heat exchange equipment with more efficient heat transfer performance, the helical heat transfer element can be changed from the traditional round section to the oval section. When the section is oval ($e \neq 1$), the heat transfer performance of the helical heat transfer elements is improved in varying degrees. Especially, when $e=2$, heat transfer performance of the helical heat transfer element is enhanced to the greatest extent, at the same time, the integrated enhanced heat transfer performance of the hollow helical deflector heat exchanger is the strongest.

The above studies all indicate that adding baffles or deflectors in the heat exchanger improves the overall heat transfer performance to varying degrees, which has important theoretical significance and practical value for the design and development of new heat exchangers.

1.3 The Main Content of This Book

1.3.1 Analysis of Existing Problems and Development Trends

Based on previous research conclusions[158-164], the design of ETB heat exchangers should follow the principle: within the parameter range that meets the requirements for enhanced heat transfer, the flow-induced vibration should be reasonably induced and appropriately controlled, while taking into account the fatigue life of ETBs to ensure that it does not undergo fatigue damage while enhancing heat transfer.

However, after investigation by the authors of this book and their research team, it has been found that there are still some shortcomings in the current research on flow-induced vibration, vibration enhanced heat transfer, vibration/heat transfer control, etc., as follows.

(1) Due to the influence of the internal elastic heat transfer element structure,

the ETB heat exchangers have a small heat transfer area per unit volume and low overall heat transfer performance, resulting in poor overall efficiency of the heat exchanger and thus affecting its market competition.

(2) Flow-induced vibration in heat exchangers is a complex FSI problem involving multi field fluid. At present, most of the relevant research is limited to experimental means, and there is a lack of systematic research on numerical analysis of flow-induced vibration response of ETBs under complex flow conditions.

(3) In practical engineering applications, there is a phenomenon of uneven vibration for ETBs in ETB heat exchangers. In this way, ETBs with severe vibration are prone to vibration damage, while ETBs with less obvious vibration have poor heat transfer efficiency, which affects the service life and heat transfer efficiency of the heat exchanger. At present, there is no good solution to this problem.

In response to the above shortcomings, this book starts from the vibration response of single/multi row ETBs induced by uniform shell-side fluid, and the vibration response of each row of ETBs in heat exchanger induced by actual shell-side fluid was studied. Then, based on the shortcomings of independent pulsating flow generator, a distributed pulsating flow generator was designed to conduct numerical analysis and experimental research on the vibration response of ETBs induced by the coupling of shell-side fluid and distributed pulsating fluid. In addition, based on the structural improvement of the ETB and the installation of baffles, the vibration and heat transfer performances of ETBs in heat exchangers were studied. The research work in this book has important theoretical and engineering significance for improving the heat transfer performance of ETB heat exchangers and effectively stimulating and controlling the vibration of ETBs.

1.3.2 The Research Content of This Work

The main research work of this work is summarized as follows.

(1) In Chapter 1, started from the overview of the research status of vibration enhanced heat transfer, this book provided a detailed introduction to the structure and working principle of ETB heat exchangers, as well as the application of flow-induced vibration in ETB heat exchangers. Then, started from the excitation and control of ETB vibration in heat exchangers, the existing pulsating flow generation devices and the application of baffles and deflectors in the heat exchanger were introduced. Finally, based on the existing problems in the research and application of ETB heat exchangers, the main research content of this book were proposed.

(2) In Chapter 2, the governing equations for fluid and structural domains were introduced first. Then, a sequential solution method for bi-directional FSI and a step-by-step calculation strategy of rough calculation plus actuarial calculation suitable for the complex flow field in the ETB heat exchanger were proposed. And also, the calculation formulas of basic parameters for evaluating the performance of the ETB heat exchanger were introduced. Finally, the numerical calculation results were compared with previous experimental data and simulation results, and the accuracy of the vibration and heat transfer calculation results had been verified.

(3) In Chapter 3, the geometric model of the structural domain and the shell-side and tube-side fluid domains of the single-row ETB were established, and the structural domain and the shell and tube-side fluid domains were meshed respectively. Based on the sequential solution method for bi-directional FSI calculation, the effects of flow velocity of uniform shell-side fluid, the structural parameters of the ETB and the tube-side fluid on the vibration responses of the single-row ETB were investigated. In addition, based on the geometric modes and meshes of the single-row of ETBs and their shell-side fluid domain, the geometric model and mesh of multi-row ETBs and their shell-side fluid domain were formed. Then, numerical analysis was conducted on the vibration responses of ETBs under uniform shell-side fluid induction.

(4) In Chapter 4, the overall shell-side fluid domain mesh of a ETB heat exchanger was formed through the basic process of fluid domain partitioning, segmented fluid domain meshing and shell-side fluid domain mesh assembly. Then, the vibration responses of each row of ETBs under different shell-side water inlet velocities, different tube row spacing and different number of ETBs were numerically analyzed. And also, the heat transfer characteristics of each row of ETBs were studied. On the other hand, based on the traditional ETB, an improved ETB (IETB) was proposed. In addition, the influence of IETB installation angle on the vibration and heat transfer characteristics of multi-row IETBs were studied, and the optimal installation angle of IETB was obtained.

(5) In Chapter 5, through the study of a two-dimensional fluid flow around vortex element with different shapes, the appropriate vortex element was selected. Based on the study of two-dimensional fluid flow around vortex element in the branch tube, the flow passage structure was designed, and the influence of inlet velocity, inlet fluid flow direction, vortex element structural parameters and external fluid domain on the pulsating flow formed by the vortex element was studied. Combined with the selected vortex element and the designed flow channel for branch tube, a distributed pulsating

flow generator and its installation were designed. Then, effects of the structural parameters of the distributed pulsating flow generator on the fluid flow at the outlet of each branch tube and the frequency and intensity of the pulsating flow were studied. And also, the suitable structural parameters of distributed pulsating flow generator in the ETB heat exchanger were selected.

(6) In Chapter 6, the overall shell-side fluid domain of the heat exchanger installed with the distributed pulsating flow generator was established firstly, and the mesh of the overall shell-side fluid domain is formed. Then, numerical analysis was conducted on the vibration and heat transfer characteristics of each row of ETBs under different inlet velocity conditions induced by coupling fluids. Additionally, based on the ETB heat exchanger used in numerical analysis, a test bench was built for the coupling of shell-side fluid and pulsating fluid to induce the vibration of ETBs. Finally, experimental research was conducted on the vibration response of each row of ETBs under the coupling of shell-side fluid and pulsating fluid.

(7) In Chapter 7, for improving the performance of heat exchangers by effectively utilizing the flow-induced vibration, based on the limitations of pulsating flow generation devices in practical applications, two types of heat exchangers: traditional ETB heat exchanger with baffles and traditional ETB heat exchanger without baffles were proposed. Based on mesh division and boundary condition settings, by comparing and analyzing the vibration and heat transfer characteristics of the inner traditional ETBs, the effects of baffles on the ETB vibration responses and vibration-enhanced heat transfer performances under different conditions were investigated. The necessity of adding baffles to improve the overall heat transfer performance of the ETB heat exchangers had been verified.

(8) In Chapter 8, considering the advantage of improved ETB that is easy to achieve vibration under low fluid velocity, based on the traditional ETB heat exchangers with or without baffles, two types of heat exchangers—improved ETB heat exchanger with baffles and improved ETB heat exchanger without baffles were proposed. The influence of inlet velocity and baffles installation on the IETB vibration and heat transfer features was studied. In addition, the heat transfer performances of the four heat exchangers (contains two types of heat exchangers in Chapter 7) were compared under different conditions. Research had shown that structural optimization of ETB can further improve the comprehensive heat transfer performance of the ETB heat exchanger.

(9) In Chapter 9, based on the improved ETB heat exchanger with baffles, the

influences of the baffle structure parameters (height and curvature) on the vibration-enhanced heat transfer performance of the improved ETB heat exchanger were systematically studied under different conditions. The purpose is to further improve the comprehensive heat transfer performance of the ETB heat exchangers.

References

[1] CHENG L. Principle and application of elastic tube bundle heat exchanger [M]. Beijing: Science Press, 2014.

[2] CAO Y, ZHANG C W. Application and development prospects of heat exchangers [J]. Machine China, 2014, 3: 116-117.

[3] DU W J, ZHAO J Z, ZHANG L X, et al. Review and prospect of the development of heat exchanger structure [J]. Journal of Shandong University, 2021, 51(5): 76-83.

[4] JI J D. Study on flow-induced vibration of elastic tube bundle with shell-side distributed pulsating flow in heat exchanger [D]. Jinan: Shandong University, 2016.

[5] WANG X T, ZHENG N B, LIU Z C, et al. Numerical analysis and optimization study on shell-side performances of a shell and tube heat exchanger with staggered baffles [J]. International journal of heat and mass transfer, 2018, 124: 247-259.

[6] BAI Y B. Optimization and application of spray heat exchanger system for waste heat utilization of mine return air [D]. Hefei: University of Science and Technology of China, 2022.

[7] ZHANG B, YANG Z T, LIU L, et al. Thermal interference of backfill heat exchangers in heat storage/release processes in deep mines [J]. Journal of China coal society, 2023, 48(3): 1155-1168.

[8] JI J D, NI X W, ZHAO J H, et al. Numerical investigation of the effect of the structure and number of spiral baffles on the vibration and heat transfer coefficient of spiral tubes within spiral tube heat exchangers [J]. Applied thermal engineering, 2014, 258: 124545.

[9] DUAN D R. Study on flow-induced vibration of elastic tube bundle in heat transfer enhancement and fatigue strength [D]. Jinan: Shandong University, 2017.

[10] CHENG L, QIU Y. Complex heat transfer enhancement by fluid-induced

vibration [J]. Journal of hydrodynamics, Ser, B. 2003, 15(1): 84-89.

[11] XU G X. Research on heat transfer enhancement in complex channels [M]. Nanjing: Nanjing University Press, 2018.

[12] LU P, LIANG Z, LUO X L, et al. Design and optimization of organic rankine cycle based on heat transfer enhancement and novel heat exchanger: A review [J]. Energies, 2023,16(3): 1380.

[13] PETROVIC M, FUKUI K, KOMINAMI K. Numerical and experimental performance investigation of a heat exchanger designed using topologically optimized fins [J]. Applied thermal engineering, 2022,218: 119232.

[14] YAN K. A study on the vibration and heattransfer characteristics of conical spiral tube bundle in heat exchanger [D]. Jinan: Shandong University, 2012.

[15] JI J D, CHEN W Q, GAO R M. Research on vibration and heat transfer in heat exchanger with vortex generator [J]. Journal of thermophysics and heat transfer, 2021, 35(1): 164-170.

[16] JI J D, GE P Q, BI W B. Numerical investigation on the flow and heat transfer performances of horizontal spiral-coil pipes [J]. Journal of hydrodynamics, 2016, 28(4): 576-584.

[17] GUO Z Y, HUANG S Y. The principle of field synergy and new technologies for enhanced heat transfer [M]. Beijing: China Electric Power Press, 2004.

[18] YAN K, GE P Q, SU Y C, et al. Numerical simulation on heat transfer characteristic of conical spiral tube bundle [J]. Applied thermal engineering, 2011, 31: 284-292.

[19] YAN K, GE P Q, BI W B, et al. Vibration characteristics of fluid-structure interaction of conical spiral tube bundle with FEM [J]. Journal of hydrodynamics, 2010, 22(1): 121-128.

[20] JI J D, GE P Q, BI W B. Numerical analysis of shell-side flow-induced vibration of elastic tube bundle in heat exchanger [J]. Journal of hydrodynamics, 2018, 30(2): 249-257.

[21] BERGLES A E. Heat transfer enhancement—the maturing of the second-generation heat transfer technology [J]. Heat transfer engineering, 1997, 18: 47-55.

[22] BERGLES A E. Heat transfer enhancement—the encouragement and accommodation of high heat flux [J]. ASME journal of heat transfer, 1997, 119: 8-19.

[23] GU W Z, SHEN J R, MA C F, et al. Heat transfer enhancement [M]. Beijing:

Science Press, 1990.

[24] DUAN D R, GE P Q, BI W B, Numerical investigation on heat transfer performance of planar elastic tube bundle by flow-induced vibration in heat exchanger [J]. International journal of heat and mess transfer, 2016, 103: 868-878.

[25] JI J D, GAO R M, CHEN W Q. Analysis of vortex flow in fluid domain with variable cross-section and design of a new vortex generator [J]. International communications in heat and mass transfer, 2020, 116: 104695.

[26] BERGLES A E. Application of heat transfer augmentation [M]. Washington D. C.: Hemisphere Publishing Corporation, 1981.

[27] YAKUT K, SAHIN B. Flow-induced vibration analysis of conical rings used for heat transfer enhancement in heat exchangers [J]. Applied energy, 2004, 78 (3): 273-288.

[28] LI Y, DING Y M, YANG W M. Numerical simulation of turbulent flow and heat transfer in heat exchange tube inserted with rotors-assembled strand [J]. Petrochemical technology, 2009, 38(9): 979-983.

[29] KIM H Y, KIM Y G, KANG B H. Enhancement of natural convection and pool boiling heat transfer via ultrasonic vibration [J]. International journal of heat and mass transfer, 2004, 47(2-13): 2831-2840.

[30] HYUN S, LEE D R, LOH B G. Investigation of convective heat transfer augmentation using acoustic streaming generated by ultrasonic vibrations [J]. International journal of heat and mass transfer, 2005, 48(3-4): 703-718.

[31] CHENG L, LUAN T, DU W. Heat transfer enhancement by flow-induced vibration in heat exchangers [J]. International journal of heat and mass transfer, 2009, 52: 1053-1057.

[32] OHADI M M, NELSON D A, ZIA A S. Heat transfer enhancement of laminar and turbulent pipe flow via corona discharge [J]. International journal of heat and mass transfer, 1991, 34: 1175-1187.

[33] WANG P, LEWIN P L, SWAFFIELD D J, et al. Electric field effects on boiling heat transfer of liquid nitrogen [J]. Cryogenics, 2009, 49(8): 379-389.

[34] MARYAMALSADAT L, JAFAR M R, IRAJ H, et al. Experimental investigation for enhanced ferrofluid heat transfer under magnetic field effect [J]. Journal of magnetism and magnetic materials, 2010, 322(21): 3508-3513.

[35] MA C F, ZHENG Q, LEE S C, et al. Impingement heat transfer and recovery

effect with submerged jets of large prandtl number liquid-i: Unconfined circular jets [J]. International journal of heat and mass transfer, 1997, 40(6): 1481-1490.

[36] AL-SANEA S A. Mixed convection heat transfer along a continuously moving heated vertical plate with suction or injection [J]. International journal of heat and mass transfer, 2004, 47(6-7): 1445-1465.

[37] BHARADWAJ P, KHONDGE A D, DATE A W. Heat transfer and pressure drop in a spirally grooved tube with twisted tape insert [J]. International journal of heat and mass transfer, 2009, 52(7-8): 1938-1944.

[38] LIEBENBERG L, MEYER J D. In-tube passive heat transfer enhancement in the process industry [J]. Applied thermal engineering, 2007, 27(16): 2713-2726.

[39] LAN Z, FANG Z, ZHANG C F, et al. Experimental investigation on coating surface on the convection-condensation heat transfer of hot humidity air [J]. Proceedings of the CSEE, 2011, 31(11): 51-56.

[40] FORREST E, WILLIAMSON E, BUONGIOMO J. Augmentation of nucleate boiling heat transfer and critical heat flux using nanoparticle thin-film coatings [J]. International journal of heat and mass transfer, 2010, 53: 58-67.

[41] KATOH K, CHOI K S, AZUMA T. Heat-transfer enhancement and pressure loss by surface roughness in turbulent channel flows [J]. International journal of heat and mass transfer, 2000, 43(21): 4009-4017.

[42] JIANG P X, LI M, LU T J. Experimental research on convection heat transfer in sintered porous plate channels [J]. International journal of heat and mass transfer, 2004, 47: 2085-2096.

[43] RIN P, HEO J, KIM Y. Film condensation heat transfer characteristics of r134a on horizontal stainless steel integral-fin tubes at low heat transfer rate [J]. International journal of refrigeration, 2009, 32: 865-873.

[44] DALKILIC A S, WONGWISES S. Intensive literature review of condensation inside smooth and enhanced tubes [J]. International journal of heat and mass transfer, 2009, 52(15-16): 3409-3426.

[45] BILIR L, LIKEN Z, EREK A. Numerical optimization of a fm-tube gas to liquid heat exchanger [J]. International journal of thermal sciences, 2012, 52: 59-72.

[46] HE F J, CAO W W, YAN P. Experimental investigation of heat transfer and flowing resistance for air flow cross over spiral finned tube heat exchanger [J].

Energy procedia, 2012, 17: 741-749.

[47] PROMVONGE P, PETHKOOL S, PIMSARN M, et al. heat transfer augmentation in a helical-ribbed tube with double twisted tape inserts [J]. International communications in heat and mass transfer, 2012, 39: 953-959.

[48] ABU-KHADER M M. Further understanding of twisted tape effects as tube insert for heat transfer enhancement [J]. Heat mass transfer, 2006, 43: 123-134.

[49] LEI Y G, HE Y L, TIAN L T, et al. Hydrodynamics and heat transfer characteristics of a novel heat exchanger with delta-winglet vortex generators [J]. Chemical engineering science, 2010, 65: 1551-1562.

[50] HENZE M, WOLFERSDORF J, WEIGAND B, et al. Flow and heat transfer characteristics behind vortex generators—a benchmark dataset [J]. International journal of heat and fluid flow, 2011, 32: 318-328.

[51] MISCEVIC M, RAHLI O, TADRIST L, et al. Experiments on flows, boiling and heat transfer in porous media: Emphasis on bottom injection [J]. Nuclear engineering and design, 2006, 236(19-21): 2084-2103.

[52] MCGILLI W R, CAREY V P. On the role of marangoni effects on the critical heat flux for pool boiling of binary mixtures [J]. ASME journal of heat transfer, 1996, 118: 103-109.

[53] SHIVKUMAR C, RAO M R. Studies on compound augmentation of laminar flow heat transfer to genernalised power law fluids in spirally corrugated tubes by means of twisted tape inserts [J]. ASNE proceedings of the 1988 national heat transfer conference, 1988, 1: 685-692.

[54] ESEN E B, OBOT N T, RABAS T J. Enhancement: Part 1. Heat transfer and pressure drop results for air flow through passages with spirally-shaped roughness [J]. Journal of enhanced heat transfer, 1994, 1(2): 145-156.

[55] LI X Y, DENG X H, DENG Q H. Compound heat transfer enhancement in converging-diverging tube with delaying self-sustaining swirl flow [J]. Chemical engineering, 2011, 39(2): 14-17.

[56] VAN-ROOYEN R S, KROGER D G. Laminar flow heat transfer in internally finned tubes with twisted-tape insets [J]. Heat transfer, 1978, 2: 577-581.

[57] DEAVER F K, PENNY W R, JEFFERSON T B. Heat transfer from an oscillating horizontal wire to water [J]. Journal of heat transfer, 1962, 84(8): 251-256.

[58] LEMLICH R, RAO M A. The effect of transverse vibration on free convection from a horizontal cylinder [J]. International journal of heat and mass transfer,

1965, 8: 27-33.

[59] PENNY W R, JEFFERSON T B. Heat transfer from an oscillating horizontal wire to water and ethylene glycol [J]. Journal of heat transfer, 1966, 88: 359-366.

[60] HSIEH R, MARSTERS G F. Heat transfer from a vibrating vertical array of horizontal cylinders [J]. Canadian journal of chemistry (engineering), 1973, 51: 302-306.

[61] DAWOOD A S, MANOCHA B L, ALI S M J. The effect of vertical vibrations on natural convection heat transfer from a horizontal cylinder [J]. International journal of heat and mass transfer, 1981, 24(3): 491-496.

[62] SAXENA U C, LAIRD A D K. Heat transfer from a cylinder oscillating in a cross-flow [J]. Journal of heat transfer, 1978, 100: 684-689.

[63] LEUNG CT, KO N W M, MA K H. Heat transfer from a vibrating cylinder [J]. Journal of sound and vibration, 1981, 75(4): 581-582.

[64] KATINAS V I, MARKERICIUS A A, ZUKAUSKAS A A. Heat transfer behavior of vibrating tubes operating in cross flow-1: Temperature and velocity fluctuations [J]. Heat transfer-soviet research, 1986, 18(2): 1-9.

[65] KATINAS V I, MARKERICIUS A A, ZUKAUSKAS A A. Heat transfer behavior of vibrating tubes operating in cross flow-2: Local and average heat transfer coefficients [J]. Heat transfer-soviet research, 1986, 18(2): 10-17.

[66] TAKAHASHI K, ENDOH K A. New correlation method for the effect of vibration on forced-convection heat transfer [J]. Journal of chemical engineering of japan, 1990, 23(1): 45-50.

[67] KARANTH D, RANKIN G W, SRIDHAR K A. Finite difference calculation of forced convective heat transfer from an oscillating tube [J]. International journal of heat and mass transfer, 1994, 37(11): 1619-1630.

[68] CHENG C H, CHEN H N, AUNG W. Experimental study of the effect of transverse oscillation on convection heat transfer from a circular cylinder [J]. Journal of heat transfer, 1997, 119: 474-482.

[69] KLACZAK A. Report from experiments on heat transfer by forced vibrations of exchangers [J]. International journal of heat and mass transfer, 1997, 32(6): 477-480.

[70] GAU C, WU J M, LIANG C Y. Heat transfer enhancement and vortex flow structure over a heated cylinder oscillating in the cross flow direction [J]. Journal of heat transfer, 1999, 121: 789-795.

[71] BRONFENBRENER L, GRINIS L, KORIN E. Experimental study of heat transfer intensification under vibration condition [J]. Chemical engineering and technology, 2001, 24(4): 367-371.

[72] FU W S, TONG B H. Numerical investigation of heat transfer from a heated oscillating cylinder in a cross flow [J]. International journal of heat and mass transfer, 2002, 45(11): 3033-3043.

[73] LENG X L, CHENG L, DU W J. Heat transfer properties of the vibrational pipe when fluid passes by it slowly [J]. Journal of engineering thermophysics, 2003,24(2):328-330.

[74] LEE Y H, KIM D H, CHANG S H. An experimental investigation on the critical heat flux enhancement by mechanical vibration in vertical round tube [J]. Nuclear engineering and design, 2004, 229(1): 47-58.

[75] BLEVINS R D. Flow-induced vibration [M]. New York: New York Van Nostrand Reinhold, 1990.

[76] WEAVER D S, Fitzpatrick J A. A review of cross-flow induced vibration in heat exchanger tube arrays [J]. Journal of fluids and structures, 1988, 2(1): 73-93.

[77] PETTIGREW M P, SYVESTRE Y, CAMPAGNA A O. Vibration analysis of heat exchangers and steam generator designs [J]. Nuclear engineering and design, 1978, 48: 97-115.

[78] BLECINS R D. Flow-induced vibration in nuclear reactors [J]. Progress in nuclear energy, 1979, 4: 24-49.

[79] AXISA F, ANTUNES J, VILLARD B. Overview of numerical methods for predicting of cylinder arrays in cross-flow [J]. ASME journal of pressure vessel technology, 1988, 110: 6-14.

[80] EISINGER F L, SULLIVAN R E, FRANCIS J T. A review of acoustic vibration criteria compared to inline tube banks [J]. ASME journal of vessel technology, 1994, 116: 17-23.

[81] JI J D, GE P Q, BI W B. Numerical analysis on shell-side flow-induced vibration and heat transfer characteristics of elastic tube bundle in heat exchanger [J]. Applied thermal engineering, 2016, 107: 544-551.

[82] JI J D, GE P Q, LIU P, et al. Design and application of a new distributed pulsating flow generator in elastic tube bundle heat exchanger [J]. International journal of thermal sciences, 2018, 130: 216-226.

[83] JI J D, GAO R M, SHI B J, et al. Investigation of fluid-induced vibration and

heat transfer of helical elastic coiled tube [J]. Journal of thermophysics and heat transfer, 2021, 35(3): 634-643.

[84] ZHENG J Z, CHENG L, DU W J. Dynamic characteristics of elastic tube bundles with component mode synthesis method [J]. Chinese journal of mechanical engineering, 2007, 43(7): 202-206.

[85] CHENG F, TIAN M C. Experimental research on heat transfer and resistance characteristics of heat exchanger of elastic tube bundle [J]. Research and exploration in laboratory, 2007, 26(11): 23-25.

[86] SU Y C. A study on the characteristics of the flow-induced vibration and heat transfer of elastic tube bundle [D]. Jinan: Shandong University, 2012.

[87] SU Y C, LI M L, LIU M L, et al. A study of the enhanced heat transfer of flow-induced vibration of a new type of heat transfer tube bundle—the planar bending elastic tube bundle [J]. Nuclear engineering and design, 2016, 309: 294-302.

[88] DUAN D R, GE P Q, BI W B, et al. An empirical correlation for the heat transfer enhancement of planar elastic tube bundle by flow-induced vibration [J]. International journal of thermal sciences, 2020, 155: 106405.

[89] DUAN D R, GE P Q, BI W B, et al. Numerical investigation on the heat transfer enhancement mechanism of planar elastic tube bundle by flow-induced vibration [J]. International journal of thermal sciences, 2017, 112: 450-459.

[90] DUAN D R, GE P Q, BI W B, et al. Numerical investigation on synthetical performance of heat transfer of planar elastic tube bundle heat exchanger [J]. Applied thermal engineering, 2016, 109: 295-303.

[91] CHENG L, TIAN M C, ZHANG G M, et al. Theoretical analysis of complex heat transfer enhancement by flow-induced vibration [J]. Journal of engineering thermophysics, 2002, 23(3): 330-332.

[92] TIAN M C, CHENG L, LIN Y Q, et al. Experimental investigation of heat transfer enhancement by crossflow induced vibration [J]. Journal of engineering thermophysics, 2002, 23(4): 485-487.

[93] JIANG B. Analysis on mechanism of heat transfer enhancement by vibration and experimental research on a new type of vibrational heat transfer component [D]. Jinan: Shandong University, 2010.

[94] YAN K, GE P Q, SU Y C, et al. Mathematical analysis on transverse vibration of conical spiral tube bundle with external fluid flow [J]. Journal of hydrodynamics, 2010, 22(6): 816-822.

[95] YAN K, GE P Q, HU R R, et al. Heat transfer and resistance characteristics of conical spiral tube bundle based on field synergy principle [J]. Chinese journal of mechanical engineering, 2012, 25(2): 370-376.

[96] YAN K, GE P Q, BI W B. Study on vibration characteristic and stress intensity of planar elastic tube bundles [J]. Material science forum, 2009, 628-629: 227-232.

[97] YAN K, GE P Q, ZHANG L, et al. Finite element analysis of vibration characteristics of planer elastic tube bundle conveying fluid [J]. Journal of mechanical engineering, 2010, 46(18): 145-149.

[98] JI J D, GAO R M, CHEN W Q, et al. Study on shell-side vibration-enhanced heat transfer of helical elastic tube heat exchanger[J]. Journal of engineering thermophysics, 2021, 42(10): 2692-2699.

[99] JI J D, GAO R M, CHENG Q H, et al. Analysis on fluid-induced vibration and heat transfer of helical elastic tube bundles[J]. Journal of thermophysics and heat transfer, 2021, 35(1): 171-178.

[100] JI J D, ZHOU R, GAO R M, et al. Analysis on heat transfer characteristic of spiral elastic tube bundle heat exchangers with different cross section shapes [J]. Journal of vibration and shock, 2022, 41(22): 219-225.

[101] JI J D, LI F Y, SHI B J, et al. Vibration-enhanced heat transfer of double-array helical elastic tube bundle heat exchanger[J]. Journal of thermophysics and heat transfer, 2022, 36(2): 351-357.

[102] SUN Y R, JI J D, HUA Z S, et al. Vibration and heat transfer performances analysis of a hollow helical baffle heat exchanger [J]. Journal of vibration and shock, 2023, 42(8): 256-263.

[103] JI J D, DENG R Y, LU Y, et al. Analysis of heat transfer characteristic of elliptical section spiral copper tube heat exchanger [J]. Journal of Xi'an Jiaotong University, 2023, 57(6): 124-133.

[104] DANESHMAND F, NIROOMANDI S T. Natural neighbour Galerkin computation of the vibration modes of fluid-structure systems [J]. Engineering computations, 2007, 8: 125-127.

[105] SEGURA S C, ZHANG P, PORFIRI M. Three-dimensional exact solution of free vibrations of a simply supported rectangular plate in contact with a fluid [J]. Journal of sound and vibration, 2022, 534: 117007.

[106] KADAPA C. A unified simulation framework for fluid-structure-control interaction problems with rigid and flexible structures [J]. International journal of

computational methods, 2022, 19(1): 2150052.

[107] YUE Q B, LIU G R, LIU J B, et al. Modelling techniques for fluid-solid coupling dynamics of bundle tubes vibrating and colliding in fluids [J]. International journal of computational fluid dynamics, 2018, 32(1): 35-48.

[108] TANDIS E, ASHRAFIZADEH A. A numerical study on the fluid compressibility effects in strongly coupled fluid-solid interaction problems [J]. Engineering with computers, 2019, 37(2): 1205-1217.

[109] PENG X, RAO G N, LI B, et al. Investigation on the gas-solid two-phase flow in the interaction between plane shock wave and quartz sand particles [J]. Applied sciences-basel, 2020, 10(24): 8859.

[110] YUN G, TAN N, XIONG Y M, et al. Numerical study of response of vortex-induced vibration on a flexible cylinder using a modified wake oscillator model [J]. Journal of vibration and shock, 2017, 36(22): 86-92.

[111] XU W H, WU Y H, ZENG X H, et al. A new wake oscillator model for predicting vortex induced vibration of a circular cylinder [J]. Journal of hydrodynamics, 2010, 22(3): 381-386.

[112] JACQUELIN E, LAINE J P, BENNANI A, et al. A modelling of an impacted structure based on constraint modes [J]. Journal of sound and vibration, 2007, 301(3-5): 789-802.

[113] ZHENG R H, ZHENG H H, ANGELI A, et al. A simplified modelling and analysis of six degree of freedom random vibration test [J]. Mechanical systems and signal processing, 2020, 150: 107304.

[114] WANG X Q, SO R M C, CHAN K T. A non-linear fluid force model for vortex-induced vibration of an elastic cylinder [J]. Journal of sound and vibration, 2003, 260(2): 287-305.

[115] FENG C C. The measurement of vortex induced effects in flow past stationary and oscillating, circular and d-section cylinders [D]. Vancouver: University of British Columbia, 1968.

[116] AHMED N A, WAGNER D J. Vortex shedding and transition frequencies associated with flow around a circular cylinder [J]. AIAA journal, 2003, 4: 542-544.

[117] GRIFFITHS I, EVANS C, GRIFFITHS N. Tracking the flight of a spinning football in three dimensions [J]. Measurement science and technology, 2005, 16 (10): 2056-2065.

[118] GABBAI R D, BENAROYA H. An overview of modeling and experiments of

vortex-induced vibration of circular cylinders [J]. Journal of sound and vibration, 2005, 282 (3-5): 575-616.

[119] GABBAI R D, BENAROYA H. A first-principles derivation procedure for wake-body models in vortex-induced vibration: Proof-of-concept [J]. Journal of sound and vibration. 2008, 312: 19-38.

[120] SO R M C, WANG X Q, XIE W C, et al. Free-stream turbulence effects on vortex-induced vibration and flow-induced force of an elastic cylinder [J]. Journal of fluids and structures, 2008, 24(4): 481-495.

[121] IVANOV O, VEDENEEV V. Vortex-induced vibrations of an elastic cylinder near a finite-length plate [J]. Journal of fluids and structures, 2021, 107: 103393.

[122] DING L, HE H Y, SONG T. Vortex-induced vibration and heat dissipation of multiple cylinders under opposed thermal buoyancy [J]. Ocean engineering, 2023, 270: 112669.

[123] MENG H T. Study on the pulsating flow generating device in elastic tube bundle heat exchanger [D]. Jinan: Shandong University, 2012.

[124] LENHARD J H. Synopsis of lift, drag, and vortex frequency data for rigid circular cylinders [M]. Pullman, WA: Technical Extension Service, Washington State University, 1966.

[125] JIANG R J. Research on vortex-induced vibrations in the flow around circular cylinders [D] Hangzhou: Zhejiang University, 2013.

[126] UEMURA T, KOMURO A, ONO R. Flow control around a pitching oscillation circular cylinder using a dielectric barrier discharge plasma actuator [J]. Journal of physics D-applied physics, 2023, 56(12): 125202.

[127] BAIKOV N D, PETROV A G. On a flow around a cylinder over uneven bottom [J]. Computational mathematics and mathematical physics, 2023, 63 (3): 401-412.

[128] RINOSHIKA H. Three-dimensional wake structures over a short cylinder having a fore angled hole [J]. Ocean engineering, 2023, 271: 113713.

[129] AFROZ F, SHARIF M A R. Numerical study of cross-flow around a circular cylinder with differently shaped span-wise surface grooves at low Reynolds number [J]. European journal of mechanics B-fluids, 2022, 91: 203-218.

[130] YAN Z Y, ZHU X L, BAI Y, et al. Characteristics of the wake of the flow around a circular cylinder in a centrifugal field [J]. Journal of the Taiwan institute of chemical engineers, 2022, 134: 104348.

[131] SEOL C, HONG J, KIM T. Flow around porous square cylinders with a periodic and scalable structure [J]. Experimental thermal and fluid science, 2023, 144: 110864.

[132] SONG T, LIU X, XU F. Moving surface boundary-layer control on the wake of flow around a square cylinder [J]. Applied sciences-basel, 2022, 12(3): 1632.

[133] BASOHBATNOVINZAD M, SHAMS M, POURYOUSSEFI S G, et al. Experimental and numerical investigation of flow around an inline square cylinder array at a high Reynolds number [J]. Archive of applied mechanics, 2022, 92(12): 3433-3446.

[134] DAI L F, WU H Y. Rarefaction effect, heat transfer, and drag coefficient for gas flow around square cylinder in transition flow regime [J]. Journal of heat transfer-transactions of the ASME, 2022, 144(7): 071801.

[135] LIU J H, YANG Z H, LIU Y, et al. Effect of inclined angle on a wall-mounted finite-height square cylinder with a passive vertical suction [J]. Ocean engineering, 2022, 257: 111654.

[136] HAFFNER Y, LI R Y, MELDI M, et al. Drag reduction of a square-back bluff body under constant cross-wind conditions using asymmetric shear layer forcing [J]. International journal of heat and fluid flow, 2022, 96: 109003.

[137] ALI M, ZEITOUN O, NUHAIT A. Forced convection heat transfer over horizontal triangular [J]. International journal of thermal sciences, 2011, 50(1): 106-114.

[138] ZEITOUN O, ALI M, NUHAIT A. Convective heat transfer around a triangular cylinder in an air cross flow [J]. International journal of thermal sciences, 2011, 50: 1685-1697.

[139] AFROZ F, SHARIF M A R. Numerical study of cross-flow around a circular cylinder with differently shaped span-wise surface grooves at low Reynolds number [J]. European journal of mechanics B-fluids, 2022, 91: 203-218.

[140] WANG R Y, LU B, ZUO X Q, et al. Numerical computations of flow around three equilateral-triangular square cylinders with rounded corners using momentum exchange-based IB-LBM at low Reynolds numbers [J]. Ocean engineering, 2022, 263: 112373.

[141] WANG R Y, LU B, ZUO X Q, et al. Numerical study on the flow past three cylinders in equilateral-triangular arrangement at $Re = 3 \times 10^6$ [J]. Applied sciences-basel, 2022, 12(2): 11835.

[142] FARUQUEE Z, TING D S K, FARTAJ A, et al. The effects of axis ratio on laminar fluid flow around an elliptical cylinder [J]. International journal of heat and fluid flow, 2007, 28(15): 1178-1189.

[143] DAOUD S, NEHARI D, AICHOUNI M, et al. Numerical simulations of an oscillating flow past an elliptic cylinder [J]. Journal of offshore mechanics and arctic engineering, 2016, 138(1): 011802.

[144] PALEI V, SEIFERT A. Effects of periodic excitation on the flow around a d-shaped cylinder at low reynolds numbers [J]. Flow turbulence combust, 2007, 78(3-4): 409-428.

[145] JI J D, ZHANG J W, GAO R M, et al. Tests for pulsating flow generator-induced vibration of elastic tube bundle[J]. Journal of vibration and shock, 2021, 40(3): 291-296.

[146] HEMMAT ESFE M, BAHIRAEI M, TORABI A, et al. A critical review on pulsating flow in conventional fluids and nanofluids: Thermo-hydraulic characteristics [J]. International communications in heat and mass transfer, 2021, 120: 104859.

[147] XU C, XU S L, WANG Z Y, et al. Experimental investigation of flow and heat transfer characteristics of pulsating flows driven by wave signals in a microchannel heat sink [J]. International communications in heat and mass transfer, 2021, 125: 105343.

[148] SAILOR D J, ROHLI D J, FU Q L. Effect of variable duty cycle flow pulsations on heat transfer enhancement for an impinging air jet [J]. International journal of heat and fluid flow, 1999, 20(6): 574-580.

[149] LI H, ZHANG X M, XU Z, et al. Design and experimental study on vane-type pulsating flow generator [J]. Transactions of the chinese society of agricultural machinery, 2008, 39(2): 63-66.

[150] GAO H, ZENG D L. Experimental study on heat transfer enhancement by self-oscillation pulsed jet [J]. Journal of engineering for thermal energy and power, 2003, 18(4): 349-360.

[151] GAO H, LIU J F. Experimental analysis of heat transfer enhancement by using Helmholtz oscillator [J]. Energy technology, 2009, 30(3): 141-144.

[152] DUAN D R, CHENG Y J, GE M R, et al. Experimental and numerical study on heat transfer enhancement by flow-induced vibration in pulsating flow [J]. Applied thermal engineering, 2022, 207: 118171.

[153] LIU J Q, TIAN M C, WANG H X, et al. Experimental study on flow-induced

vibration of compound bent beam under pulsating flow [J]. Journal of hydrodynamics, 1998, 13(4): 467-472.

[154] JI J D, LU Y, SHI B J, et al. Numerical research on vibration and heat transfer performance of a conical spiral elastic bundle heat exchanger with baffles [J]. Applied thermal engineering, 2023, 232: 121036.

[155] CHEN J, LI N Q, DING Y, et al. Experimental thermal-hydraulic performances of heat exchangers with different baffle patterns [J]. Energy, 2020, 205: 118066.

[156] WANG K, LIU J Q, LIU Z C, et al. Fluid flow and heat transfer characteristics investigation in the shell side of the branch baffle heat exchanger [J]. Journal of applied fluid mechanics, 2021, 14: 1775-1786.

[157] SUN Y R, JI J D, HUA Z S, et al. Flow-induced vibration and heat transfer analysis for a novel hollow heat exchanger[J]. Journal of thermophysics and heat transfer, 2023, 37(1): 94-103.

[158] JI J D, ZHANG J W, LI F Y, et al. Numerical research on vibration-enhanced heat transfer of improved elastic tube bundle heat exchanger[J]. Case studies in thermal engineering, 2022, 33: 101936.

[159] MOHAMMED A M, KAPAN S, SEN M, et al. Effect of vibration on heat transfer and pressure drop in a heat exchanger with turbulator [J]. Case studies in thermal engineering, 2022, 28: 101680.

[160] WANG J G, JI J D, GAO R M, et al. Vibration-enhanced heat transfer of helical elastic tube with different number of tubes[J]. Journal of thermophysics and heat transfer, 2022, 36(4): 1015-1024.

[161] JI J D, GAO R M, ZHANG J W, et al. Improved tube structure and segmental baffle to enhance heat transfer performance of elastic tube bundle heat exchanger [J]. Applied thermal engineering, 2022, 200: 117703.

[162] JI J D, ZHANG J W, GAO R M, et al. Numerical research on vibration-enhanced heat transfer of elastic scroll tube bundle [J]. Journal of thermophysics and heat transfer, 2022, 36(1): 61-68.

[163] JI J D, DENG X, ZHANG J W, et al. Study on vibration and heat transfer performances of a modified elastic tube bundle heat exchanger [J]. Journal of physics: conference series, 2022, 2181: 012043

[164] ZHANG J W, JI J D, PAN Y L, et al. Effect of baffles on heat transfer characteristics of improved plan elastic tube bundle heat exchanger [J]. Journal of vibration and shock, 2023, 42(10): 51-57.

Chapter 2 Numerical Calculation Method

In practical engineering applications of heat exchangers, both the shell-side and tube-side of ETBs are filled with flowing fluid media. Through the coupling-induced effect of the two fields of fluid in the shell-side and tube-side, the ETBs in heat exchanger are vibrated, thus the vibration-enhanced heat transfer has been achieved. Compared with the tube-side domain and its internal fluid, the structure and internal fluid flow characteristics of the shell-side domain are more complex.

When the fluid impacts the structure, the structure is subjected to stress and strain and thus undergoes deformation or motion. The flow field will also change due to the changes of the structure. The phenomenon of FSI arises from this interaction. Therefore, solving FSI problems requires consideration of both the fluid and structural domains. With the rapid development of computer technology and computational fluid dynamics software, research on FSI vibration based on numerical simulation has gradually attracted attention from academia[1-5].

In this chapter, the governing equations for fluid and structural domains were introduced first. Then, a sequential solution method for bi-directional FSI calculation and a step-by-step calculation strategy of rough calculation plus actuarial calculation suitable for the complex flow field in the ETB heat exchanger were proposed. And, the calculation formulas of basic parameters for evaluating the performance of the ETB heat exchanger were introduced. Finally, the numerical calculation results were compared with previous experimental data and simulation results, the accuracy of the vibration and heat transfer calculation results have been verified.

2.1 Fundamental Governing Equations

2.1.1 Governing Equations for Fluid and Structural Domains

The fluid flow should follow the basic physical conservation law. The conservation law based on incompressible fluid can be described by the following governing equations[6-7].

Continuity equation is

$$\frac{\partial u_x}{\partial x}+\frac{\partial u_y}{\partial y}+\frac{\partial u_z}{\partial z}=0 \tag{2.1}$$

where x, y and z are the axes of the Cartesian coordinate system; u_x, u_y and u_z are the fluid velocities in the x, y and z directions.

Momentum conservation equations are

$$\rho\left(\frac{\partial u_x}{\partial t}+u_x\frac{\partial u_x}{\partial x}+u_y\frac{\partial u_x}{\partial y}+u_z\frac{\partial u_x}{\partial z}\right)=-u_x\frac{\partial p}{\partial x}+\mu\left(\frac{\partial^2 u_x}{\partial x^2}+\frac{\partial^2 u_x}{\partial y^2}+\frac{\partial^2 u_x}{\partial z^2}\right) \tag{2.2}$$

$$\rho\left(\frac{\partial u_y}{\partial t}+u_x\frac{\partial u_y}{\partial x}+u_y\frac{\partial u_y}{\partial y}+u_z\frac{\partial u_y}{\partial z}\right)=-u_x\frac{\partial p}{\partial y}+\mu\left(\frac{\partial^2 u_y}{\partial x^2}+\frac{\partial^2 u_y}{\partial y^2}+\frac{\partial^2 u_y}{\partial z^2}\right) \tag{2.3}$$

$$\rho\left(\frac{\partial u_z}{\partial t}+u_x\frac{\partial u_z}{\partial x}+u_y\frac{\partial u_z}{\partial y}+u_z\frac{\partial u_z}{\partial z}\right)=-u_x\frac{\partial p}{\partial z}+\mu\left(\frac{\partial^2 u_z}{\partial x^2}+\frac{\partial^2 u_z}{\partial y^2}+\frac{\partial^2 u_z}{\partial z^2}\right) \tag{2.4}$$

where ρ is the density of fluid; t is the time; μ is the dynamic viscosity of fluid; p is the pressure of fluid.

Energy conservation equation is

$$\rho C_{\mathrm{p}}\left(\frac{\partial T}{\partial t}\right)+\rho C_{\mathrm{p}}\left(u_x\frac{\partial T}{\partial x}+u_y\frac{\partial T}{\partial y}+u_z\frac{\partial T}{\partial z}\right)=\lambda\left(\frac{\partial^2 T}{\partial x^2}+\frac{\partial^2 T}{\partial y^2}+\frac{\partial^2 T}{\partial z^2}\right) \tag{2.5}$$

where λ is the thermal conductivity of fluid; T is the temperature; C_{p} is the specific heat at constant pressure.

In the calculations of this book, both the shell-side and tube-side fluids are incompressible water. Even if the inlet velocity is not high, the vibration of the ETBs changes the turbulence characteristics of the fluid, making *Re* increase sharply. Therefore, the standard $k-\varepsilon$ turbulence model is adopted in this book, which can be given by

$$\frac{\mathrm{d}k}{\mathrm{d}t}=\frac{1}{\rho}\frac{\partial}{\partial y}\left[\left(\frac{\mu_{\mathrm{t}}}{\sigma_k}+\mu_{\mathrm{l}}\right)\frac{\partial k}{\partial y}\right]+\frac{\mu_{\mathrm{t}}}{\rho}\left(\frac{\partial u_x}{\partial y}+\frac{\partial u_y}{\partial x}\right)\frac{\partial u_x}{\partial y}-\varepsilon \tag{2.6}$$

$$\frac{\mathrm{d}\varepsilon}{\mathrm{d}t}=\frac{1}{\rho}\frac{\partial}{\partial y}\left[\left(\frac{\mu_{\mathrm{t}}}{\sigma_\varepsilon}+\mu_{\mathrm{l}}\right)\frac{\partial \varepsilon}{\partial y}\right]+\frac{C_{\varepsilon 1}\mu_{\mathrm{t}}}{\rho}\frac{\varepsilon}{k}\left(\frac{\partial u_x}{\partial y}+\frac{\partial u_y}{\partial x}\right)\frac{\partial u_x}{\partial y}-C_{\varepsilon 2}\frac{\varepsilon^2}{k} \tag{2.7}$$

In the above equations, the turbulent viscosity (μ_t) could be obtained by the combination of Eq. (2.6) and Eq. (2.7) as

$$\mu_{\mathrm{t}}=C_\mu\frac{\rho k^2}{\varepsilon} \tag{2.8}$$

where μ_{l} and μ_{t} are the dynamic viscosities of laminar and turbulent fluids; σ_k and σ_ε

are the diffusion Prandtl numbers; C_μ, $C_{\varepsilon 1}$ and $C_{\varepsilon 2}$ are the turbulence model constants. Based on references [8-12], in the subsequent calculations, the parameter values are $C_{\varepsilon 1}=1.44$, $C_{\varepsilon 2}=1.92$, $\sigma_k=1.0$, $\sigma_\varepsilon=1.3$, $C_\mu=0.09$.

In terms of structure, the global equation of motion of the ETB is established by using Hamilton principle. Considering the effect of shell-side fluid and tube-side fluid, the structural mechanics equation of the ETB after dispersion is expressed as

$$M_s\ddot{S} + C_s\dot{S} + K_sS = F_s + F_f \tag{2.9}$$

where the subscripts "s" and "f" respectively represent the structural and the fluid; M_s, C_s and K_s are the mass matrix, damping matrix and stiffness matrix of the structural domain; ; S is the displacement of the structural domain F_s is the fluid force acting on the structure; F_f is the force within the structural domain.

2.1.2 Basic Calculation Process

According to different solution methods of control equations, FSI problems can be divided into strong coupling method (or direct method) and weak coupling method (or separated method). According to the different data transfers, FSI problems can also be divided into uni-directional coupling analysis and bi-directional coupling analysis. Among them, bi-directional coupling analysis can be divided into sequential solution method and simultaneous solution method[13].

The sequential solution method uses a partitioned method where the fluid and structure are solved separately by their respective solvers. Then, through interface information exchange, the fluid transfers stress and viscous forces to the structure, and the structure feeds back deformation and velocity to the fluid. The simultaneous solution method uses a monolithic method where the control equations of the fluid and structure are integrated into a set of equations for the overall coupled solution. The overall calculation method is more accurate than the step-by-step calculation method, but it requires the development of new matrix solution methods and has a large amount of calculation. The distributed calculation method is relatively flexible and can be used by integrating existing fluid and structure solvers.

Based on the research question of this book, a sequential solution method for bi-directional FSI is adopted to numerically analyze the vibration response and heat transfer performance of ETB under shell and/or tube fluid induction. The modeling of both the structural domain and the fluid domain is carried out using the Design Modeler modeling module in the Workbench platform; the mesh division of the structural domain is carried out using the Meshing module in the Workbench platform;

the Transient Structure module is selected for transient dynamic analysis. The mesh division of the fluid domain is carried out using ICEM mesh division software and the general CFD analysis software ANSYS CFX is selected for computational fluid dynamics analysis.

The overall planning process of FSI calculation is shown in Fig. 2. 1 and the specific implementation is as follows.

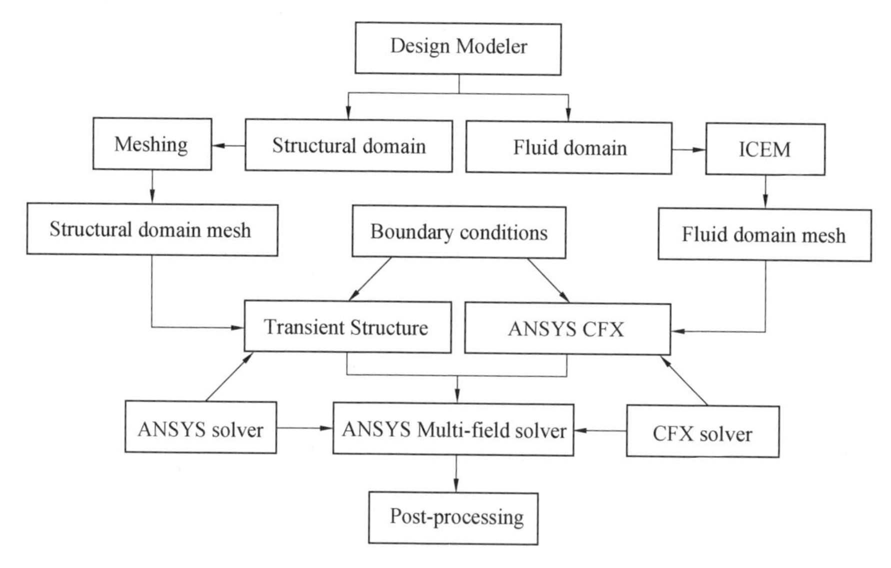

Fig. 2. 1 Flow chart of the FSI calculation

(1) Through the Design Modeler modeling module in the Workbench platform, geometric models of the structural domain and fluid domain are established respectively.

(2) Import the geometric model of the structural domain into the Meshing module, and use appropriate mesh division strategies to divide the structural domain.

(3) Import the geometric model of the structural domain into the mesh division software ICEM, use appropriate mesh division strategies to divide the fluid domain, and export the fluid domain meshes after the mesh division is completed.

(4) Set the boundary conditions of the structural domain for transient dynamic analysis, including: applying fixed constraints, setting the gravity and direction, defining the FSI interfaces, setting the calculation time and time step, etc.

(5) Based on the Workbench platform, establish a connection between the Transient Structure module and the general CFD analysis software ANSYS CFX preprocessing to prepare for data transfer during FSI calculations.

(6) Import the fluid domain meshes into the pre-processing module of ANSYS CFX, and carry out the relevant settings of FSI calculation, including: selecting the solving equation, defining the FSI interfaces, setting the velocity and temperature conditions of fluid inlet and outlet, setting the solution time and time step, defining the FSI calculation order, etc.

(7) Based on the ANSYS Multi-field solver, the FSI calculation is carried out by the sequential solution method of bi-directional FSI, and the data transfer between the fluid domain and the structural domain is completed through the FSI interface.

(8) Perform post-processing of the results.

Due to the large number of elements in the fluid and structural domains, FSI calculations are computationally intensive, with long computation times and high requirements in terms of computer hardware. In order to obtain the vibration response of the ETB under the fully developed fluid domain conditions, the calculation process needs to be planned rationally. For this reason, a step-by-step calculation strategy of rough calculation plus actuarial calculation is proposed. The specific process of the step-by-step calculation strategy is shown in Fig. 2. 2.

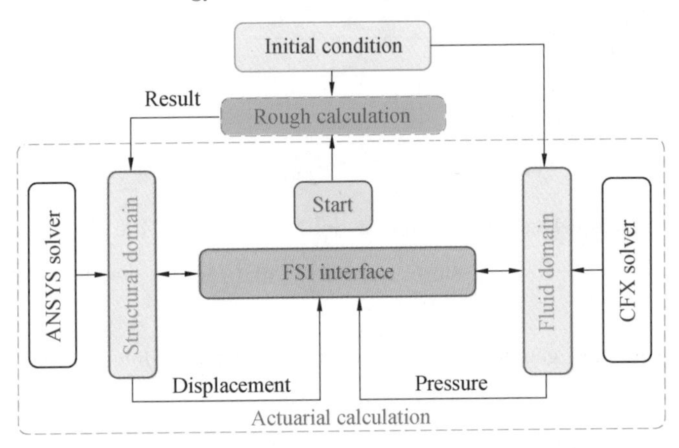

Fig. 2. 2 The specific process of the step-by-step calculation strategy

Step 1: Rough calculation, where only the ANSYS CFX software is used to calculate the fluid domain until the internal fluid domain is fully developed.

Step 2: Actuarial calculation, using the calculation results at the end of the rough calculation as the initial conditions, FSI calculations are performed for the fluid domain and the structural domain to obtain the vibration response of the ETBs under fully developed fluid domain conditions.

The specific implementation of the actuarial calculation is as follows.

(1) Based on developed fluid domain conditions calculated by the rough calculation, the CFX solver is used to calculate the fluid domain and obtain the pressure data at the FSI interface of the fluid domain.

(2) The pressure data at the FSI interface of the fluid domain is transferred to the FSI interface of the structural domain. Then, using this pressure data, combined with the initial conditions of the structural domain, the ANSYS solver is used to perform transient dynamic calculations on the structural domain and obtain the displacement data at the FSI interface of the structural domain.

(3) The displacement data at the FSI interface of the structural domain is transferred to the FSI interface of the fluid domain. Based on the new boundary conditions, the calculation of the fluid domain within the next time step is performed.

(4) Repeat the iteration alternately until the calculation is completed.

Additionally, it should be noted that, to ensure that the fluid domain gets fully developed and the total calculation time of the rough calculation is obtained, the average outlet temperature is monitored. When this temperature reaches stability, it indicates that the flow characteristics of the internal fluid reach stability and the fluid domain obtains full development. The specific implementation process can refer to "4.1.3 Boundary conditions" and "4.2.1 Shell-side outlet temperature distribution".

2.1.3 Data Transfer Between FSI Interfaces

In the actuarial calculation process, FSI interfaces are used to transfer the pressure data from the fluid domain and the displacement data from the structural domain.

The following conservation criteria should be followed for the calculation:

$$n_f \tau_f = n_s \tau_s \tag{2.10}$$

$$g_f = g_s \tag{2.11}$$

where n_f and n_s are the normal vectors at the FSI interface of the fluid domain and structural domain; τ_f and τ_s are the stress at the FSI interface of the fluid domain and structural domain; g_f and g_s are the displacements at the FSI interface of the fluid domain and structural domain.

If heat transfer is considered during the FSI calculation, the conservation of heat flux and temperature must also be considered. The expressions are

$$q_f = q_s \tag{2.12}$$

$$T_f = T_s \tag{2.13}$$

where q_f and q_s are the heat flux densities at the FSI interface of the fluid domain and structural domain; T_f and T_s are the temperatures at the FSI interface of the fluid domain and structural domain.

Generally speaking, the mesh distribution of the fluid domain and the structural domain are different. Therefore, the node positions of the two models at the FSI interface are not consistent.

During the FSI calculation, the structural domain transfers displacement data to the fluid domain through its FSI interface. And also, the fluid domain transfers pressure data to the structural domain through its FSI interface. The schematic diagram of data transfer at the FSI interface is shown in Fig. 2. 3.

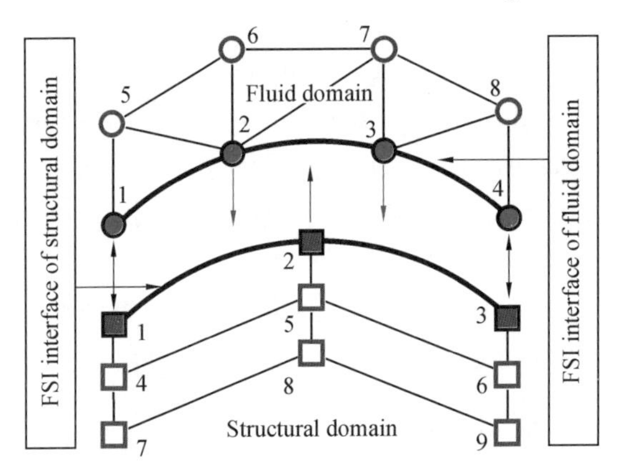

Fig. 2. 3 Data transfer at the FSI interfaces

The displacement data of nodes at the FSI interface of the fluid domain is obtained by interpolating the displacement data of nodes at the FSI interface of the surrounding structural domain. The displacement data of nodes in the fluid domain that are not at the FSI interface is calculated using ANSYS CFX. As shown in Fig. 2. 3, the displacement data of node 2 at the FSI interface of the fluid domain is obtained by interpolating the displacement data of nodes 1 and 2 at the FSI interface of the structural domain.

Based on dynamic conditions, the fluid stress on the FSI interface is first integrated by

$$F(t) = \int h^{s} \tau_{f} \mathrm{d}S \tag{2.14}$$

where h^s is the node displacement of the structural domain.

Then, the concentrated force is applied to the nodes of the FSI interface of the structure domain.

The stress data of nodes at the FSI interface of the structural domain is obtained by interpolating the stress data of nodes at the FSI interface of the surrounding fluid domain.

As shown in Fig. 2. 3, the stress data of node 2 at the FSI interface of the structural domain is obtained by interpolating the stress data of nodes 2 and 3 at the FSI interface of the fluid domain. The stress data of nodes 1 and 3 at the FSI interface of the structural domain are equal to those of nodes 1 and 4 at the FSI interface of the fluid domain. According to Eq. (2. 14), the force of fluid acting on node 2 of the structural domain includes the stress of nodes 1, 2 and 3 of the structural domain, and the stress of fluid is the sum of compressive stress and shear stress, so the solution of the nodes in fluid domain and the solution of the nodes in structural domain and are fully coupled.

2.2 Data Processing

In this book, the heat transfer coefficient of each row of ETBs is calculated based on the steady-state heat flow method[10, 14]. Based on the Newtonian cooling equation, we have

$$\mathrm{d}Q = h_x(T_{\mathrm{w}x} - T_{\mathrm{f}x})\mathrm{d}A \tag{2.15}$$

where Q is the convection heat exchange; h is the local heat transfer coefficient; T is the fluid temperature; A is the heat transfer area; the subscript "x" represents the coordinate.

For the constant heat flow condition, the heat flow density expression is expressed as

$$q = \frac{1}{A}\iint_A q(A)\mathrm{d}A = \frac{Q}{A} = \frac{\mathrm{d}Q}{\mathrm{d}A} \tag{2.16}$$

where q is the heat flow density.

The expression for the local heat transfer coefficient of the ETB is

$$h_x = \frac{\mathrm{d}Q}{\mathrm{d}A(T_{\mathrm{w}x} - T_{\mathrm{f}x})} = \frac{q}{T_{\mathrm{w}x} - T_{\mathrm{f}x}} \tag{2.17}$$

The expression for the heat transfer coefficient of the ETB is

$$h = \frac{q}{\Delta T} \tag{2.18}$$

where the expression of ΔT is

$$\Delta T = \frac{1}{A}\int_A (T_{wx} - T_{fx})\,dA = \bar{T}_w - \bar{T}_f \tag{2.19}$$

$$\bar{T}_w = \frac{1}{n}\sum_{i=1}^{m} t_{w,i} \tag{2.20}$$

$$\bar{T}_f = \frac{1}{2}(T_{in} + T_{out}) \tag{2.21}$$

where m is the number of nodes; T_{in} is the inlet temperature of the fluid; T_{out} is the outlet temperature of the fluid; the subscripts "in" and "out" represent inlet and outlet.

The expression for the Nu of the ETB is

$$Nu = h\frac{dT}{\lambda} \tag{2.22}$$

The average heat transfer performance of the ETBs is reflected by the average h or average Nu, their expressions are

$$Nu_a = \sum_{i=1}^{n} \frac{Nu_i}{n} \tag{2.23}$$

$$h_a = \sum_{i=1}^{n} \frac{h_i}{n} \tag{2.24}$$

where n is the number of ETBs; Nu_a is the average Nusselt number; h_a is the average heat transfer coefficient; the subscript "a" represents average.

To measure the vibration-enhanced heat transfer performance of the ETBs, the performance evaluation criteria PEC has been adopted, which is expressed as[1, 15]

$$\text{PEC} = \frac{Nu_{av}/Nu_a}{(f_v/f)^{1/3}} \tag{2.25}$$

where the expression of f is

$$f = \frac{2\Delta p d}{L_f \rho u_a^2} \tag{2.26}$$

$$\Delta p = p_{in} - p_{out} \tag{2.27}$$

where L_f is the length of the fluid domain; u_a is the average flow velocity; p_{in} and p_{out} respectively represent the inlet and outlet pressures of the fluid domain; the subscript "v" represents vibration.

To compare the overall heat transfer performance of two types of heat exchangers, the factor JF (dimensionless numbers related to Colburn factor J and Fanning friction factor F) is used, and the expressions are[16]

$$\text{JF} = \left(\frac{J_{\text{I}}}{J_{\text{II}}}\right) / \left(\frac{F_{\text{I}}}{F_{\text{II}}}\right)^{1/3} \tag{2.28}$$

$$J = \frac{Nu_a}{RePr^{1/3}} \tag{2.29}$$

$$F = \frac{2\Delta P}{\rho u_a^2} \times \frac{dT}{L_f} \tag{2.30}$$

where J is a dimensionless number related to the Colburn factor; F is a dimensionless number related to the Fanning friction factor; the subscripts " Ⅰ " and " Ⅱ " represent two types of heat exchangers Ⅰ and Ⅱ.

The JF factor involves the factors J (reflecting the heat transfer performance of the ETBs) and F (reflecting the resistance of the fluid), which is used to compare the overall heat transfer performance of two types of heat exchangers. Based on Eq. (2.27), if JF>1, it indicates that the overall heat transfer performance of the heat exchanger Ⅰ is bigger than that of the heat exchanger Ⅱ. On the contrary, if JF <1, it indicates that the overall heat transfer performance of the heat exchanger Ⅰ is smaller than that of the heat exchanger Ⅱ.

2.3 Numerical Method Validation

2.3.1 Verification of Vibration Calculation Results

In order to verify the correctness of the numerical calculation method and the accuracy of the vibration calculation results in this book, based on the structural parameters of the experimental ETB in the literature [17], the corresponding numerical analysis model is established and the vibration frequency and acceleration of the two monitoring points in the x-direction (vertical direction) are compared.

It should be noted that the ETB used for the study in this book is an improvement on the ETB tested in the literature [17], and a comparison between the two is shown in Fig. 2. 4. As shown in Fig. 2. 4, the structural dimensions of both ETBs are identical; compared with the ETB in the literature [17], the ETB used in this book differs only in the bends of the innermost and outermost copper tubes. In addition, compared with the ETB in the literature [17], the ETB used in this book has the following characteristics.

(1) Reduced bending of the innermost bundle and outermost copper tube, which reduces the stress concentration and makes processing and manufacturing easier[18].

(2) The inherent frequency of the ETB is reduced, which makes it easier to

achieve vibration under the induction of low velocity fluid, thus improving its heat transfer characteristics under low velocity flow conditions[19-20].

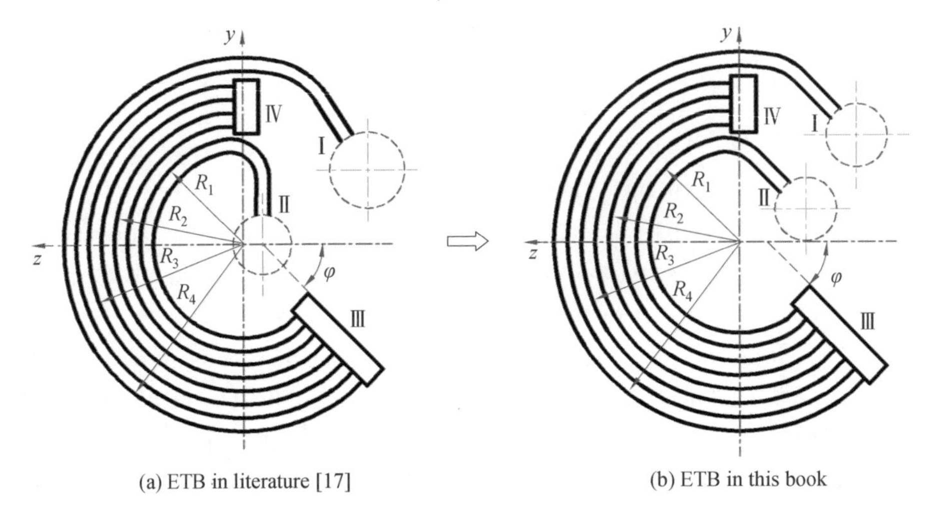

Fig. 2.4 Comparison of two ETBs

Based on the ETB shown in Fig. 2.4(a) for the test, a vibration test rig for a single-row ETB induced by the shell-side fluid was constructed in the literature [21]. The vibration response signals of the two stainless steel connectors were collected using accelerometers, and the vibration response of the single-row ETB was tested at different shell-side inlet velocities (u_{in} = 0.2–0.4 m/s). The dimensions (including shell-side fluid inlet and outlet tubes, tube-side fluid inlet and outlet tubes, cylinder, upper and bottom shell covers, etc.) are consistent with the heat exchanger studied in this book. The ETB under test is located in the middle of the heat exchange, and the sensors are mounted on the two stainless steel connectors Ⅲ and Ⅳ, labeled as A and B.

Based on the shell-side fluid domain of the single-row ETB for the test, the corresponding geometric model is established, and the fluid domain partitioning, segmented fluid domain meshing and mesh assembly are carried out sequentially to form the actual shell-side fluid domain mesh of the single-row ETB, as shown in Fig. 2.5. For the specific process, please refer to "4.1.2 Mesh division strategy".

It should be noted that because this model is different from the six-row ETB shell-side fluid domain model, a different mesh partition is used. The tetrahedral mesh is used for the bottom shell cover domain, and the hexahedral mesh is used for the rest of the partitioned domains, with a total of 782,486 nodes for the overall shell-side fluid domain mesh. The meshing of the structural domain of the single-row ETB also includes a hexahedral mesh (copper bend tubes) and a tetrahedral mesh (stainless steel connectors), with a total of 54,792 nodes.

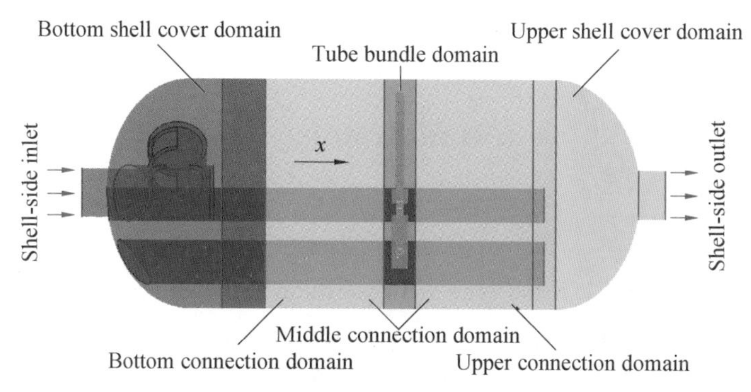

Fig. 2.5 Shell-side fluid domain model for a single row of ETB

During the calculation, for the settings related to the structural domain, the fluid domain, as well as the computation time and time step, refer to "4.2.1 Shell-side Outlet Temperature Distribution". In addition, after the trial calculation, the mesh of the single-row ETB shell-side fluid domain meets the independence requirement.

In order to verify the correctness of the numerical calculation method and the accuracy of the vibration calculation results in this book, the vibration frequency (f_x) and acceleration (a_x) of the monitoring points on the two stainless steel connectors of the ETB in the x-direction (vertical direction) are calculated for different shell-side inlet velocities (u_{in} = 0.2 m/s and 0.4 m/s), and the calculated results are compared with the experimental data, as shown in Table 2.1.

Table 2.1 Comparison of numerical results with experimental data

u_{in} /(m·s^{-1})	Monitoring points	f_x			a_x		
		Numerical results/Hz	Experimental data/Hz	Relative Errors	Numerical results /(m·s^{-2})	Experimental data /(m·s^{-2})	Relative Errors
0.2	A	28.8	28.0	2.78%	0.097	0.088	9.28%
	B	29.5	29.0	1.69%	0.090	0.082	8.89%
0.4	A	28.6	28.0	2.10%	0.377	0.340	9.81%
	B	29.8	29.0	2.68%	0.406	0.375	7.58%

Note: Relative error = |Numerical result − Experimental data|/Numerical result ×100%.

As can be seen from Table 2.1, the numerical calculation results are basically consistent with the experimental data, and the relative error of a_x is larger compared with the relative error of f_x, but the maximum relative error is less than 10%. Therefore, it is reasonable to use the numerical method in this book to calculate the vibration results.

2.3.2 Verification of Heat Transfer Calculation Results

Considering the accuracy of the heat transfer calculation results, a quantitative comparison was made with the experimental data of Salimpour[3] at the same thermal and geometric properties. If the error between the heat transfer calculation results and the experimental results is very small, it indicates that the heat transfer calculation results calculated by using the numerical calculation method in this book are reasonable.

During the comparison, the geometric parameters of "Heat exchanger 1" in literature [3] are selected. In the range of u_{in} from 0.019 kg/s to 0.136 kg/s, the shell and coiled tube heat exchanger was simulated by the numerical calculation method of this study. The corresponding results are shown by 10% error bars in Fig. 2.6(a). Compared with experimental data in literature [3], the deviation of Nu is 1.37%–8.66%. These small differences are attributed to the model simplification and the inevitable measurement errors (the measurement error is the error between the data extracted from the graph in literature [3] and the real data in the graph).

In addition to comparing with experimental data, this study also compared with the simulation data of Duan et al.[22] The structure, material and flow field of the ETB studied in this book are consistent with those simulated in the literature [22]. The heat exchanger with tube row spacing of 50 mm in literature [22] is selected as the research object. The corresponding results are shown by 10% error bars in Fig. 2.6(b). Compared to the simulated data, the deviation of heat transfer coefficients is 1.33%–4.87%.

The calculation needs to be carried out under the condition that the numerical calculation method is accurate. The above comparisons indicate that the heat transfer calculation results calculated using the numerical method in this book are effective.

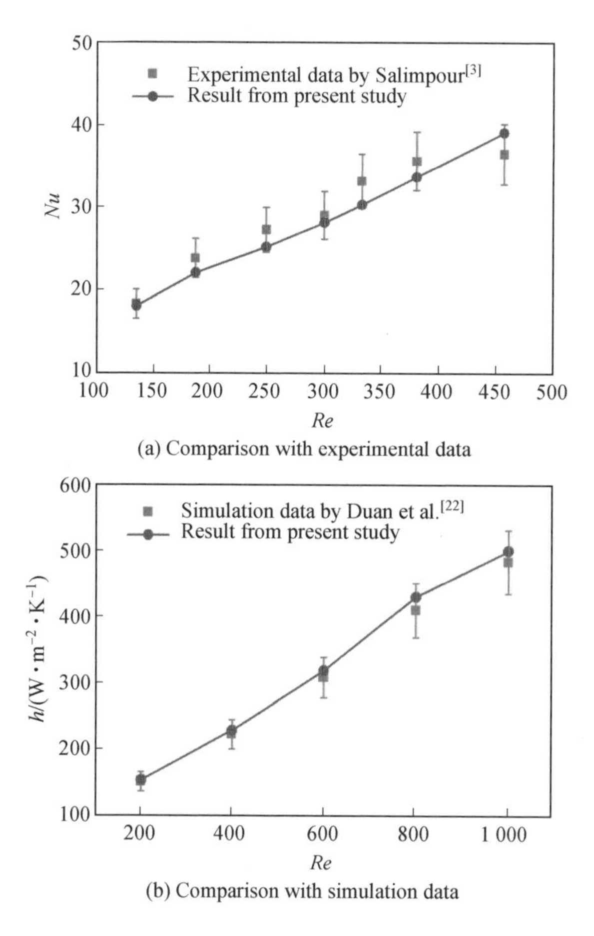

Fig. 2.6 Method validation of the present simulation with literatures

References

[1] LU G F, ZHOU G B. Numerical simulation on performances of plane and curved winglet type vortex generator pairs with punched holes [J]. International journal of heat and mass transfer, 2016, 102: 679-690.

[2] JI J D, PAN Y L, DENG X, et al. Research on the heat transfer performance of an improved elastic tube bundle heat exchanger under fluid-induced vibration [J]. Case studies in thermal engineering, 2023, 49: 103184.

[3] SALIMPOUR M R. Heat transfer coefficients of shell and coiled tube heat exchangers [J]. Experimental thermal and fluid science, 2009, 33(2): 203-207.

[4] JI J D, LI F Y, SHI B J, et al. Analysis of the effect of baffles on the vibration and heat transfer characteristics of elastic tube bundles [J]. International communications in heat and mass transfer, 2022, 136: 106206.

[5] ZOU Q F, DING L, HE H Y, et al. Investigation on heat transfer and flow-induced vibration of three cylinders in equilateral triangle arrangement [J]. International communications in heat and mass transfer, 2022, 136: 106177.

[6] JI J D, NI X W, SHI B J, et al. Influence of deflector direction on heat transfer capacity of spiral elastic tube heat exchanger [J]. Applied thermal engineering, 2024, 236: 121754.

[7] JI J D, LU Y, SHI B J, et al. Numerical research on vibration and heat transfer performance of a conical spiral elastic bundle heat exchanger with baffles [J]. Applied thermal engineering, 2023, 232: 121036.

[8] JI J D. Study on flow-induced vibration of elastic tube bundle with shell-side distributed pulsating flow in heat exchanger [D]. Jinan: Shandong University, 2016.

[9] DUAN D R. Study on flow-induced vibration of elastic tube bundle in heat transfer enhancement and fatigue strength [D]. Jinan: Shandong University, 2017.

[10] YAN K. A study on the vibration and heat transfer characteristics of conical spiral tube bundle in heat exchanger [D]. Jinan: Shandong University, 2012.

[11] JI J D, GE P Q, BI W B. Numerical analysis of shell-side flow-induced vibration of elastic tube bundle in heat exchanger [J]. Journal of hydrodynamics, 2018, 30(2): 249-257.

[12] GAO R M. Study on fluid-induced vibration-enhanced heat transfer characteristics of helical elastic tube bundle in heat exchanger [D]. Huainan: Anhui University of Science & Technology, 2022.

[13] HAN Z Z, WANG J, LAN X P. FLUENT-Fluid engineering simulation calculation example and application [M]. Beijing: Beijing Institute of Technology Press, 2010.

[14] CHENG L. Principle and application of elastic tube bundle heat exchanger [M]. Beijing: Science Press, 2014.

[15] WEBB R L. Performance evaluation criteria for use of enhanced heat transfer surfaces in heat exchanger design [J]. International journal of heat and mass transfer, 1981, 24: 715-726.

[16] HSIEH C T, JANG J Y. Parametric study and optimization of louver finned-tube

heat exchangers by Taguchi method [J]. Applied thermal engineering, 2012, 42: 101-110.

[17] CHENG F, TIAN M C. Experimental research on heat transfer and resistance characteristics of heat exchanger of elastic tube bundle [J]. Research and exploration in laboratory, 2007, 26(11): 23-25.

[18] JI J D, GE P Q, BI W B, et al. Shell-side flow-induced vibration of elastic tube bundle based on different tube combinations [J]. Journal of Xi'an Jiaotong University, 2018, 52(3): 69-75.

[19] JI J D, GE P Q, BI W B. Numerical analysis on flow-induced vibration responses of elastic tube bundle [J]. Journal of vibration and shock, 2016, 35 (6): 80-84.

[20] JI J D, GE P Q, BI W B. Numerical analysis on the combined flow induced vibration responses of elastic tube bundle in heat exchanger [J]. Journal of Xi'an Jiaotong University, 2015, 49(9): 24-29.

[21] ALI M, ZEITOUN O, NUHAIT A. Forced convection heat transfer over horizontal triangular [J]. International journal of thermal sciences, 2011, 50 (1): 106-114.

[22] DUAN D R, GE P Q, BI W B, et al. Numerical investigation on synthetical performance of heat transfer of planar elastic tube bundle heat exchanger [J]. Applied thermal engineering, 2016, 109: 295-303.

Chapter 3 Vibration of ETBs Induced by Uniform Shell-side Fluid

The vibration responses of a single-row ETB induced by the uniform shell-side fluid is the basis for conducting the vibration responses of a multi-row ETBs induced by the actual shell-side fluid and distributed pulsating fluid. Changes in the flow parameters of the uniform shell-side fluid, the structure parameters of the ETB and the number of tube rows affect the vibration responses of the ETB induced by the uniform shell-side fluid.

In this chapter, the geometric model of the structural domain and the shell-side and tube-side fluid domains of the single-row ETB were established, and the structural domain and the shell and tube-side fluid domains were meshed respectively. Based on the sequential solution method for bi-directional FSI calculation, the effects of flow velocity of uniform shell-side fluid, the structural parameters of the ETB (outer section diameter and wall thickness) and the tube-side fluid on the vibration responses of the single-row ETB were investigated. In addition, based on the geometric modes and meshes of the single-row of ETBs and their shell-side fluid domain, the geometric model and mesh of multi-row ETBs and their shell-side fluid domain were formed. Furthermore, numerical analysis was conducted on the vibration responses of ETBs with different tube rows under uniform shell-side fluid induction.

3.1 Geometric Model of Single-row ETB and Its Mesh

3.1.1 Single-row ETB and Its Shell-side Fluid Domain

The schematic diagram of the structure of the single-row ETB and its shell-side and tube-side fluid domains are shown in Fig. 3.1.

The ETB consists of four copper bend tubes (bend radius—R_1, R_2, R_3, R_4; outer section diameter—d; wall thickness—δ) and two stainless steel connectors (Ⅲ, Ⅳ). Ⅰ and Ⅱ are fixed ends, Ⅲ and Ⅳ are free ends, and φ is the position angle of connector Ⅲ. The tube-side fluid flows in from the port at Ⅰ and out from the port

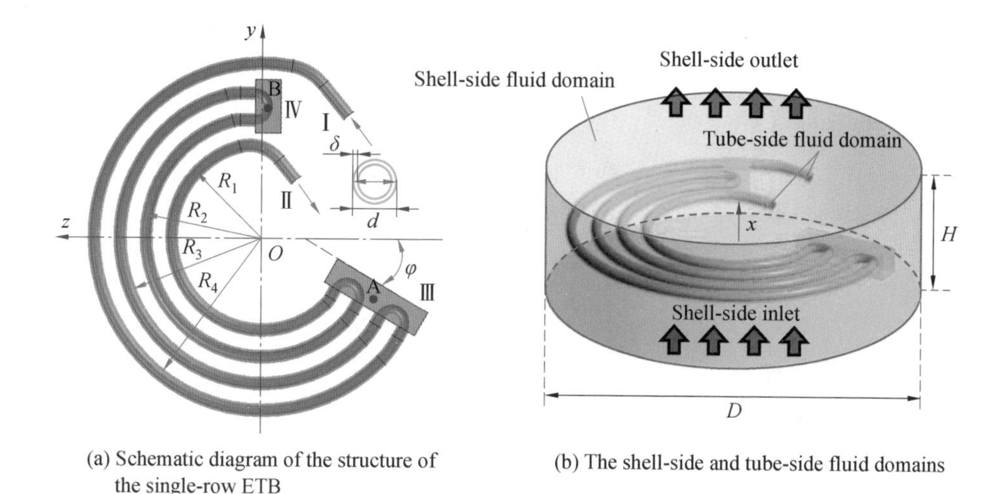

(a) Schematic diagram of the structure of the single-row ETB

(b) The shell-side and tube-side fluid domains

Fig. 3. 1 Schematic diagram of the structure of the single-row ETB and its shell-side and tube-side fluid domains

at Ⅱ; the fluid in the uniform shell-side fluid domain flows in from the bottom and out from the top, and x is the flow direction[1-2]. During the calculation, the diameter of the shell-side fluid domain $D=300$ mm and the height $H=100$ mm. In Fig. 3. 1, points A and B are two monitoring points established on the two stainless steel connectors Ⅲ and Ⅳ to monitor the vibration response of the ETB under fluid induction.

In order to analyze the vibration responses of the ETB under the uniform shell-side and/or tube-side fluid domain induction, the specific structural parameters of the ETB in the calculation process are shown in Table 3. 1, and the specific material properties of the copper bend tubes and the stainless steel connectors are shown in Table 3. 2.

Table 3. 1 Specific structural parameters of the ETB in the calculation process

Structural parameters	Value
Bend radius R_1, R_2, R_3, R_4/mm	70, 90, 110, 130
Stainless steel connector Ⅲ /mm	80×20×20
Stainless steel connector Ⅳ /mm	40×20×20
Tube outer diameter d /mm	7–12.0
Tube wall thickness δ /mm	1.0–2.0
Position angle φ /(°)	30

Table 3.2 Specific material properties of the copper bend tubes and the stainless steel connectors

Structure	Properties	Value
Copper bend tube	Density ρ /(kg · m^{-3})	8,900
	Elastic modulus E /Pa	1.29×10^{11}
	Poisson's ratio υ	0.33
Stainless steel connector	Density ρ/(kg · m^{-3})	7,800
	Elastic modulus E/Pa	2.10×10^{11}
	Poisson's ratio υ	0.30
	Stainless steel connector Ⅲ/mm	80×20×20
	Stainless steel connector Ⅳ/mm	40×20×20

3.1.2 Boundary Conditions

The boundary conditions of the structural domain are set as follows.

(1) The cross sections at the two fixed ends Ⅰ and Ⅱ are set as "Fixed Support".

(2) The outer and inner surfaces of the ETB are set as "Fluid Solid Interface".

(3) The direction of gravitational acceleration (Standard Earth Gravity) is set as the x-direction, and its value is 9.806 6 m/s^2.

The boundary conditions of the fluid domain are set as follows.

1. For the Uniform Shell-side Fluid Domain

(1) Set the bottom inlet boundary type as "Inlet", given the inlet water velocity.

(2) Set the top exit boundary type as "Outlet", given the relative static pressure of 0 Pa.

(3) The inner surface of the shell-side fluid domain is set as "Fluid Solid Interface", which corresponds to the outer surface of the ETB.

(4) The side wall of the shell-side fluid domain is set a set to "Wall", and the boundary details included: the "Mesh Motion" option is set as "Stationary", and the "Heat Transfer" option is set as "Adiabatic".

2. For the Tube-side Fluid domain

(1) The inlet boundary type is "Inlet" at Ⅰ, and the inlet water velocity is

given.

(2) The outlet boundary type is "Outlet" at Ⅱ, and the relative static pressure is given as 0 Pa.

(3) The outer surface of the tube-side fluid domain is set as "Fluid Solid Interface", which corresponds to the inner surface of the ETB.

In this calculation, the total calculation time of the fluid domain is 1.2 s and the time step is 0.001 s.[3-4] The total calculation time and time step settings of the structural domain are consistent with those of the fluid domain.

3.1.3 Mesh and Independence Analysis

Fig. 3.2 shows the structural domain mesh of the ETB, which is divided using the Workbench platform's mesh division module. It can be seen that the structural domain mesh contains a tetrahedral mesh (stainless steel connector) and a hexahedral mesh (copper bend tube) with a minimum mesh quality of 0.42, a total of 9,366 elements and 54,777 nodes.

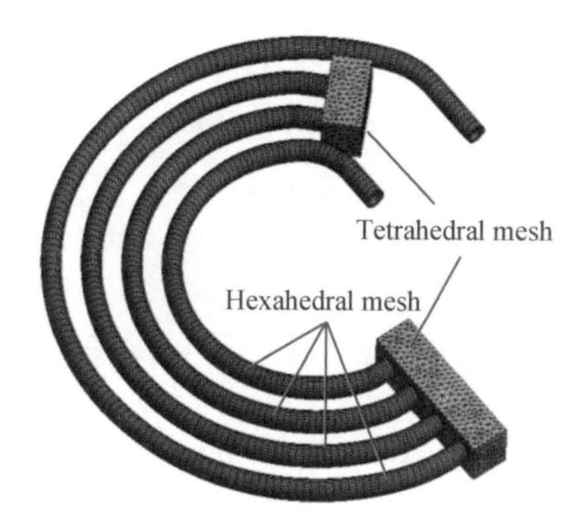

Fig. 3.2 Schematic diagram of the structural domain mesh

Fig. 3.3 shows the meshes of the tube-side fluid domain and the uniform shell-side fluid domain, both of which are divided by the meshing software ICEM, and both are hexahedral meshes.

As shown in Fig. 3.3, for the tube-side fluid domain, the minimum mesh quality is 0.74, the element number is 52,940, the node number is 47,880. For the uniform shell-side fluid domain, the minimum mesh quality is 0.50, the element number is 329,708, the node number is 309,305. In order to increase the mesh density in the area around the ETB, boundary layer meshes are applied at the near wall of both shell-side and tube-side fluid domains. Among them, a 4-layer boundary layer mesh

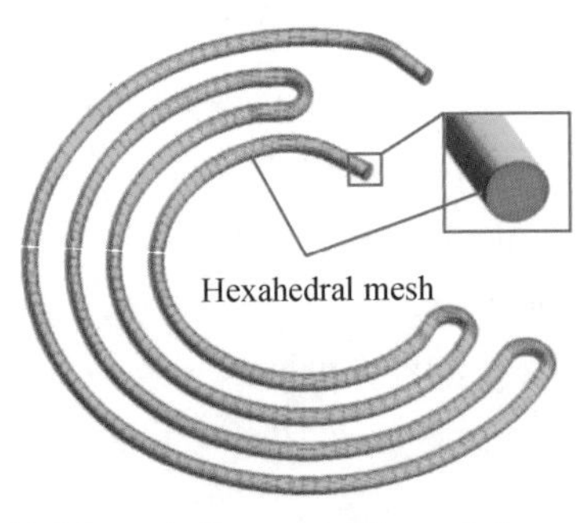

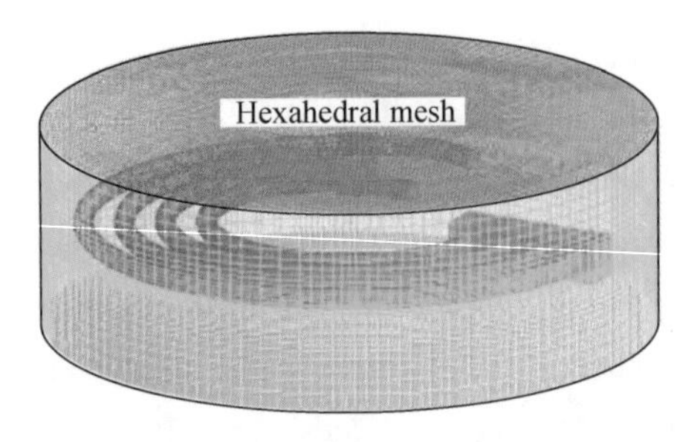

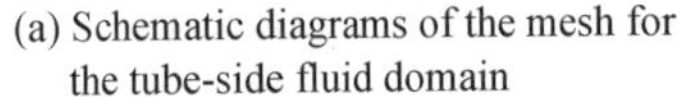
(a) Schematic diagrams of the mesh for the tube-side fluid domain

(b) Uniform shell-side fluid domain

Fig. 3.3 Schematic diagrams of the mesh for the tube-side fluid domain and the uniform shell-side fluid domain

is applied at the near wall of the tube-side, and a 6-layer boundary layer mesh is applied at the near wall of the shell-side[5].

The mesh independence analysis is carried out for the dominant frequency ($f_{x\text{-main}}$) and amplitude (A_x) of the vibration at monitoring point B in the x-direction (longitudinal direction) when the coupling between the shell-side and tube-side fluids induces the ETB vibration. For the trial calculation, the ETB cross-sectional outer diameter $d = 10$ mm, wall thickness $\delta = 1.5$ mm, shell-side and tube-side fluid medium are water, and the flow velocity is 0.8 m/s. The standard $k-\varepsilon$ model is used for the calculation, and the HP Z800 workstation with 8 cores is used for parallel operation.

Table 3.3 shows the comparison of vibration dominant frequency ($f_{x\text{-main}}$) and amplitude (A_x) of monitoring point B in the x-direction (longitudinal direction) for the three mesh division schemes. In Table 3.3, the number of nodes refers to the total number of nodes in the shell-side and tube-side fluid domains and the structural domain. Case 2 refers to the mesh strategy mentioned above, and cases 1 and 3 are obtained by reducing or increasing the mesh density of the shell-side, tube-side fluid domain and structural domain based on case 2.

Table 3.3 Comparison of calculation results under different mesh schemes

Case	Nodes	Calculation results		Relative errors/%		Calculation
		$f_{x\text{-main}}$/Hz	A_x/mm	$f_{x\text{-main}}$	A_x	time/h
1	301,826	17.6	0.022	3.83	15.38	18.92
2	432,365	18.3	0.026	—	—	22.65
3	976,209	18.8	0.024	2.73	7.69	48.08

Note: The relative error is calculated based on the calculation results of case 2.

As can be seen from Table 3.3, when the number of meshes is increased (case 3), the impact on the calculation results is not significant, and the maximum relative error is only 7.69%, but the calculation time is about 2.12 times of the calculation time of case 2. In addition, when the number of meshes is decreased (case 1), the impact on the calculation results is greater, the maximum relative error is up to 15.38%, and the calculation time is about 0.84 times of the calculation time of case 2. Based on the consideration of both computational accuracy and computational efficiency, the mesh of case 2 is chosen as the mesh for calculation.

3.2 Vibration Analysis of Single-row ETB

3.2.1 Modal Analysis

In order to facilitate the analysis of the vibration responses of the ETB under the induction of uniform shell-side fluid, the inherent frequencies and vibration modes of the ETB are calculated first. Table 3.4 shows the first six orders of inherent frequencies and the corresponding vibration modes of the ETB.

Table 3.4 Inherent frequency and vibration pattern of the ETB

Order	1	2	3	4	5	6
Inherent frequency	23.10	24.05	24.98	33.18	40.72	54.75
Vibration type	In-plane	Out-plane	Out-plane	In-plane	Out-plane	In-plane

In Table 3.4, the "in-plane" vibration refers to the transverse vibration in the plane of the ETB, and the "out-plane" vibration refers to the longitudinal vibration perpendicular to the plane of the ETB, which is described in detail in the literatures [6–8]. From Table 3.4, it can be seen that the vibration model of the ETB is divided into in-plane vibration and out-plane vibration. The inherent frequencies corresponding to in-plane vibration are 23.10 Hz, 33.18 Hz and 54.75 Hz, i.e., the first, fourth and sixth order inherent frequencies; the inherent frequencies corresponding to out-plane vibration are 24.05 Hz, 24.98 Hz and 40.72 Hz, i.e., the second, third and fifth order inherent frequencies.

3.2.2 Effect of Velocity

Based on different inlet velocities (u_{in}), the vibration response of a single-row ETB induced by a uniform shell-side fluid domain is investigated. Fig. 3.4 shows the variation of vibration displacements (S) of monitoring points A and B in the x, y and z directions with the calculated time (t) for u_{in} = 0.6 m/s, 0.8 m/s and 1.0 m/s. During the calculation, the ETB cross-sectional outside diameter d = 10 mm, wall thickness δ = 1.5 mm, tube-side no fluid medium.

From Fig. 3.4, it can be concluded the followings.

(1) The velocity of uniform shell-side fluid domain has a large influence on the vibration response of the ETB, and the fluctuation of vibration displacement increases with the increase of inlet velocity.

(2) Due to the influence of gravity of the ETB, when the inlet velocity is low (0.6 m/s, 0.8 m/s), the displacement curve in the x-direction of the monitoring point has an obvious "double-peak" phenomenon. When the inlet velocity is higher (1.0 m/s), the "double-peak" phenomenon disappears, indicating that the impact of fluid is enhanced, which weakens the influence of gravity on the vibration response.

(3) Due to the influence of fluid impact and gravity of the ETB, the vibration equilibrium position in the x-direction of the monitoring point shifts upward with the increase of the inlet velocity. Take the equilibrium position of monitoring point A in the x-direction as an example: the equilibrium position is −0.115 mm when u_{in} = 0.6 m/s, and the equilibrium position is 0.465 mm when u_{in} = 1.0 m/s.

In order to further analyze the vibration response of the ETB at different velocities, the fast Fourier transform (FFT) is performed on the displacement time curves shown in Fig. 3.4 for different inlet velocities. Fig. 3.5, Fig. 3.6 and Fig. 3.7 show the vibration displacement spectra of monitoring points A and B on the two stainless steel connections under different inlet velocities, respectively.

From Fig. 3.5, Fig. 3.6 and Fig. 3.7, it can be seen as follows.

(1) When the inlet velocity is low (0.6 m/s, 0.8 m/s), the vibration of the two monitoring points in the x-direction has obvious second harmonic frequency, which echoes the "double-peak" phenomenon observed in Fig. 3.4.

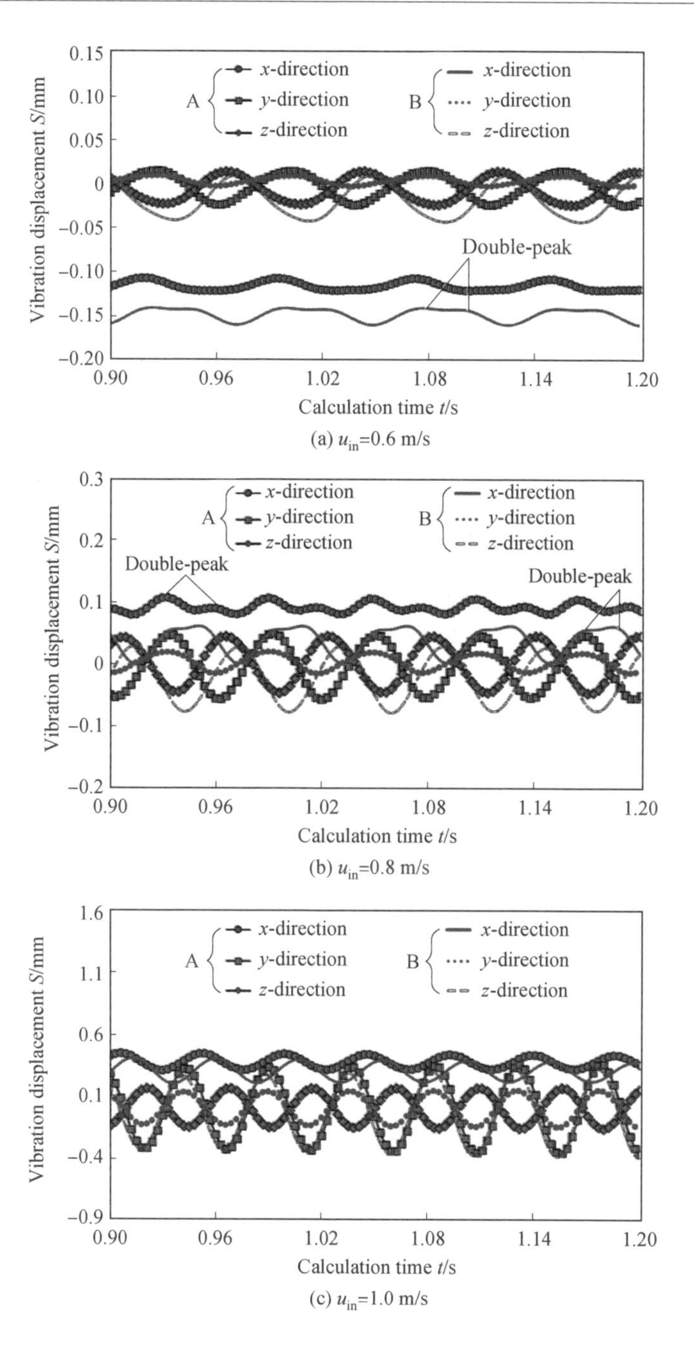

(a) u_{in}=0.6 m/s

(b) u_{in}=0.8 m/s

(c) u_{in}=1.0 m/s

Fig. 3. 4 Variation of vibration displacement of monitoring point with calculation time at different inlet velocities (d = 10 mm, δ = 1.5 mm)

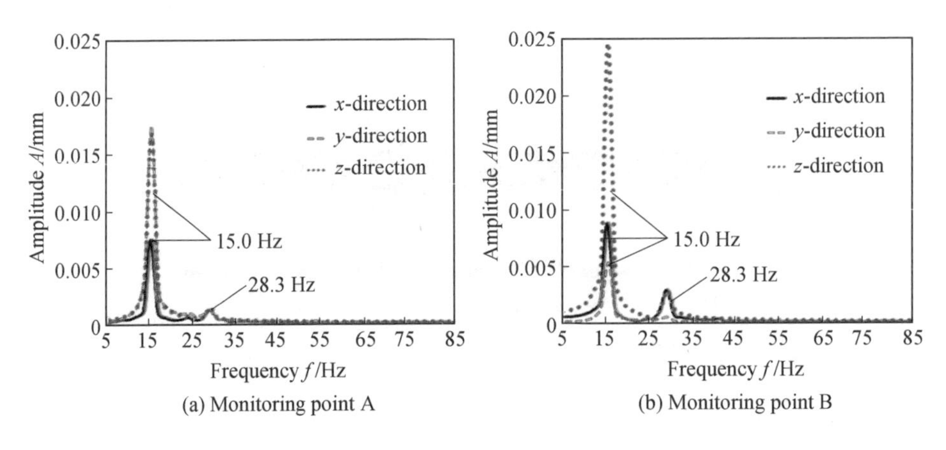

Fig. 3.5 Vibration displacement spectrum of the monitoring point (u_{in} = 0.6 m/s)

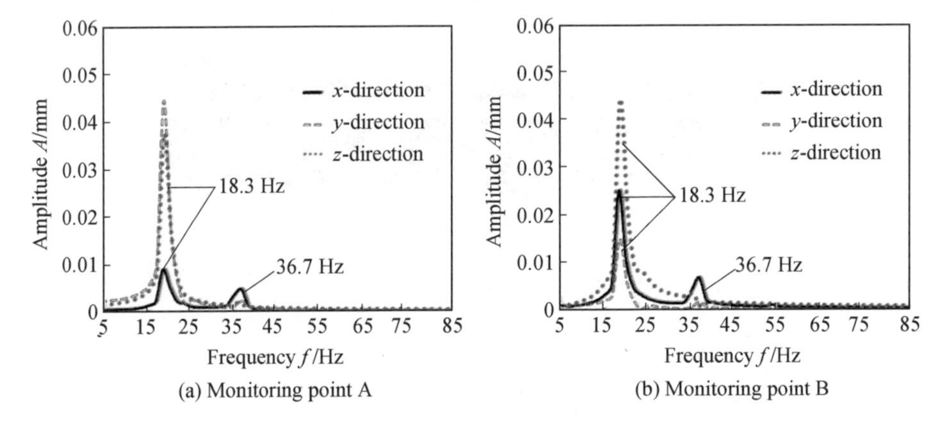

Fig. 3.6 Vibration displacement spectrum of the monitoring point (u_{in} = 0.8 m/s)

(2) Due to the induction of the shell-side fluid, the magnitudes of the main frequency (f_{main}) and harmonic frequency (f_{harm}) are the same in all directions at the two monitoring points at the same velocity.

(3) The amplitude of monitoring point B is higher when u_{in} = 0.6 m/s and 0.8 m/s, and the amplitude of monitoring point A is slightly higher when u_{in} = 1.0 m/s. The amplitude of monitoring point B is higher when u_{in} = 1.0 m/s. The amplitude of monitoring point A is higher when u_{in} = 1.0 m/s.

(4) As the inlet velocity increases, the amplitude and frequency of the monitoring point in the x-, y- and z-directions increase. The peak of amplitude occurs when u_{in} = 1.0 m/s, which is due to the fact that the principal frequency of flow-induced ETB vibration (23.3 Hz) is close to the 1st in-plane vibration intrinsic frequency of the ETB (23.1 Hz), i.e., the 1st order intrinsic frequency.

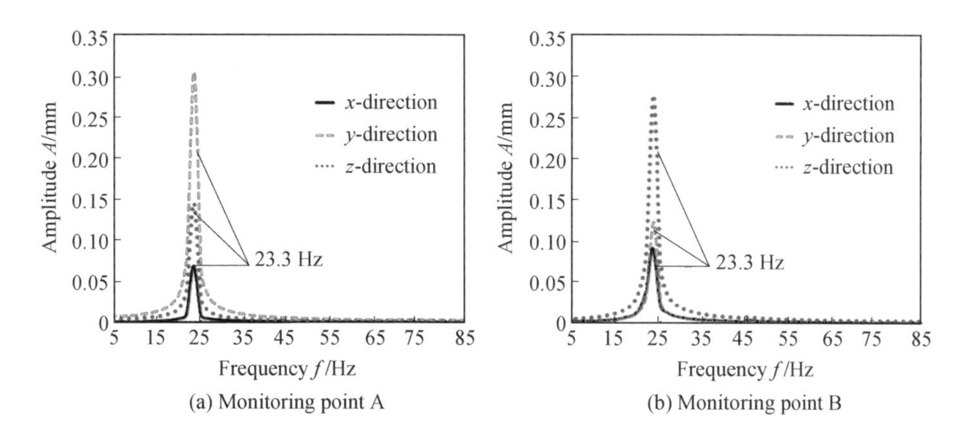

Fig. 3.7 Vibration displacement spectrum of the monitoring point (u_{in} = 1.0 m/s)

(5) The vibration at monitoring point A is dominated by the vibration in the y-direction (highest amplitude), and the vibration at monitoring point B is dominated by the vibration in the z-direction (highest amplitude), which indicates that the vibration is mainly manifested as in-plane vibration.

To further analyze the effect of uniform inlet velocity on the vibration response of the ETB, the vibration response of the ETB is studied for inlet velocities of 0.4 m/s to 1.6 m/s. Fig. 3.8 shows the variation of the main frequency of vibration at the monitoring point with the inlet velocity. Fig. 3.9 shows the variation of the amplitude of the two monitoring points in each direction with the inlet velocity.

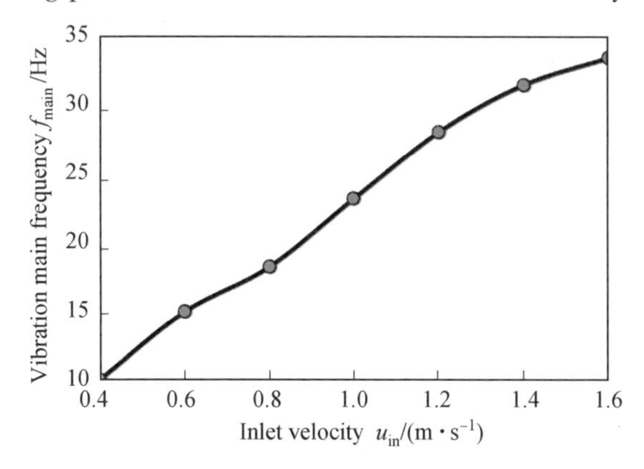

Fig. 3.8 Variation of the main frequency of vibration at the monitoring point with the inlet velocity

From Fig. 3.8, it can be seen that the main frequency of vibration increases with the increase of inlet velocity in the range of parameters calculated in this chapter, and the relationship is approximately linear.

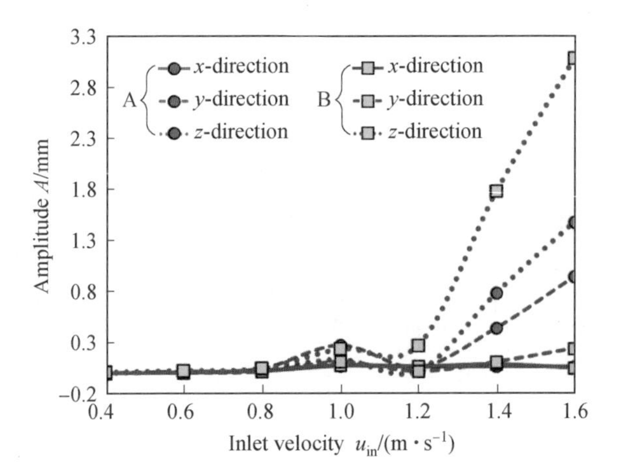

Fig. 3.9 Variation of vibration amplitude with inlet velocity at monitoring points

From Fig. 3.9, the following conclusions can be drawn as follows.

(1) In the parameter range calculated in this book, when the inlet velocity is lower than 0.8 m/s, the amplitude is basically constant with the increase of the inlet velocity, and when the inlet velocity is higher than 0.8 m/s, the amplitude oscillates with the increase of the inlet velocity.

(2) For a certain determined flow velocity, the amplitude in the *z*-direction is significantly higher than that in the *x*- and *y*-directions, which means that the vibration of the monitoring point of the ETB is mainly in the *z*-direction under the induction of uniform shell-side fluid, and the ETB mainly shows in-plane vibration at this time.

(3) When $u_{in} = 1.0$ m/s, there is a peak of amplitude with the change of inlet velocity. This is because at this flow velocity, the dominant frequency of flow-induced vibration (23.3 Hz) is close to the first in-plane vibration intrinsic frequency (23.1 Hz) of the ETB, i.e., the first-order intrinsic frequency.

(4) At $u_{in} = 1.2$ m/s, a trough in the variation of the amplitude occurs. This is because at this shell flow velocity, the principal frequency of flow-induced vibration is far from the intrinsic frequency corresponding to the vibration in the face of the ETB.

(5) When $u_{in} = 1.4$ m/s, the main frequency of flow-induced vibration (31.7 Hz) approaches the 2nd in-plane vibration intrinsic frequency of the ETB (33.18 Hz), which is the 4th order intrinsic frequency, causing the amplitude to increase further.

3.2.3 Effect of Structural Parameters

The influence of the structural parameters of the ETB: the cross-sectional outer diameter d and the wall thickness δ, on the vibration response of the ETB induced by a uniform shell-side fluid domain cannot be ignored. In order to study the effects of these two structural parameters, the vibration response of the ETB is calculated for different cross-sectional outer diameters ($d = 7-12$ mm) and wall thicknesses ($\delta = 1.0-2.0$ mm). During the calculation, the inlet velocity of the uniform shell-side fluid is 0.8 m/s, and there is no fluid medium in the tube-side.

The time plots of the displacement of the two monitoring points in each direction with respect to the calculation time for $d = 8$ mm, $\delta = 1.5$ mm and $d = 12$ mm, $\delta = 1.5$ mm are shown in Fig. 3.10 and Fig. 3.11, respectively.

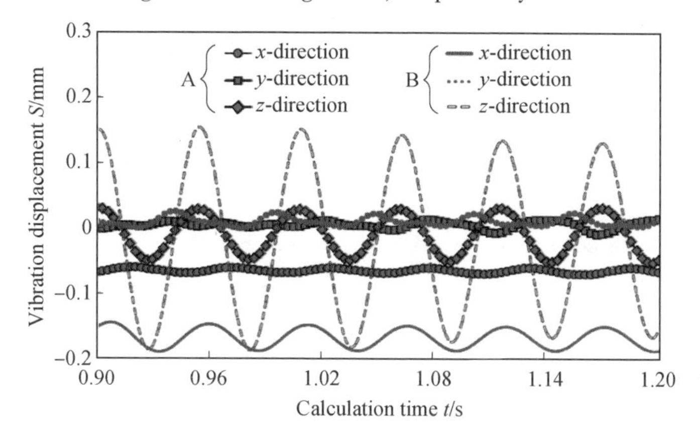

Fig. 3.10 Variation of monitoring point displacement with calculation time (d=12 mm, δ=1.5 mm)

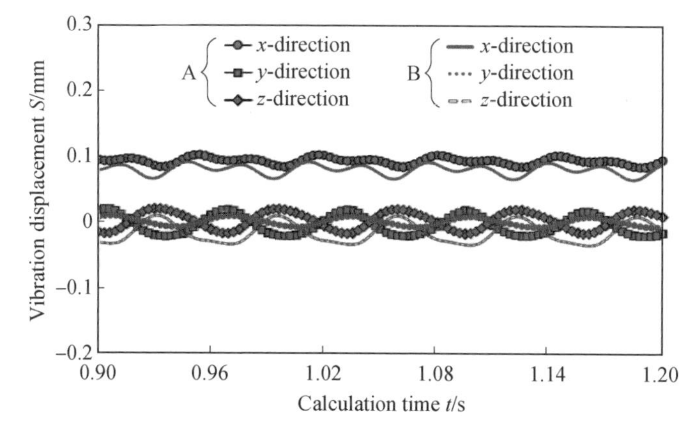

Fig. 3.11 Variation of monitoring point displacement with calculation time (d=8 mm, δ=1.5 mm)

From Fig. 3.10 and Fig. 3.11, combined with $d=10$ mm, $\delta=1.5$ mm shown in Fig. 3.4(b), the following conclusions can be drawn.

(1) The influence of the cross-sectional outer diameterd of the ETB on the vibration response of the monitoring point is significant. When the cross-sectional outer diameter is small ($d=8$ mm), the vibration intensity of the monitoring point in all directions is enhanced, especially the vibration of the monitoring point B in the z-direction is unusually intense.

(2) The cross-section outer diameter d of the ETB has a significant effect on the vibration balance position of the two monitoring points in the x-direction, and the vibration balance position in the x-direction is lower when the section outer diameter is smaller. The vibration balance position of monitoring point A in the x-direction on the stainless-steel connector Ⅲ: when the outer diameter $d=8$ mm, the balance position is -0.072 mm; when the outer diameter $d=12$ mm, the balance position is 0.095 mm.

Fig. 3.12 and Fig. 3.13 show the vibration displacement spectra of the monitoring points A and B on the two stainless steel connectors when the outer diameter of the ETB section $d=8$ mm and $d=12$ mm, respectively.

The following conclusions can be drawn from Fig. 3.12 and Fig. 3.13.

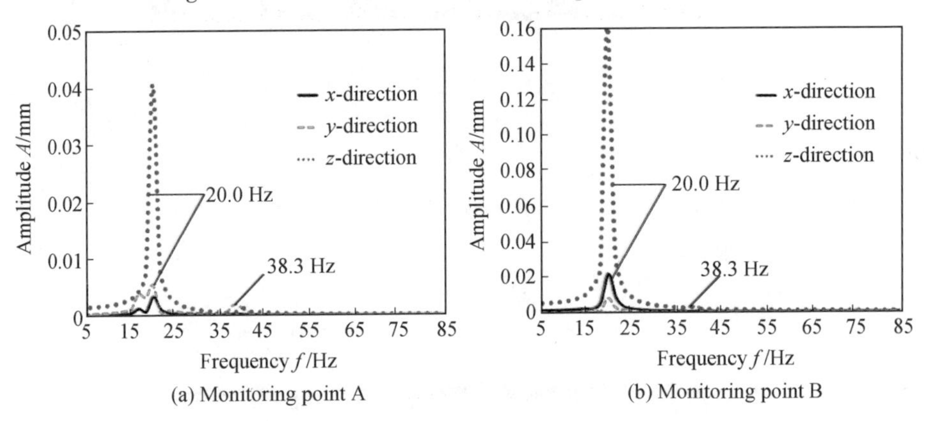

Fig. 3.12 Vibration displacement spectrum of the monitoring point ($d=8$ mm, $\delta=1.5$ mm)

(1) When the ETB cross-sectional outside diameter is small ($d=8$ mm), the monitoring point vibration frequency is larger (main frequency is 20.0 Hz; harmonic frequency is 38.3 Hz), and the amplitude is higher (monitoring points A, B in the z-direction of the main frequency amplitude are 0.041 mm and 0.159 mm), especially the monitoring point B in the z-direction of the vibration is unusually violent. It means that the vibration of stainless steel connector Ⅳ is more violent compared with the

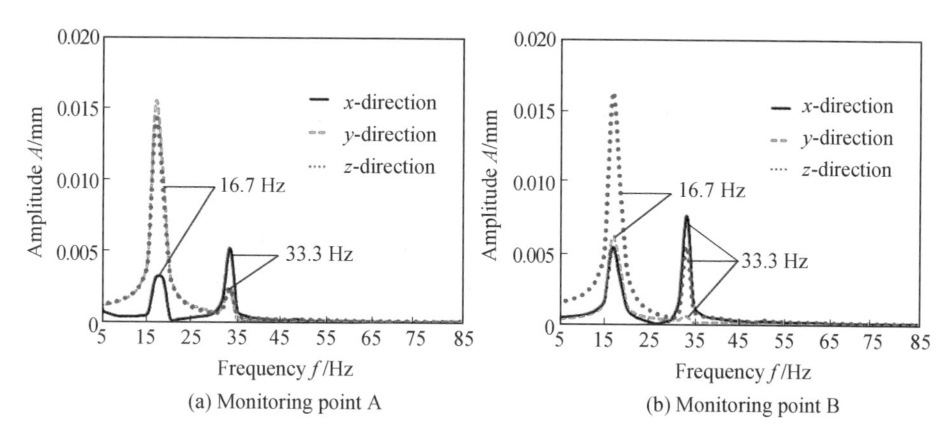

Fig. 3. 13 Vibration displacement spectrum of the monitoring point (d=12 mm, δ=1.5 mm)

vibration of stainless steel connector Ⅲ.

(2) When the ETB cross-sectional outside diameter is larger, the monitoring point vibration frequency is smaller (main frequency is 16.7 Hz; harmonic frequency is 33.3 Hz), and the vibration amplitude is lower (monitoring points A, B in the z-direction of the main frequency amplitude are 0.014 mm and 0.017 mm), the impact of vibration harmonic frequency is more significant, and the second harmonic frequency amplitude in the x-direction of the two monitoring points are higher than its main frequency amplitude (the amplitude of the main and harmonic frequencies of monitoring point A in the x-direction are 0.003 mm and 0.005 mm; the amplitude of the main and harmonic frequencies of monitoring point B in the x-direction are 0.005 mm and 0.008 mm.)

Fig. 3. 14 shows the variation of the vibration main frequency at the monitoring point with the outer diameter of the ETB section. From Fig. 3. 14, it can be seen that the vibration main frequency decreases approximately in a stepwise manner with the increase of the outer diameter of the ETB section within the calculated parameter range.

Fig. 3. 15 shows the variation of the amplitude of the two monitoring points on the stainless-steel connectors Ⅲ and Ⅳ in each direction with the outer diameter of the ETB. From Fig. 3. 15, it can be seen as follows.

(1) The overall trend shows that the vibration intensity (maximum amplitude) of the two monitoring points decreases with the increase of the outer diameter of the ETB cross-section.

(2) The peak amplitude of monitoring point B in the z-direction occurs when the section outer diameter is 8 mm, and the peak amplitude of monitoring point A in the

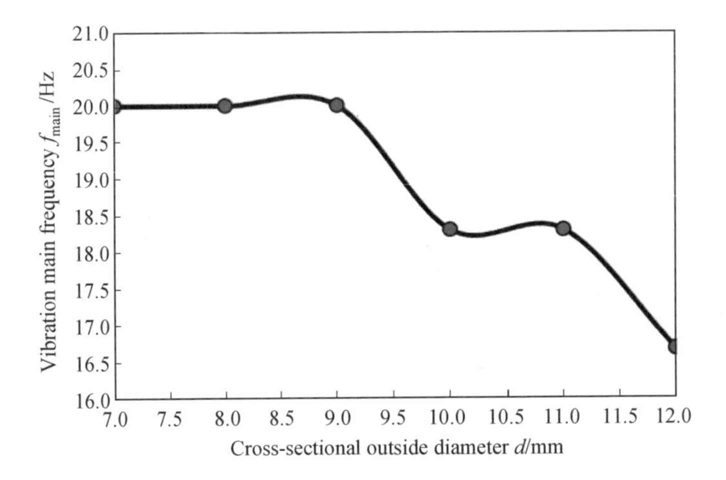

Fig. 3. 14 Variation of the vibration main frequency at the monitoring point with the outer diameter of the ETB section

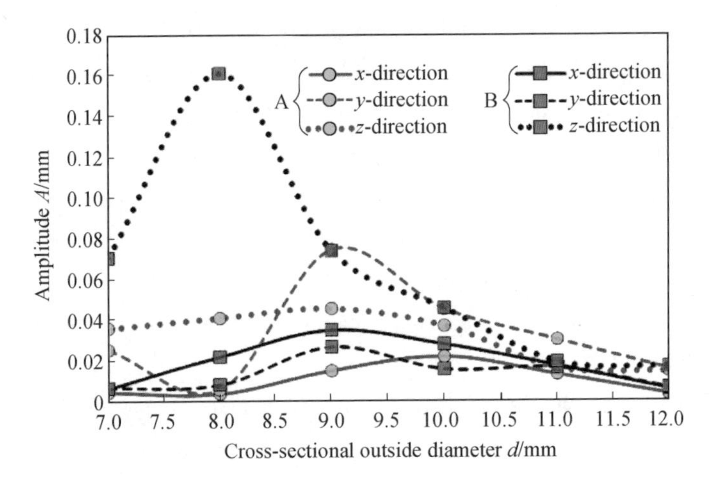

Fig. 3. 15 Variation of the amplitude of the monitoring point with the outer diameter of the ETB section

y-direction occurs when the section outer diameter is 9 mm.

Time plots of the displacement of the two monitoring points in each direction with respect to the calculation time are shown in Fig. 3. 16 and Fig. 3. 17 for $d=10$ mm, $\delta=1.0$ mm and $d=10$ mm, $\delta=2.0$ mm, respectively.

From Fig. 3. 16 and Fig. 3. 17, combined with $d=10$ mm, $\delta=1.5$ mm shown in Fig. 3.4(b), the following conclusions can be drawn.

(1) The wall thickness δ of the ETB has a significant effect on the vibration response of the monitoring point, and the vibration intensity of the monitoring point in

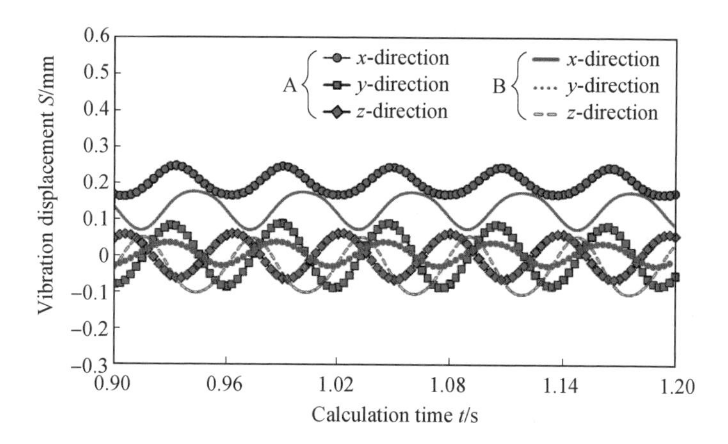

Fig. 3. 16 Variation of monitoring point displacement with calculation time (d=10 mm, δ=1.0 mm)

all directions is weakened when the wall thickness of the ETB is larger.

(2) The wall thickness δ of the ETB has a significant effect on the vibration balance position of the two monitoring points in the x-direction, and the vibration balance position in the x-direction is lower when the wall thickness is larger. The vibration balance position of monitoring point A in the x-direction on the stainless steel connector Ⅲ: when the wall thickness δ = 1.0 mm, the balance position is 0.205 mm; when the wall thickness δ=2.0 mm, the balance position is 0.039 mm.

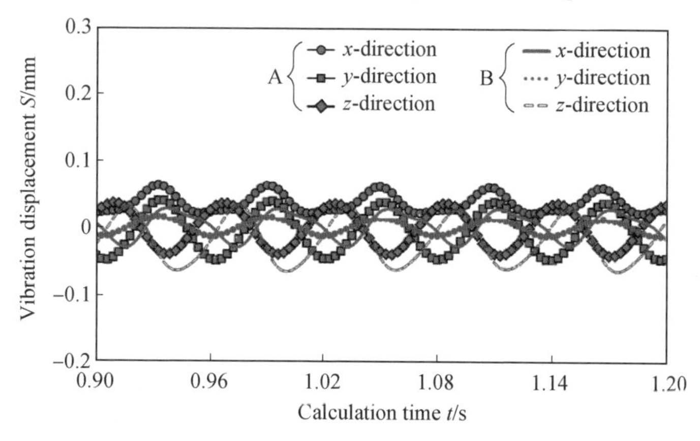

Fig. 3. 17 Variation of monitoring point displacement with calculation time (d=10 mm, δ=2.0 mm)

Fig. 3. 18 and Fig. 3. 19 show the vibration displacement spectra of monitoring points A and B on the two stainless steel connectors for the ETB wall thickness δ = 1.0 mm and δ=2.0 mm, respectively.

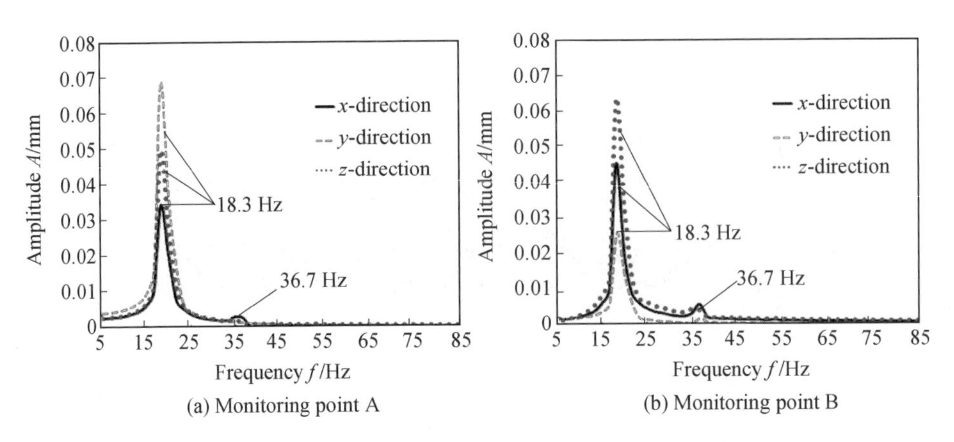

Fig. 3. 18 Vibration displacement spectrum of the monitoring point (d=10 mm, δ=1.0 mm)

From Fig. 3. 18 and Fig. 3. 19, the following conclusions can be drawn as follows.

(1) The change of wall thickness of the ETB does not affect the vibration frequencies of the two monitoring points, but the amplitude of vibration is higher when the wall thickness is smaller. For example, the main frequency of vibration at monitoring point A in the y-direction for different wall thicknesses is 18. 3 Hz. When the wall thickness δ=1.0 mm, the amplitude A=0.069 mm; when the wall thickness δ=2.0 mm, the amplitude A=0.038 mm.

(2) When the wall thickness of the ETB is larger, the effect of the second harmonic frequency, especially in the x-direction, is enhanced.

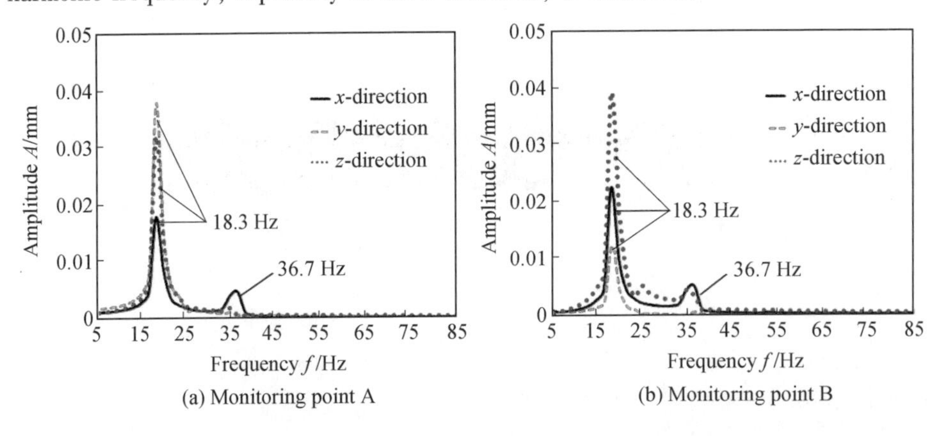

Fig. 3. 19 Vibration displacement spectrum of the monitoring point (d=10 mm, δ=2.0 mm)

Because the wall thickness of the ETB has basically no effect on the vibration frequency at the monitoring point, only the variation of amplitude with the wall thickness of the ETB is made, as shown in Fig. 3. 20.

From Fig. 3. 20, the following conclusions can be drawn.

(1) The amplitude of the monitoring points in each direction decreases with the increase of the wall thickness of the ETB.

(2) The maximum amplitude of monitoring point A appears in the y-direction and the maximum amplitude of monitoring point B appears in the z-direction, both in the plane of the ETB.

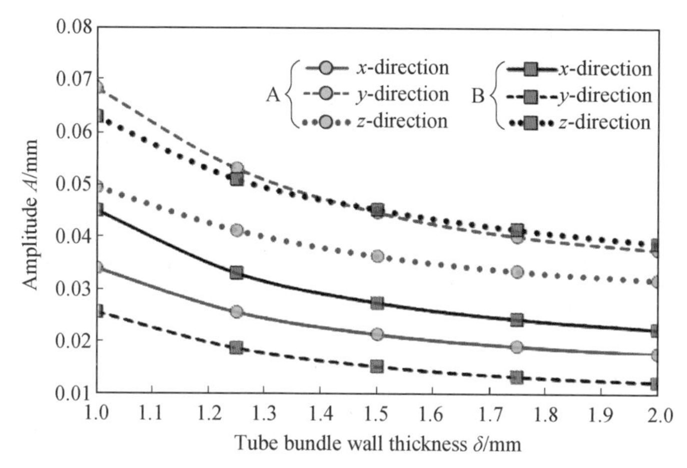

Fig. 3. 20 Variation of the amplitude of the monitoring point with the wall thickness of the ETB

3.2.4 Effect of Tube-side Fluid

In the actual working engineering of the ETB, the vibration of the ETB is caused by the coupling-induced effect of the two fields of fluid in shell-side and tube-side[9]. Therefore, it is necessary to analyze the vibration response of the ETB under the coupling induction of the two fields of the shell-side and tube-side.

Fig. 3. 21 shows the variation of vibration displacements of monitoring points A and B in the x-, y- and z-directions with the calculated time when the flow velocity of both shell-side and tube-side is 0. 8 m/s (i. e. $u_{in} = 0.8$ m/s).

During the calculation, the ETB cross-sectional outer diameter $d = 10$ mm, wall thickness $\delta = 1.5$ mm, and the shell-side and tube-side fluid medium are water.

It can be seen that: the vibration displacement curves induced by the coupling of the two fluid domains of shell-side and tube-side pass shown in Fig. 3. 21 are basically consistent with the vibration curves induced by the shell-side fluid only in Fig. 3. 4(b). This indicates that the influence of the tube-side fluid on the vibration

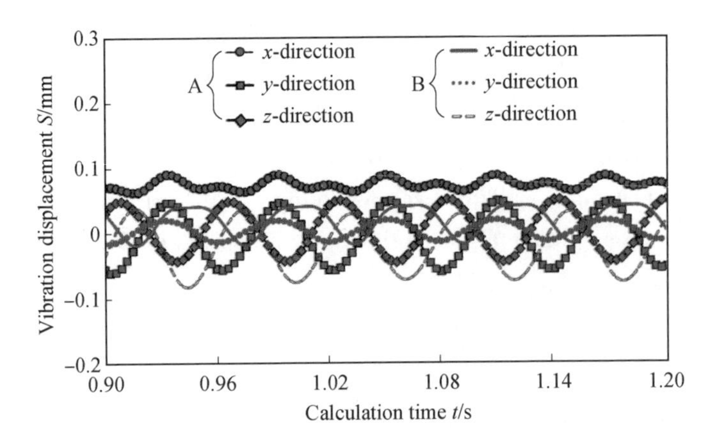

Fig. 3.21 Variation of vibration displacement with calculation time

response of the ETB is not significant in the parameter range studied in this chapter.

For further analysis, the vibration curves of monitoring point B are compared for two induction conditions (coupled induction of shell-side and tube-side fluid and induction of shell-side fluid only) at different inlet velocities, as shown in Fig. 3.22.

From Fig. 3.22, the following conclusions can be drawn.

(1) When the shell-side and tube-side fluid flow velocity is small (0.8 m/s), the vibration equilibrium position in the *x*-direction is lower when coupling is induced, and the vibration displacement curves in the *y*- and *z*-directions basically coincide.

(2) When the fluid flow velocity of shell and tube is larger (1.0 m/s), the vibration displacement curves in the *x*-direction basically coincide when coupling is induced, the vibration displacement curve in the *y*-direction is shifted to the right but the amplitude is basically unchanged, and the vibration displacement curve in the *z*-direction is shifted to the right and the amplitude becomes larger.

(3) Compared with the vibration displacement curve of the ETB induced by the shell-side fluid only, the vibration displacement curve of the ETB induced by the coupling of the shell-side and tube-side fluids does not change much, which means that the vibration of the ETB is mainly induced by the shell-side fluid, and the influence of the tube-side fluid is not significant.

Based on the above comparison of the vibration response of the ETB under the two induced conditions, the vibration displacement curves of the ETB induced by the coupled shell-side and tube-side fluids are basically consistent with those of the ETB induced by the shell-side fluid only. It shows that the shell-side fluid is the main factor to induce the vibration of ETB in the range of calculation parameters of this

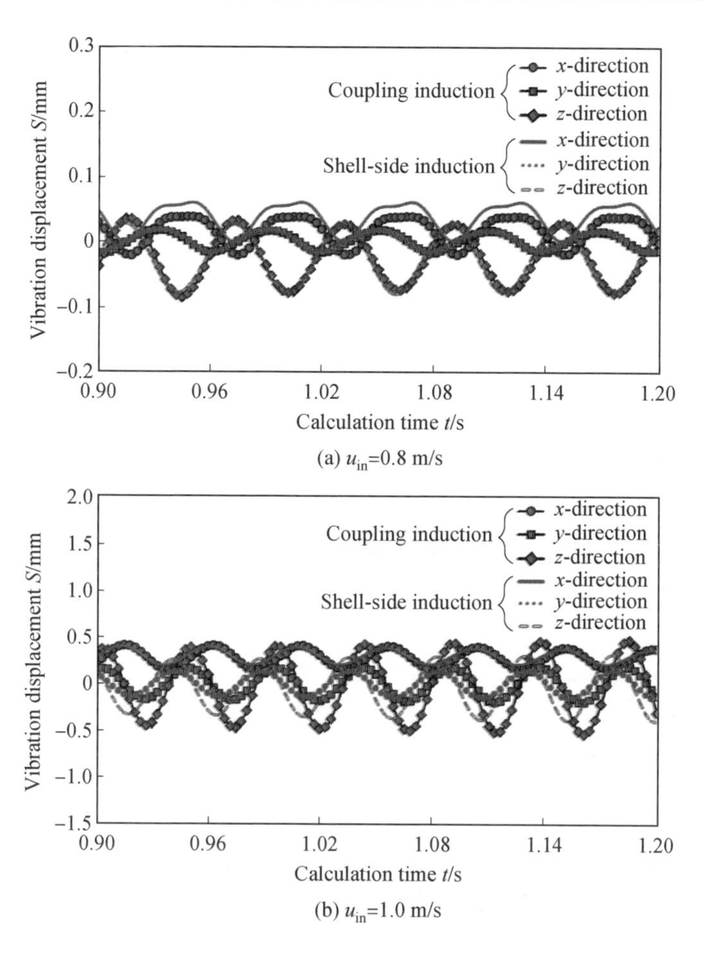

Fig. 3.22 Comparison of vibration displacement curves of monitoring point B under two induced conditions

book, and the influence of the tube-side fluid on the vibration response of ETB is smaller. Therefore, only the effect of the shell-side fluid on the vibration response of the ETB is considered in the subsequent studies.

3.3 Assembly of Shell-side Fluid Domain for Multi-row ETBs

3.3.1 Geometric Model of Multi-row ETBs and Its Mesh

The Geometric model of the multi-row ETB and its shell-side fluid domain is formed on the basis of the single-row ETB and its shell-side fluid domain (Fig. 3.1). During the calculation process, outer section diameter of the ETB $d = 10$ mm, the wall thickness $\delta = 1.5$ mm, tube row spacing $H = 60$ mm. The remaining structural parameters are shown in Table 3.1, and the material properties are shown in Table 3.2.

The structural domainfor multi-row ETBs is replicated along the x-direction (height direction) using the Design Modeler modeling module in the Workbench platform, with a spacing of 60 mm, on the basis of the single-row ETB. The mesh of each row of ETBs is consistent with the mesh shown in Fig. 3.2.

Fig. 3.23 shows the schematic diagram of the uniform shell-side fluid domain for multi-row ETBs, where taking 5 rows of ETBs as an example. The overall uniform shell-side fluid domain is formed by the docking of multiple single-row ETB fluid domains (height: H), where the shell-side fluid flows in from the bottom and out from the top. The ETBs are uniformly arranged in the uniform shell-side fluid domain, numbered 1, 2,..., 5 from bottom to top. A_n and B_n ($n = 1, 2, ..., 5$) are monitoring points located at the midpoint of the stainless steel connectors Ⅲ and Ⅳ of each row of ETBs, used to monitor the vibration of multi-row ETBs induced by uniform shell-side fluid.

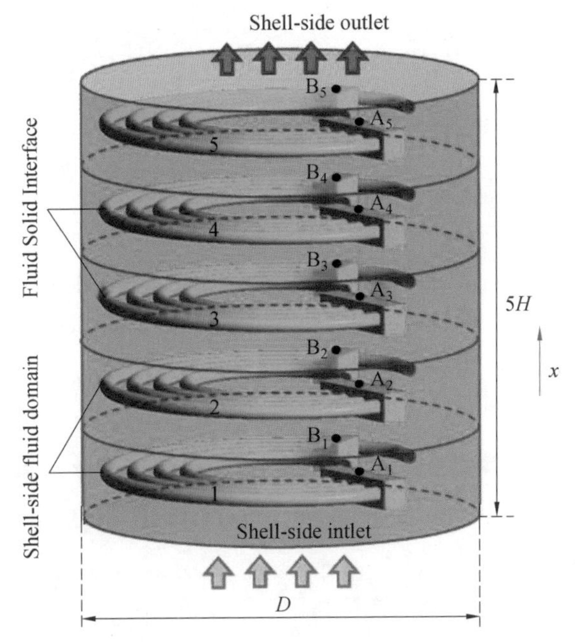

Fig. 3.23 Schematic diagram of uniform shell-side fluid domain for multi-row ETBs

In Fig. 3.23, x is the flow direction of the shell-side fluid. During the calculation process, the diameter of the uniform shell-side fluid domain is $D = 300$ mm, and the height of the uniform shell-side fluid domain for a single-row ETB (or tube row spacing) is $H = 60$ mm.

Unlike the shell-side fluid domain shown in Fig. 3.1, the height of the shell-side fluid domain for single-row ETB (H) has been reduced to a certain extent (from 100 mm to 60 mm) due to the consideration of tube row spacing. Therefore, the mesh division software ICEM was used to redivide the mesh of the shell-side fluid domain, as shown in Fig. 3.24.

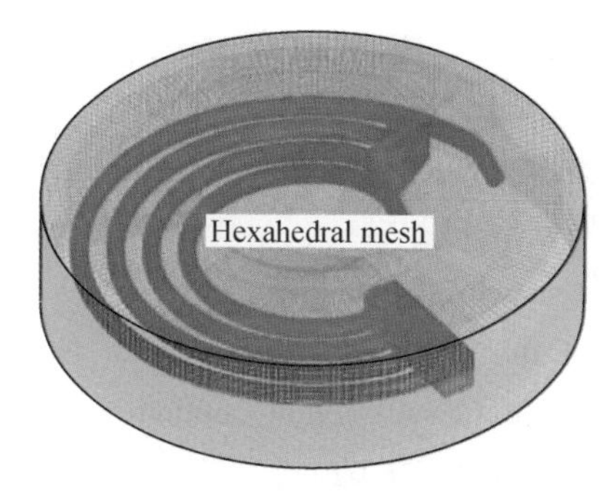

Fig. 3.24 Mesh diagram of the shell-side fluid domain for single-row ETB

Similar to the mesh shown in Fig. 3.3, the mesh of the shell-side fluid domain for single-row ETB is hexahedron mesh, with a minimum mesh quality of 0.54, a total of 181,944 elements and 166,856 nodes. Six layers of boundary layer mesh are set in the fluid domain in contact with the ETB.

The mesh of shell-side fluid domain for multi-row ETBs is formed by copying the mesh in the x-direction as shown in Fig. 3.24 using the mesh duplication function of ANSYS CFX software. The interface between fluid domains is established between the meshes of each part, and the mesh connection adopts a 1 : 1 method. In this way, the total number of elements in the overall shell-side fluid domain is 909,720, and the total number of nodes is 834,280.

3.3.2 Boundary Conditions

The boundary condition setting is similar to that of the single-row ETB and its shell-side fluid domain, but multiple FSI interfaces need to be set.

The boundary conditions of the structural domain are set as follows.

(1) The sections at the two fixed ends Ⅰ and Ⅱ of each row of ETBs are set as "Fixed Support".

(2) The outer surfaces of each row of ETBs are set as "Fluid Solid Interface", and are sequentially named FSI-1, FSI-2, . . . , FSI-5 starting from the bottom row of ETBs.

(3) The direction of gravitational acceleration (Standard Earth Gravity) is set as x-direction, and its value is 9.8066 m/s^2.

The boundary conditions of the shell-side fluid domain are set as follows.

(1) The bottom surface of the shell-side fluid domain is set as the fluid inlet, with a boundary type of "Inlet", and a given inlet flow velocity of 0.8 m/s.

(2) The top surface of the shell-side fluid domain is set as the fluid outlet, and with a boundary type of "Outlet". The relative static pressure at the outlet is set to 0 Pa.

(3) The internal surfaces of the shell-side fluid domain of each single row ETB are respectively set as "Fluid Solid Interface", and correspond one-to-one with FSI-1, FSI-2, . . . , FSI-5 of the structural domain.

(4) The side wall of the shell-side fluid domain is set as "Wall", and the boundary details included: the "Mesh Motion" option is set as "Stationary", and the "Heat Transfer" option is set as "Adiabatic".

3.3.3 Mesh Independence Analysis

For multi-row ETBs and their shell-side fluid domains, the meshes of each single-row ETB fluid domain are consistent, and the overall shell-side fluid domain is formed by connecting the shell-side fluid domains of the single-row ETB. Therefore, the mesh independence analysis of a single-row of ETB and its shell-side fluid domain can verify the mesh independence of the multi-row ETBs and their shell-side fluid domain.

When the uniform shell-side fluid is used to induce the vibration of a single-row ETB, the main frequency and amplitude of vibration at monitoring point A on stainless steel connector Ⅲ in the y-direction are analyzed for mesh independence. During the calculation process, the shell-side fluid medium is water, and the inlet velocity u_{in} = 0.8 m/s.

Table 3.5 shows the comparison of the main frequency ($f_{x\text{-main}}$) and amplitude (A_x) of vibration at monitoring point A on stainless steel connector Ⅲ in the y-direction under three mesh division schemes.

Table 3.5 Comparison of calculation results under different mesh schemes

Cases	Nodes	Calculation results		Relative errors/%		Calculation time/h
		$f_{x\text{-main}}$/Hz	A_x/mm	$f_{x\text{-main}}$	A_x	
1	176,552	17.8	0.045	2.73	11.76	12.85
2	221,633	18.3	0.051	—	—	16.31
3	534,231	18.7	0.054	2.19	5.88	37.60

Note: The relative error is calculated based on the calculation results of case 2.

In Table 3.5, the number of nodes refers to the total number of nodes in the shell-side and tube-side fluid domains and the structural domain. Case 2 refers to the mesh strategy mentioned above, and cases 1 and 3 are obtained by reducing or increasing the mesh density of the shell-side fluid domain and structural domain based on case 2.

From Table 3-5, it can be seen that increasing the number of meshes (case 3) has little impact on the calculation results, with a maximum relative error of only 5.88%, but the calculation time is about 2.31 times that of case 2. In addition, when the number of meshes decreases (case 1), it has a significant impact on the calculation results, with a maximum relative error of 11.76% and a calculation time of approximately 0.79 times that of case 2. Based on considerations of computational accuracy and efficiency, the mesh of case 2 is selected as the computational mesh.

3.4 Vibration Analysis of Multi-row ETBs

3.4.1 Vibration Response of Two Rows of ETBs

Based on the sequential solution method of bi-directional FSI, the vibration response of two rows of ETBs induced by uniform shell-side fluid is numerically analyzed. Fig. 3.25 shows the variation of vibration displacement in each direction of two monitoring points on ETB 1 with the calculation time, and the vibration displacement spectrum is shown in Fig. 3.26. In addition, Fig. 3.27 shows the variation of vibration displacement in each direction of two monitoring points on ETB 2 with the calculation time, and the vibration displacement spectrum is shown in Fig. 3.28.

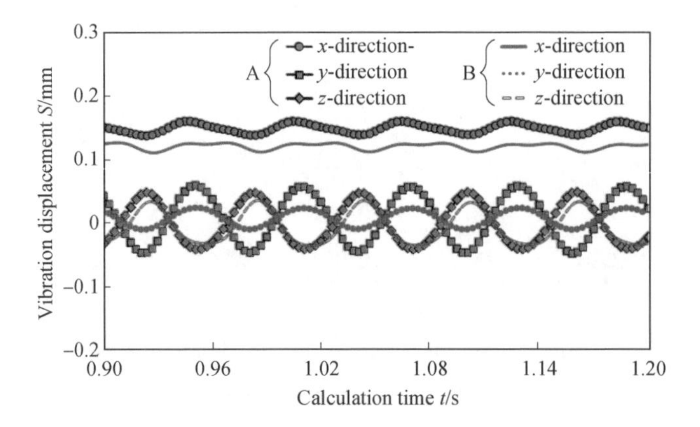

Fig. 3. 25 Variation of vibration displacement in each direction of two monitoring points on ETB 1 with calculation time

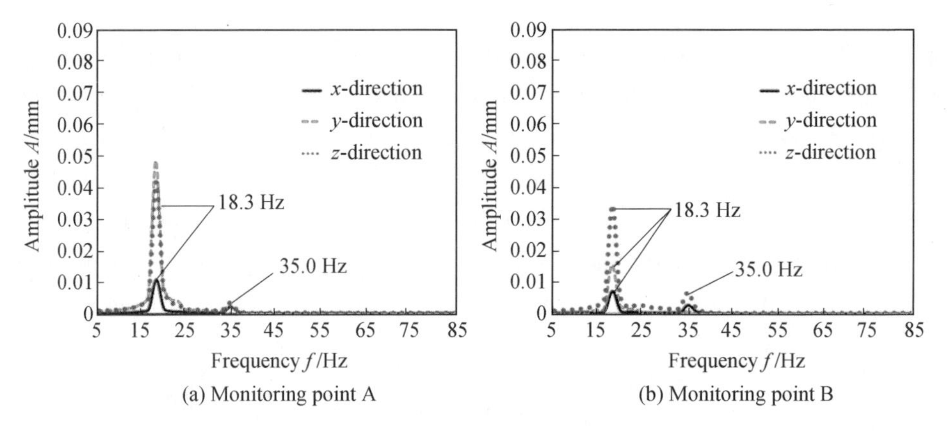

Fig. 3. 26 Vibration displacement spectrum of the monitoring point on ETB 1

From Fig. 3. 25 to Fig. 3. 28, the following conclusions can be drawn.

(1) Due to the influence of ETB 2, the vibration of stainless steel connector Ⅲ of ETB 1 (maximum amplitude: 0. 048 mm) is more severe than that of stainless steel connector Ⅳ (maximum amplitude: 0. 034 mm), and stainless steel connector Ⅲ mainly vibrates in the y-direction, while stainless steel connector Ⅳ mainly vibrates in the z-direction.

(2) Due to the influence of ETB 1, the vibration of stainless steel connector Ⅳ of ETB 2 (maximum amplitude: 0. 082 mm) is more severe than that of stainless steel connector Ⅲ (maximum amplitude: 0. 053 mm), and both stainless steel connectors mainly vibrate in the z-direction.

(3) The vibration balance positions (0. 153 mm and 0. 114 mm) of the two monitoring points in the x-direction of the ETB 1 are located above the plane of the

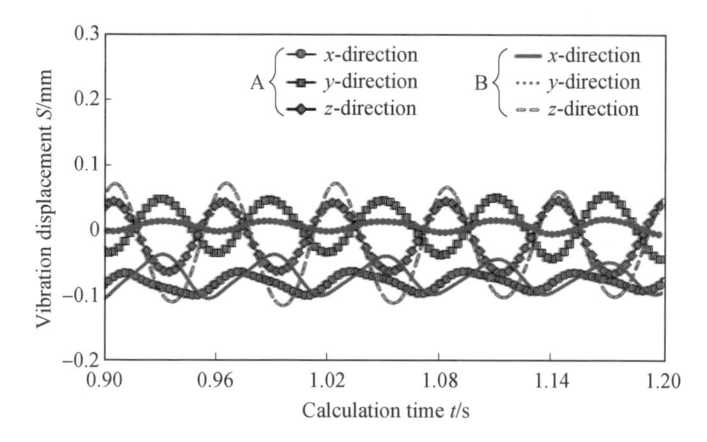

Fig. 3. 27 Variation of vibration displacement in each direction of two monitoring points on ETB 2 with calculation time

ETB (similar to the situation of the single-row of ETB). Due to the weakening of the impact force, the vibration balance positions (−0.074 mm and −0.067 mm) in the x-direction of the two monitoring points of the ETB 2 are located below the plane of the ETB.

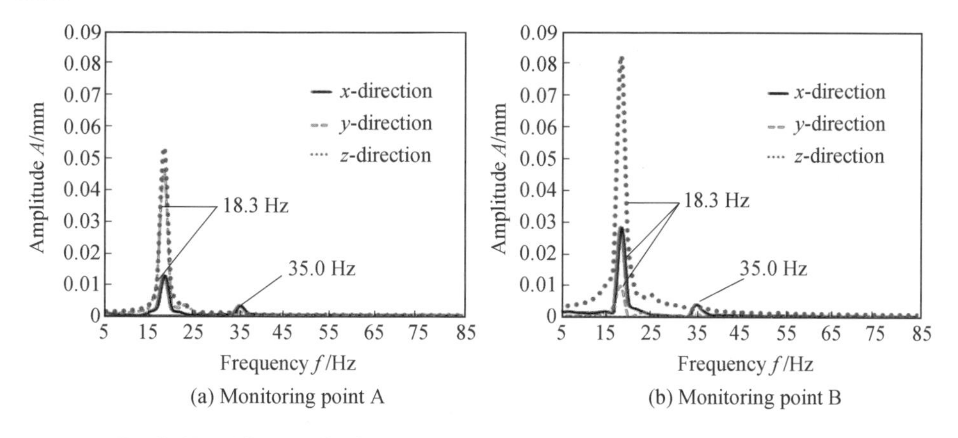

Fig. 3. 28 Vibration displacement spectrum of the monitoring point on ETB 2

(4) Compared with ETB 1, the vibration of ETB 2 is more severe. There are obvious vibration harmonics in the x- and z-directions at the two monitoring points of ETB 1, while there are obvious vibration harmonics in the x-direction at the two monitoring points of ETB 2.

(5) Compared with the case of fluid induction in the shell-side of the single row ETB (Fig. 3. 4(b) and Fig. 3. 6), the magnitude of the main frequency of vibration remains unchanged, and the harmonic frequency decreases by about 5%.

3.4.2 Vibration Response of Five Rows ETBs

In order to analyze the vibration response of multi-row ETBs induced by the uniform shell-side fluid, the flow-induced vibration response of 3–5 rows of ETBs is studied when $u_{in}=0.8$ m/s.

Fig. 3.29 shows the variation of the main frequency amplitude of vibration at the monitoring point A_n ($n=1, 2, \dots, 5$) on the stainless steel connector Ⅲ of the multi-row ETBs in various directions with the number of the ETB.

From Fig. 3.29, the following conclusions can be drawn.

(1) The monitoring pointsof connector Ⅲ on each row of ETBs have larger amplitudes in the *y*- and *z*-directions, while the amplitudes in the *x*-direction are smaller, indicating that the vibration of each row of ETBs is manifested as in-plane vibration.

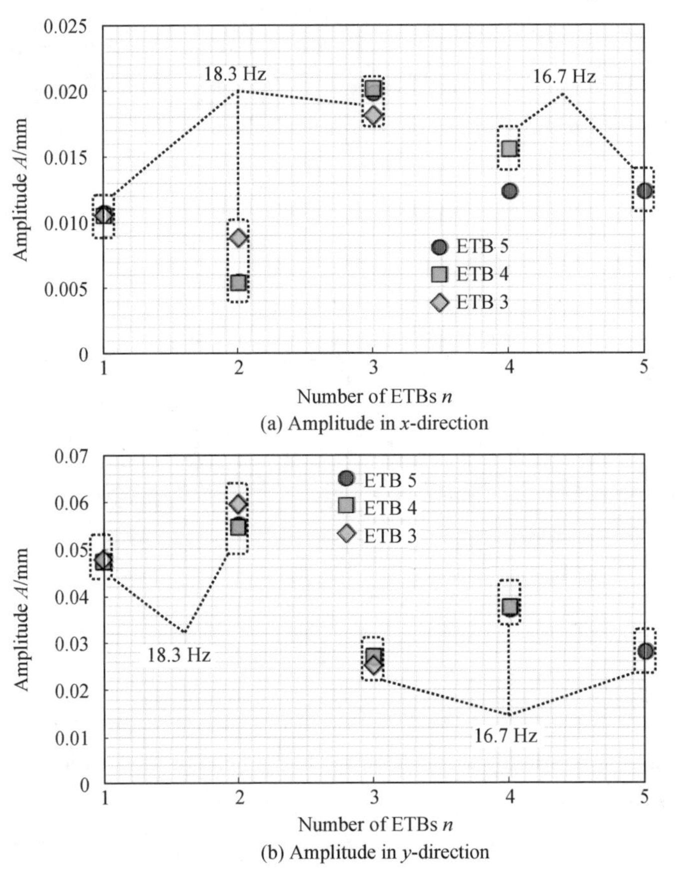

Fig. 3.29 Variation of the main frequency amplitude of vibration at the monitoring point A_n on the stainless steel connector Ⅲ of the multi-row ETBs in various directions with the number of the ETB

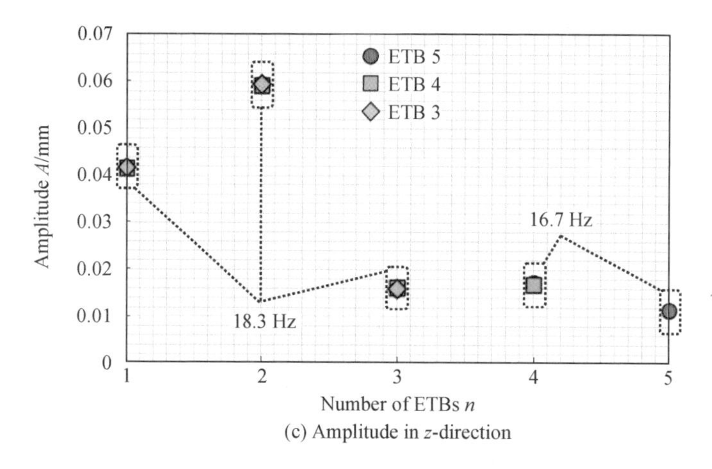

(c) Amplitude in z-direction

Fig. 3.29 (continued)

(2) The amplitude of the monitoring points on the stainless steel connector Ⅲ of each row of ETBs in the x-direction decreases first, then increases, and then decreases along the number of ETBs, and the amplitude of the stainless steel connector Ⅲ on ETB 3 is the highest in the x-direction.

(3) The amplitude of the monitoring points on thestainless steel connector Ⅲ of each row of ETBs in the y- and z-directions increases first, and then decreases with the variation of the number of ETBs. Moreover, the amplitude of the stainless steel connector Ⅲ on ETB 2 is the highest in the y- and z-directions.

(4) The vibration frequency of the monitoring points in the x- and z-directions for the first three rows of ETBs (18.3 Hz) is higher than that of the second two rows of ETBs (16.7 Hz), and the vibration frequency of the monitoring points in the y-direction for the first two rows of ETBs (18.3 Hz) is higher than that of the last three rows of ETBs (16.7 Hz).

(5) For the number of tube rows studied in this chapter, the vibration intensity of connector Ⅲ for the bottom row ETBs (ETBs 1 and 2) is relatively high, while the vibration intensity of connector Ⅲ for the top row ETBs (ETBs 3, 4 and 5) is relatively low, and the vibration intensity of connector Ⅲ for ETB 2 is the highest.

Fig. 3.30 shows the variation of the main frequency amplitude of vibration at the monitoring point B_n ($n = 1, 2, \dots, 5$) on the stainless steel connector Ⅳ of the multi-row ETBs in various directions with the number of the ETB.

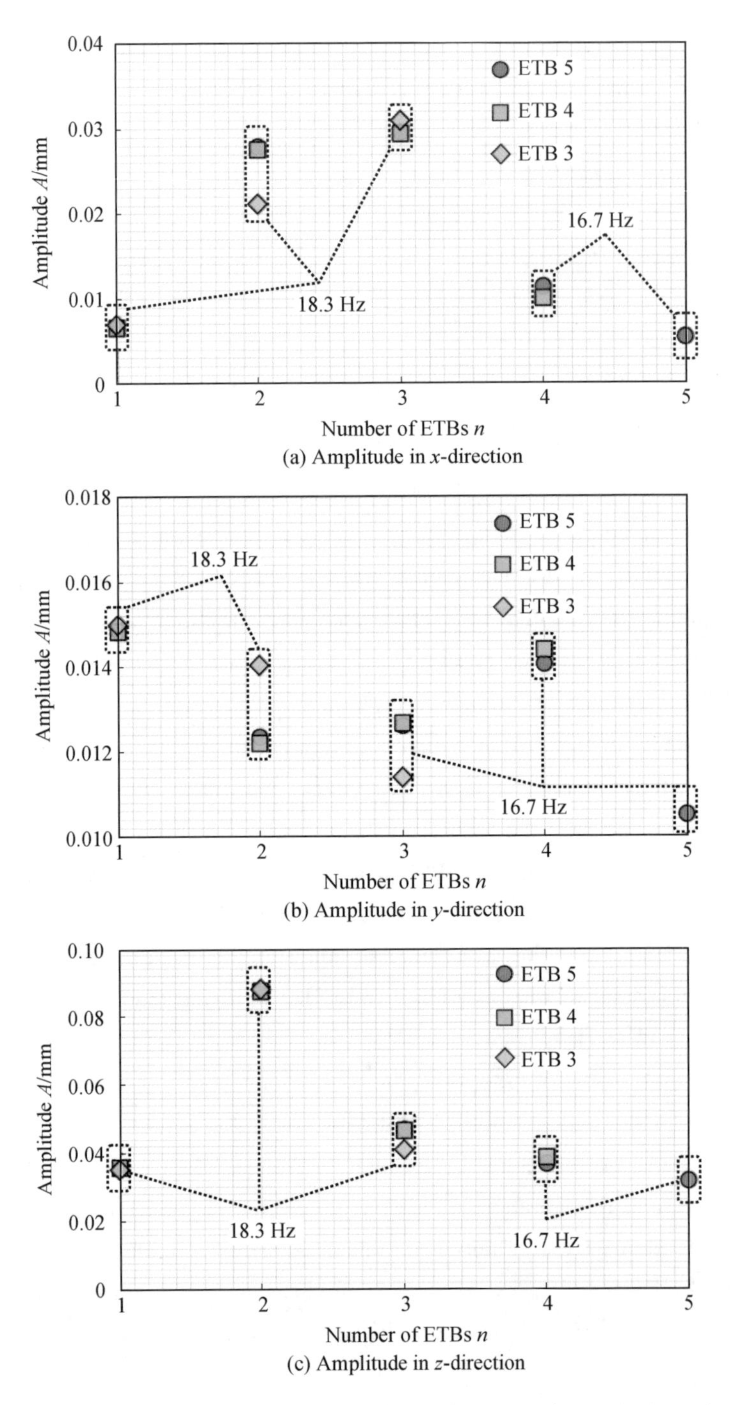

Fig. 3. 30 Variation of the main frequency amplitude of vibration at the monitoring point B_n on the stainless steel connector Ⅳ of the multi-row ETBs in various directions with the number of the ETB

From Fig. 3.30, the following conclusions can be drawn.

(1) The monitoring points on the stainless steel connector Ⅳ of each row of ETBs have the highest amplitude in the z-direction, followed by the amplitude in the x-direction, and the amplitude in the y-direction is the smallest, indicating that the vibration of the stainless steel connector Ⅳ of each row of ETBs is mainly manifested as in-plane vibration.

(2) The amplitude of the monitoring points on the stainless steel connector Ⅳ of each row of ETBs increases first and then decreases along the number of the ETB in the x- and z-directions, and the amplitude of the stainless steel connector Ⅳ on ETB 3 is the highest in the x-direction, while the amplitude of the stainless steel connector Ⅳ on ETB 2 is the highest in the z-direction.

(3) The amplitude of the monitoring points on the stainless steel connector Ⅳ of each row of ETBs in the y-direction decreases first, then increases, and then decreases along the number of the ETB, and the amplitude of the stainless steel connector Ⅳ on ETB 1 is the highest in the y-direction.

(4) Consistent with the conclusion ofstainless steel connector Ⅲ, the vibration frequency of the monitoring points in the x- and z-directions for the first three rows of ETBs (18.3 Hz) is higher than that of the second row of ETBs (16.7 Hz), and the vibration frequency of the monitoring points in the y-direction for the first two rows of ETBs (18.3 Hz) is higher than that of the last three rows of ETBs (16.7 Hz).

(5)For the number of tube rows studied in this book, the vibration intensity of the stainless steel connector Ⅳ for the first row (bottom row) ETB is the weakest, while the vibration intensity of the stainless steel connector Ⅳ for the second row ETB is the strongest, and the vibration intensity of the stainless steel connector Ⅳ for the second, third, fourth and fifth row ETBs shows a gradual weakening trend.

References

[1]JI J D, GE P Q, BI W B. Numerical analysis on flow-induced vibration responses of elastic tube bundle [J]. Journal of vibration and shock, 2016, 35(6): 80-84.

[2]JI J D, LI F Y, SHI B J, et al. Analysis of the effect of baffles on the vibration and heat transfer characteristics of elastic tube bundles [J]. International communications in heat and mass transfer, 2022, 136: 106206.

[3]JI J D, GAO R M, SHI B J, et al. Improved tube structure and segmental baffle

to enhance heat transfer performance of elastic tube bundle heat exchanger [J]. Applied thermal engineering. 2022, 200: 117703.

[4] DUAN D R, GE P Q, BI W B, et al. An empirical correlation for the heat transfer enhancement of planar elastic tube bundle by flow-induced vibration [J]. International journal of thermal sciences, 2020, 155: 106405.

[5] JI J D, GE P Q, LIU P, et al. Design and application of a new distributed pulsating flow generator in elastic tube bundle heat exchanger [J]. International journal of thermal sciences, 2018, 130: 216-226.

[6] YAN K. A study on the vibration and heat transfer characteristics of conical spiral tube bundle in heat exchanger [D]. Jinan: Shandong University, 2012.

[7] JI J D. Study on flow-induced vibration of elastic tube bundle with shell-side distributed pulsating flow in heat exchanger [D]. Jinan: Shandong University, 2016.

[8] DUAN D R, GE P Q, BI W B. Numerical investigation on heat transfer performance of planar elastic tube bundle by flow-induced vibration in heat exchanger [J]. International journal of heat and mess transfer, 2016,103: 868-878.

[9] CHENG L, LUAN T, DU W. Heat transfer enhancement by flow-induced vibration in heat exchangers [J]. International journal of heat and mass transfer, 2009, 52: 1053-1057.

Chapter 4 Vibration and Heat Transfer Analysis of ETBs Induced by Actual Shell-side Fluid

Based on the analysis of the shell-side flow-induced single-row ETB vibration response in Chapter 3 of this book, it is known that for water-water heat exchangers, the shell-side fluid is the main factor that induces ETB vibration. In an ETB heat exchanger with multiple rows of ETBs uniformly arranged, the shell-side fluid flows into the heat exchanger from the shell-side inlet, and the internal fluid flows approximately spirally upward. Changes in inlet velocity, tube row spacing, and number of ETBs will affect the lateral and longitudinal flow of the winding ETB, which in turn affects the vibration and heat transfer characteristics of each row of ETBs.

In this chapter, based on the structural characteristics of the shell-side fluid domain of the ETB heat exchanger, the overall shell-side fluid domain mesh was formed through the basic process of fluid domain partitioning, segmented fluid domain meshing and shell-side fluid domain mesh assembly. Then, the vibration responses of each row of ETBs under different shell-side inlet velocities, different tube row spacings and different number of ETBs were numerically analyzed by a step-by-step calculation strategy of rough calculation and actuarial calculation. The heat transfer characteristics of each row of ETBs were also studied based on the vibration response of each row of ETBs induced by the actual shell-side fluid domain.

In addition, based on the traditional ETB (TETB), an improved ETB (IETB) was proposed to study the influence of shell-side fluid flow direction on the vibration and heat transfer characteristics. Then, the influence of IETB installation angle on the vibration and heat transfer characteristics of multi-row IETBs in a heat exchanger were studied, and the optimal installation angle of IETB was obtained.

4.1 Computational Domain and Its Mesh

4.1.1 ETB Heat Exchanger and Its Shell-side Fluid Domain

Fig. 4.1 shows a schematic diagram of a six-row ETB heat exchanger and its shell-side fluid domain. In Fig. 4. 1, x is the coordinate system in the vertical direction of the heat exchanger. In the calculation, the heat exchanger cylinder size[1-2] is ϕ 325 mm×10 mm, the tube-side inlet tube and outlet tube size ϕ 45 mm× 3 mm, the shell-side inlet tube size is ϕ 60 mm×3 mm, and the shell-side outlet tube size is ϕ 73 mm×4 mm.

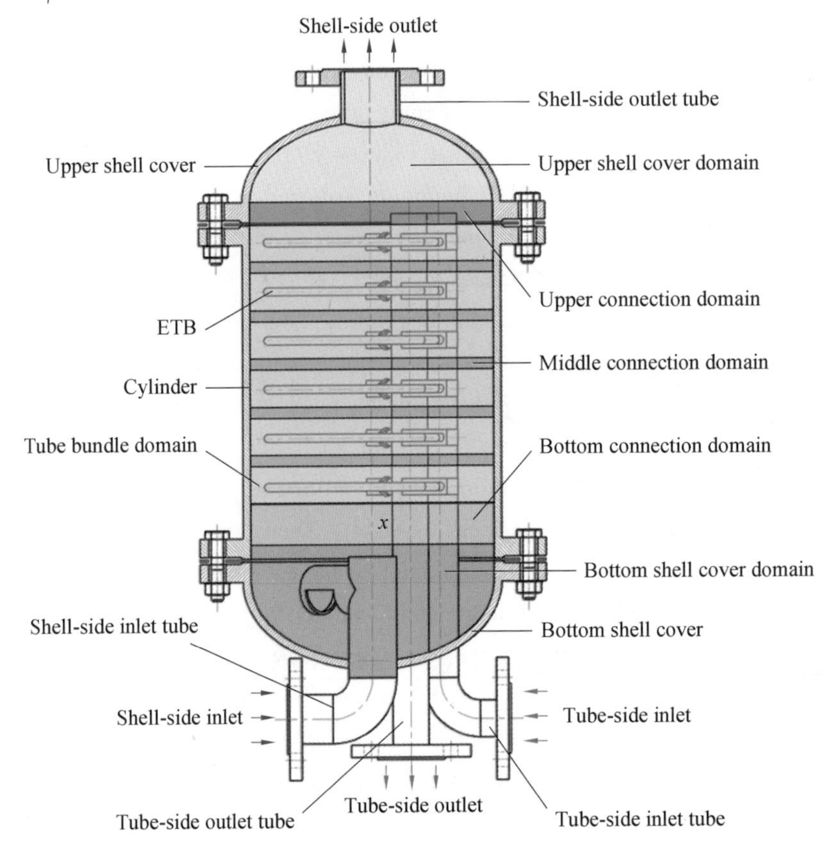

Fig. 4. 1 Schematic diagram of a six-row ETB heat exchanger and its shell-side fluid domain

As can be seen from Fig. 4. 1, the main components of the ETB heat exchanger include upper shell cover, cylinder, bottom shell cover, ETB, tube-side inlet tube, tube-side outlet tube, shell-side inlet tube, shell-side outlet tube, etc. All parts are made of high-quality ordinary steel or stainless steel, except for the elbow part of the ETB, which is made of copper tube by repeated forming. The tube-side inlet tube, tube-side outlet tube and shell-side inlet tube are welded to the bottom shell cover, and the shell-side outlet tube is welded to the upper shell cover, the rest of the parts are connected by bolts, and rubber gaskets are used to increase the sealing.

As shown in Fig. 4. 1, the ETBs are arranged uniformly in the heat exchanger. The remaining structural parameters are shown in Table 3. 1, and the material properties of the bundle are shown in Table 3. 2. For the purpose of analysis, the ETBs in the heat exchanger are numbered 1, 2, ..., 6 from bottom to top. Set monitoring points A_n, B_n ($n=1, 2, \ldots, 6$) on the two stainless steel connectors of each row of ETBs to monitor the vibration responses of the ETBs induced by the actual shell-side fluid. The placement of monitoring points is shown in Fig. 4. 1.

Fig. 4. 2 shows the installation positions of the tube-side inlet tube, tube-side outlet tube and shell-side inlet tube on the bottom shell cover. In Fig. 4. 2, y and z are the coordinate systems in the plane of the ETB, O is the coordinate center, and θ is the swing angle of the shell-side inlet tube. The position parameters are shown in Table 4. 1 if not specified in the calculation process.

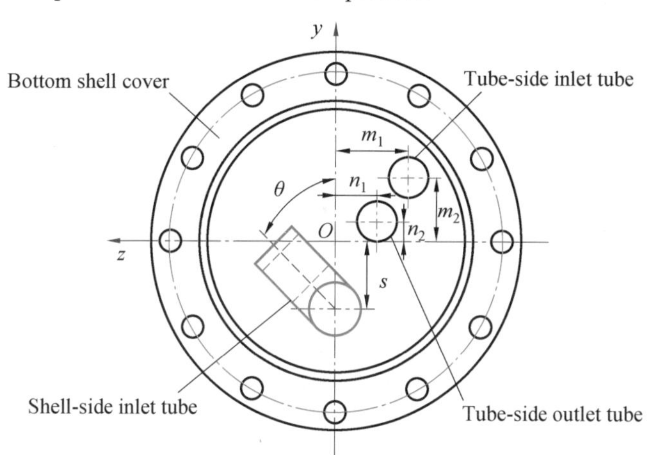

Fig. 4. 2 Installation positions of the tube-side inlet tube, tube-side outlet tube and shell-side inlet tube on the bottom shell cover

In practical engineering applications, the shell-side fluid flows in from the shell-side inlet, enters the heat exchanger through the shell-side inlet tube and impacts the bottom shell cover of the heat exchanger. After being folded, it flows approximately spirally upwards and flows out from the shell-side outlet through the shell-side outlet tube. Inside the heat exchanger, the shell-side fluid sweeps across the tube-side inlet tube, tube-side outlet tube and uniformly arranged ETBs, inducing vibration of the ETBs, thereby achieving enhanced heat transfer.

The structure of the internal shell-side fluid domain of the ETB heat exchanger is complex, the characteristic of the actual shell-side fluid is turbulent, and the vibration responses of the internal ETBs induced by the shell-side fluid is very different from that of the ETBs induced by the uniform shell-side fluid. Therefore, a reasonable planning is necessary for the meshing strategy of the shell-side fluid domain.

Table 4.1 Position parameters of the tube-side inlet tube, tube-side outlet tube and shell-side inlet tube

Parameter	Value	Parameter	Value
m_1/mm	85.0	n_2/mm	23.0
m_2/mm	75.5	s /mm	76.25
n_1/mm	48.0	θ /(°)	45

4.1.2 Mesh Division Strategy

Since both the tube-side inlet tube and the tube-side outlet tube are steel structures, they basically do not generate vibration under fluid induction. In order to improve the calculation efficiency, only the ETB of the structural domain is retained without affecting the calculation results[3-4]. The specific partitioning strategy of the ETB mesh is described in "3.1.3 Mesh and Independence Analysis", and the schematic diagram of the mesh is shown in Fig. 3.2.

Due to the complex structure of the shell-side fluid domain, it is difficult and inefficient to divide the overall mesh, and it is not easy to adjust and modify the mesh[5]. For this reason, the planned meshing strategy is as follows.

(1) The shell-side fluid domain is appropriately partitioned according to its structural characteristics.

(2) Each segmented fluid domain is meshed separately using a suitable meshing method.

(3) Based on ANSYS CFX software, the whole shell-side fluid domain mesh is assembled by using the functions of mesh import, duplication, translation and connection.

1. Fluid Domain Segmentation

The shell-side fluid domain segmentation is shown in Fig. 4.1. The segmented fluid domain consists of six parts: upper shell cover domain, bottom shell cover domain, tube bundle domain, upper connection domain, middle connection domain and bottom connection domain. Each segmented fluid domain is shown in Fig. 4.3.

In Fig. 4.3, the heights of each segmented fluid domain are: upper shell cover domain (without fluid in shell-side outlet tube)—108 mm; bottom shell cover domain (without fluid in shell-side inlet tube)—148 mm; tube bundle domain—40 mm; upper connection domain—30 mm; middle connection domain—20 mm; bottom connection domain—56. 5 mm. Thus, the tube row spacing $H = 60$ mm. When adjusting the tube row spacing, the height of the middle connection domain (Fig. 4. 3 (e)) can be appropriately increased or decreased, while the other dimensions remain unchanged. When changing the position of the shell-side inlet tube, only model and mesh the lower head region.

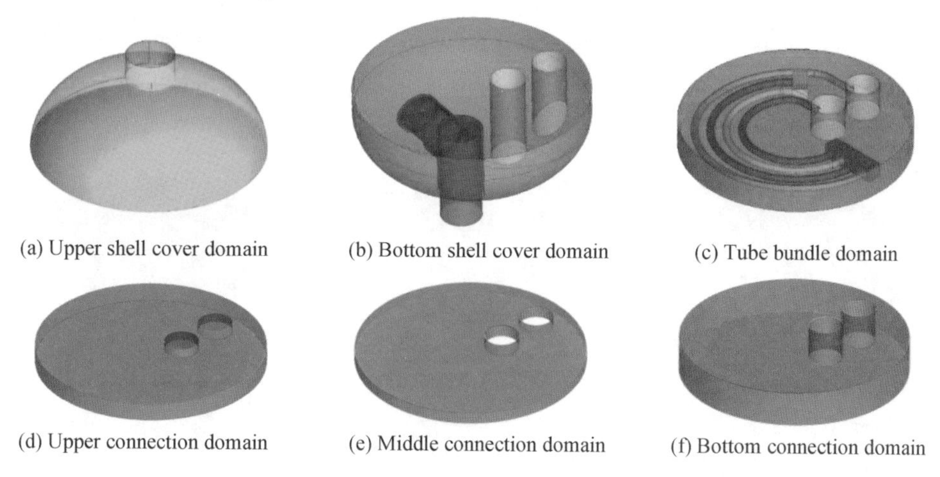

(a) Upper shell cover domain (b) Bottom shell cover domain (c) Tube bundle domain

(d) Upper connection domain (e) Middle connection domain (f) Bottom connection domain

Fig. 4. 3 Schematic diagram of each segmented fluid domain

When meshing each segmented fluid domain, the following points need to be noted.

(1) The tube bundle domain is the key region for the FSI calculation, and its mesh quality and density directly affect the accuracy of the calculation. Due to the regular structure of the tube bundle domain, a finer hexahedral mesh is used for meshing.

(2) The upper shell cover domain, the upper connection domain, the middle connection domain and the bottom connection domain have simple structures and regular shapes, and they are all non-FSI regions, so coarser hexahedral meshes are used for those domains.

(3) The bottom shell cover domain contains the shell-side inlet tube, part of the tube-side inlet tube and part of the tube-side outlet tube. The structure is complex, and it is not easy to use hexahedron mesh for division, so it is easy to use tetrahedron mesh for division.

In this way, according to the structural characteristics of each segmented fluid domain, a suitable meshing strategy is used to mesh each segmented fluid domain separately.

2. Mesh for Segmented Fluid Domains

Fig. 4. 4 shows the mesh diagram of each segmented fluid domain, and the meshes are divided by the meshing software ICEM. From Fig. 4. 4, it can be seen that only the bottom shell cover domain adopts tetrahedral mesh, and the rest of each segmented fluid domain adopts hexahedral mesh. In addition, in order to improve the accuracy of calculation, the mesh size of the contact surface between the tube bundle domain and the ETB is smaller and the density is larger, and six layers of boundary layer mesh are set.

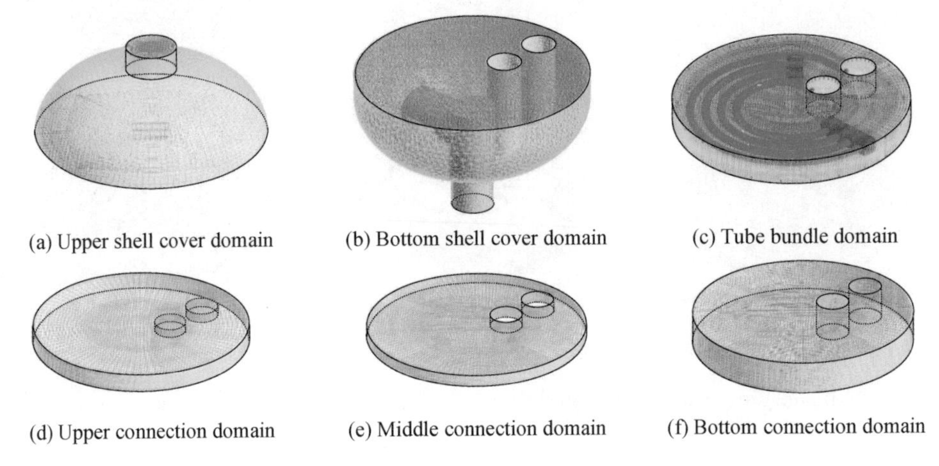

Fig. 4. 4 Mesh diagram of each segmented fluid domain

Table 4. 2 shows the mesh quantity and quality of each segmented fluid domain. Since the geometric structure of the middle connection domain and the bottom connection domain are basically the same, only the heights are different, and the same meshing strategy is used for meshing, so the meshes of both have the same quality. The tube bundle domain has more meshes because of the higher mesh density and the boundary layer mesh is used. The mesh quality of the upper connection domain, middle connection domain and bottom connection domain is relatively high, while the mesh quality of the upper shell cover domain and bottom shell cover domain is relatively low, and the lowest minimum mesh quality is 0. 36, which appears in the upper shell cover domain, a non-FSI region.

Table 4.2 Mesh quantity and quality of each segmented fluid domain

Segmented fluid domain	Number of meshes		Minimum mesh quality
	Elements	Nodes	
Upper shell cover domain	58,932	53,868	0.36
Bottom shell cover domain	175,249	30,918	0.39
Tube bundle domain	203,578	186,520	0.52
Upper connection domain	19,784	16,929	0.60
Middle connection domain	32,322	27,390	0.67
Bottom connection domain	51,872	46,370	0.67

3. Mesh Assembly of the Fluid Domain

The mesh assembly of the shell-side fluid domain is done in ANSYS CFX software, and the assembly steps of the shell-side fluid domain are shown in Fig. 4.5. It should be noted that, for the sake of observation, the mesh lines are not shown in Fig. 4.5.

The steps of the mesh assembly are as follows.

(1) Import the mesh of the bottom shell cover domain.

(2) Import the mesh of the bottom connection domain, and establish a connection between the meshes of the bottom connection and the bottom shell cover domains.

(3) Import the mesh of the tube bundle domain, and establish a connection between meshes of the the tube bundle and the bottom connection domains.

(4) Import mesh of the the middle connection domain, and establish a connection between the meshes of the middle connection and the tube bundle domains.

(5) Make 5 equally spaced copies of the tube bundle domain mesh and 4 copies of the middle connection domain mesh along the x-direction with a spacing of 60 mm (consistent with the tube row spacing H), and establish the connection between the meshes of the connected fluid segmented fluid domains.

(6) Import the mesh of the upper connection domain, and establish the connection between the meshes of the upper connection domain and the adjacent tube bundle domain.

(7) Import the mesh of the upper shell cover domain, and establish a connection between the meshes of the upper shell cover and the upper connection domains.

The mesh connections between the above segmented fluid domains are made by

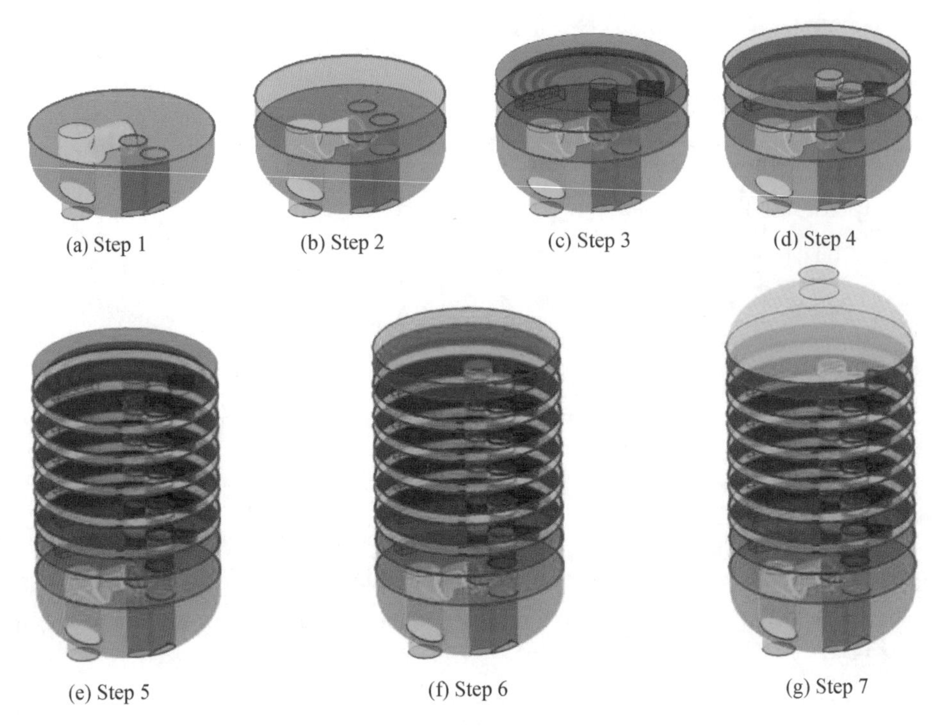

Fig. 4.5 Assembly steps of the shell-side fluid domain

the General Grid Interface (GGI) method. At this point, the overall shell-side fluid domain mesh assembly is completed. When the tube row spacing is changed, the height of the middle joint domain is changed, and the copy spacing of the tube bundle domain and the middle connection domain should be adjusted accordingly. When the upper connection domain and the upper shell cover domain are imported into CFX, they need to be shifted to a suitable position first, and then the mesh connection is made.

Thus, take the shell-side fluid domain mesh of a six-row ETB heat exchanger with tube row spacing $H=60$ mm as an example.

The number of elements in the shell-side fluid domain (n_c) is

$$n_c = 58{,}932 + 175{,}249 + 203{,}578 \times 6 + 19{,}784 + 32{,}322 \times 5 + 51{,}872 = 1{,}688{,}915 \tag{4.1}$$

The number of nodes in the shell-side fluid domain (n_n) is

$$n_n = 53{,}868 + 30{,}918 + 186{,}520 \times 6 + 16{,}929 + 27{,}390 \times 5 + 46{,}370 = 1{,}404{,}155 \tag{4.2}$$

The above meshing strategy has the following characteristics.

(1) Mesh division is carried out according to the structural characteristics of

each segmented fluid domain, which significantly reduces the quantity of elements and improves the quality of elements.

(2) Breaking the whole into parts allows for easy adjustment of the structure and mesh of each partition, avoiding the complexity of overall mesh partitioning and improving the efficiency of mesh partitioning.

(3) Based on the mesh duplication, translation and connection functions of the general CFD analysis software ANSYS CFX, the shell-side fluid domain with arbitrary number of tube rows can be formed.

(4) By adjusting the height of the middle connection domain, a shell-side fluid domain with arbitrary tube row spacing can be assembled. In addition, by adjusting the height of the upper and bottom connection domains, the distance between the ETB and the shell-side fluid inlet and outlet can be controlled.

(5) Due to the complex structure of the bottom shell cover domain, an easy-to-generate unstructured mesh is used to easily adjust parameters such as the size, position and swing angle of the shell-side fluid inlet tube.

4.1.3 Boundary Conditions

The boundary condition settings are similar to those used in Chapter 3 of this book to calculate the vibration responses of a single-row ETB induced by a uniform shell-side fluid. It should be noted that multiple FSI interfaces need to be set up.

The boundary conditions of the structural domain are set as follows.

(1) The cross sections at the two fixed ends Ⅰ and Ⅱ of each row of ETBs are set as "Fixed Support".

(2) The outer surfaces of each row of ETBs are set as "Fluid Solid Interface", and are sequentially named from the bottom row of ETBs as FSI-1, FSI-2, ..., FSI-6.

(3) The direction of gravitational acceleration (Standard Earth Gravity) is set as x-direction, and its value is 9.8066 m/s^2.

The boundary conditions of thefluid domain are set as follows.

(1) Set the shell-side fluid inlet boundary type to "Inlet", and give the inlet velocity and temperature.

(2) Set the shell-side fluid outlet boundary type to "Outlet", and give the outlet relative static pressure to 0 Pa.

(3) Set the inner surfaces of all tube bundle domains to "Fluid Solid Interface", and give the wall temperature. The FSI interfaces in the fluid domain correspond one-

to-one with the FSI interfaces in the structural domain.

(4) The boundary types of the other surfaces are set as "Wall", and the boundary details included: the "Mesh Motion" option is set as "Stationary", and the "Heat Transfer" option is set as "Adiabatic".

In this calculation, the total calculation time for rough calculation is 300 s and the time step is 0.1 s. In addition, the total calculation time for precise calculation is 1.2 s and the time step is 0.001s. The structural domain only participates in the precise calculation, and its total calculation time and time step settings are consistent with those of the precise calculation.

In the subsequent analysis, it can be seen that such time and time step settings can ensure that the vibration of the ETB is stable, and the vibration state at this time is the vibration response of the ETB under the condition of fully developed fluid domain induced by the actual shell-side fluid.

4.1.4 Mesh Independence Analysis

In order to analyze the mesh independence of the overall shell-side fluid domain andstructural domain, the vibration responses of the ETB at the shell-side inlet velocity $u_{in} = 0.4$ m/s is tested based on three different meshing schemes. Among them, the fluid medium is water and the numerical calculation method refers to Chapter 2.

Table 4.3 shows the comparison of vibration frequency f_x and amplitude A_x in x-direction (vertical direction) for monitoring point A_1 on the bottom row of the ETB under different mesh division schemes.

Table 4.3 Mesh independence analysis

Case	Number of nodes	Calculation results		Relative Errors/%		Calculation time/h
		f_x/Hz	A_x/mm	f_x	A_x	
1	1,072,939	21.24	0.0270	8.96	12.05	60
2	1,732,817	23.33	0.0307	—	—	78
3	3,891,534	23.82	0.0321	2.10	4.56	152

Note: The relative error is calculated based on the calculation results of case 2.

In Table 4.3, the number of nodes refers to the total number of nodes for the overall shell-side fluid domain and the six rows of ETBs. Case 2 refers to the meshing strategy mentioned above, and cases 1 and 3 are obtained by decreasing or increasing the mesh density of the structural domain and each segmented fluid domain on the basis of case 2.

As can be seen from Table 4.3, when the number of nodes is increased (case 3), the impact on the calculation results is not significant, and the maximum relative error is only 4.56%, but the calculation time is about 1.95 times of the calculation time of case 2. When the number of nodes is decreased (case 1), the impact on the calculation results is greater, the maximum relative error can reach 12.05%, and the calculation time is about 0.77 times of the calculation time of case 2. Based on the consideration of both computational accuracy and efficiency, the mesh of case 2 is chosen for calculation.

4.2 Vibration and Heat Transfer Analyses

4.2.1 Shell-side Outlet Temperature Distribution

Based on the step-by-step calculation strategy proposed in "2.1.2 Basic Calculation Process", in order to ensure the full development of the shell-side fluid domain, the average temperature of the shell-side outlet was monitored. Fig. 4.6 shows the variation of the average shell-side outlet temperature (T_{out}) with the rough calculation time (t_r) for shell-side inlet velocity u_{in} = 0.1 m/s and 0.4 m/s, shell-side inlet fluid temperature T_{in} = 293.15 K, and ETB wall temperature T_w = 353.15 K.

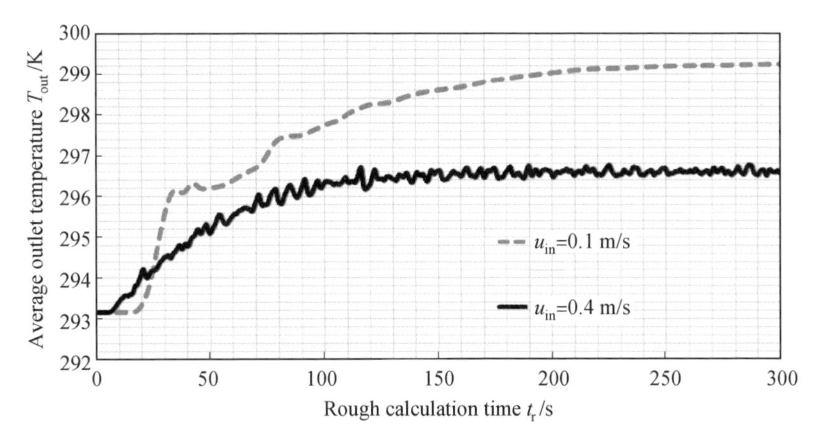

Fig. 4.6 Variation of the average shell-side outlet temperature with the rough calculation time

From Fig. 4. 6, it can be seen as follows.

(1) The average temperature of the shell-side outlet increases with the calculation time showing such a trend: first basically unchanged, then increasing, and then gradually reaching stability.

(2) The larger the shell-side inlet velocity, the lower the average shell-side outlet temperature, and the shorter the calculation time to reach stability.

(3) The average shell-side outlet temperature at different shell-side inlet velocities reaches stability, which indicates that the fluid domain in the heat exchanger has been fully developed.

4.2.2 Effect of Shell-side Inlet Velocity on Vibration

Fig. 4. 7 shows the variation of the displacement of monitoring point A_1 in the x-direction (S_x) with the FSI calculation time (actuarial calculation time t_a) for the shell-side inlet flow velocity $u_{in}=0.1$ m/s and 0.4 m/s, where the tube row spacing $H=60$ mm, and the number of tube rows $n=6$.

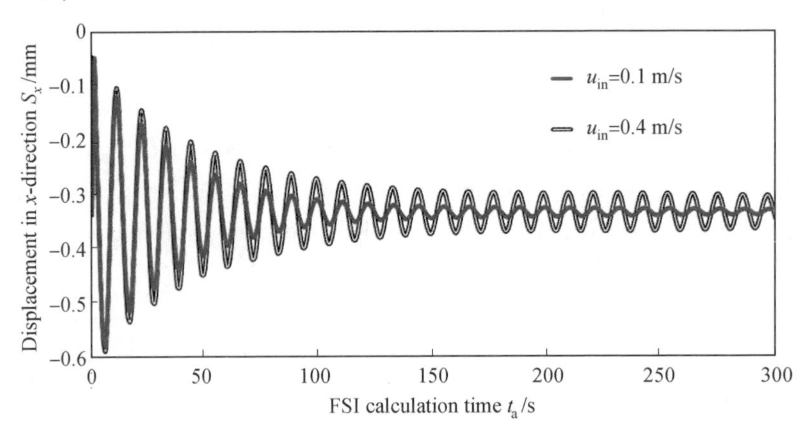

Fig. 4. 7 Variation of the displacement of monitoring point A_1 in x-direction with the FSI calculation time

Based on Fig. 4. 7, the following conclusions can be found.

(1) As the FSI calculation time increases, the vibration amplitude gradually decreases and becomes stable, and the vibration basically reaches stability when the FSI calculation time $t_a=0.6$ s.

(2) The larger the inlet velocity of the shell-side fluid, the higher the vibration amplitude of the monitoring point after stabilization.

At the same time, this also indicates that the time and time step settings of the

FSI calculation can ensure that the vibration of the ETB reaches stability, and the vibration response after stabilization is the vibration response of the ETB under the fully developed fluid domain conditions induced by the actual shell-side fluid.

In order to facilitate further analysis of the amplitude and frequency of the monitoring points, the FFT is performed on the displacement time curve after vibration stabilization (t_a = 0.6 – 1.2 s). Table 4.4 shows the vibration frequency and amplitude of the monitoring points on the two stainless steel connectors of the bottom row ETB (ETB 1) and the top row ETB (ETB 6) in all directions when u_{in} = 0.4 m/s.

Table 4.4 Vibration frequency and amplitude of the monitoring points on the two stainless steel connectors of the bottom row ETB and top row ETB in all directions

Monitoring points	Direction	Calculation results		Monitoring points	Direction	Calculation results	
		f/Hz	A/mm			f/Hz	A/mm
A_1	x	23.33	0.0307	A_6	x	23.33	0.0336
	y	23.33	0.0014		y	23.33	0.0019
	z	23.33	0.0012		z	23.33	0.0011
B_1	x	24.17	0.0220	B_6	x	24.17	0.0263
	y	24.17	0.0013		y	24.17	0.0020
	z	24.17	0.0018		z	24.17	0.0013

It can be obtained from Table 4.4 as follows.

(1) The vibration frequency of the monitoring point on the stainless-steel connector Ⅲ is lower and the vibration amplitude is higher. On the other hand, the vibration frequency of the monitoring point on the stainless-steel connector Ⅳ is higher and the vibration amplitude is lower.

(2) The vibration of each monitoring point is mainly in the x-direction, indicating that the vibration is mainly manifested as out-plane vibration, which is different from the vibration response of each row of ETBs when the uniform shell-side fluid induces multiple rows of ETBs.

(3) Through further comparison and analysis, the vibration frequency of the monitoring points on the same stainless-steel connector of each row of ETBs is the same in all directions, but the amplitude is different.

Because each row of ETBs in the heat exchanger exhibits out-plane vibration induced by the actual shell-side fluid, and the vibration amplitude in the out-plane direction (x-direction) is one order of magnitude higher than the vibration amplitude in the in-plane direction (y- and z-directions). Therefore, in the subsequent analysis, only the vibration in the out-direction (x-direction) will be discussed.

Table 4.5 shows the vibration frequency (f_x) and amplitude (A_x) of the monitoring points in the x-direction on the two stainless steel connectors of the bottom row ETB (ETB 1) and the top row ETB (ETB 6) under different inlet flow velocity (u_{in} = 0.1 m/s to 0.4 m/s).

Table 4.5 Vibration frequency and amplitude of the monitoring points in the x-direction on the two stainless steel connectors of the bottom row ETB and top row ETB

Monitoring points	u_{in}	Calculation results		Monitoring points	u_{in}	Calculation results	
		f_x/Hz	A_x/mm			f_x/Hz	A_x/mm
A_1	0.1 m/s	23.12	0.0064	A_6	0.1 m/s	23.12	0.0065
	0.2 m/s	23.15	0.0155		0.2 m/s	23.15	0.0154
	0.3 m/s	23.26	0.0237		0.3 m/s	23.26	0.0243
	0.4 m/s	23.33	0.0307		0.4 m/s	23.33	0.0306
B_1	0.1 m/s	23.76	0.0064	B_6	0.1 m/s	23.76	0.0067
	0.2 m/s	23.85	0.0144		0.2 m/s	23.85	0.0158
	0.3 m/s	24.02	0.0194		0.3 m/s	24.02	0.0225
	0.4 m/s	24.17	0.0220		0.4 m/s	24.17	0.0263

It can be obtained from Table 4.5 as follows.

(1) As the inlet velocity of the shell-side increases, the frequency and amplitude of the monitoring points on both rows of ETBs increase.

(2) In terms of frequency, the increase in monitoring point A_n on stainless steel connector Ⅲ (0.90%) is lower than the increase in monitoring point B_n on stainless steel connector Ⅳ (1.73%), which is basically consistent with the experimental conclusion in reference[6].

(3) For amplitude, the increase in monitoring point A_i on stainless steel connector Ⅲ (A_1 is 379.7%, A_6 is 370.8%) is higher than the increase in monitoring point B_i on stainless steel connector Ⅳ (B_1 is 243.8%, B_6 is 292.5%).

(4) The amplitude of monitoring points on stainless steel connector Ⅲ is greatly affected by the shell-side inlet velocity, while the frequency of monitoring points on stainless steel connector Ⅳ is greatly affected by the shell-side inlet velocity.

Fig. 4.8 shows the variation of the amplitude in x-directior of the two monitoring points on each row of ETBs in x-direction with the number of the ETBs at different shell-side inlet velocities (u_{in} = 0.1 m/s and 0.4 m/s).

From Fig. 4.8, it can be seen as follows.

(1) When the shell-side inlet velocity is low, the amplitude of the monitoring points on the two stainless steel connectors of each row of ETBs is basically the same. When u_{in} = 0.1 m/s, the vibration amplitude of stainless steel connector Ⅲ of ETB 2 is higher, and that of stainless steel connector Ⅳ of ETBs 4–6 is higher.

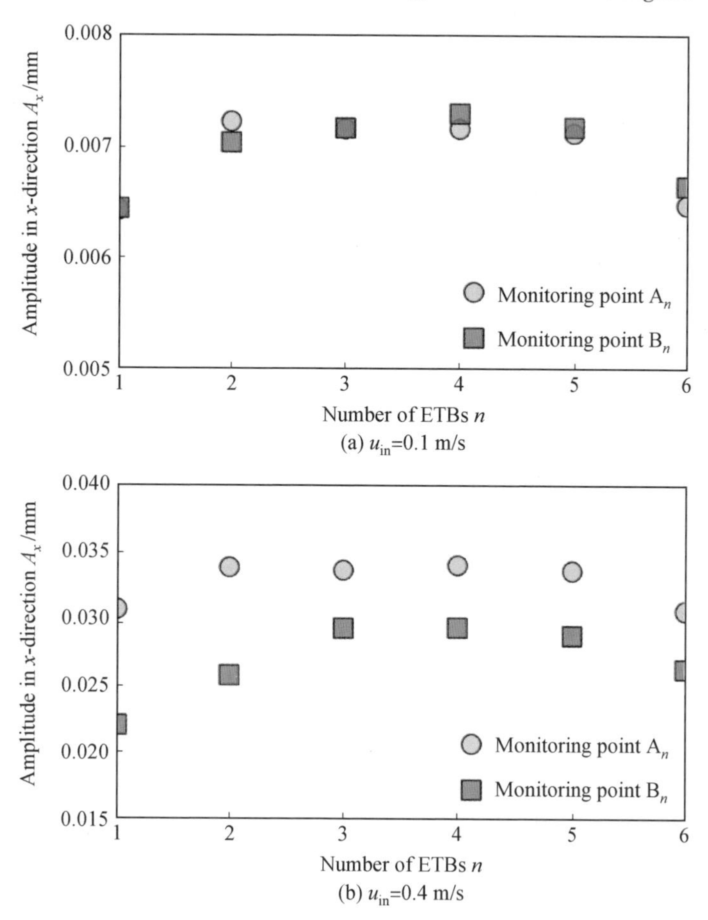

Fig. 4.8 Variation of the amplitude in x-direction of the two monitoring points on each row of ETBs in x-direction with the number of the ETBs at different shell-side inlet velocities

(2) When the shell-side inlet velocity is high (u_{in} = 0.4 m/s), the amplitude of the stainless steel connector Ⅲ of each row of ETBs is higher than that of the stainless steel connector Ⅳ.

(3) As the shell-side inlet velocity increases, the amplitude of the two stainless steel connectors in each row of ETBs increases, but the amplitude of the stainless steel connector Ⅲ increases more significantly.

(4) Under different shell-side inlet velocity conditions, the amplitude of the two monitoring point changes with the number of the ETB, with a trend of first increasing and then decreasing, which is basically consistent with the variation law when the uniform shell-side fluid induces vibration of multiple rows of ETBs.

(5) When the shell-side inlet velocity is low (u_{in} = 0.1 m/s), the maximum relative errors of the amplitudes of the two monitoring points in each row of ETBs are: A_n is 11.09% and B_n is 11.78%. When the shell-side inlet velocity is high (u_{in} = 0.4 m/s), the maximum relative errors of the amplitudes of the two monitoring points in each row of ETBs are: A_n is 10.00% and B_n is 25.17%. This indicates that the higher the shell-side inlet velocity, the lower the uniformity of vibration at each monitoring point (especially on the stainless steel connector Ⅲ).

4.2.3 Effect of Tube Row Spacing on Vibration

By changing the height of the middle connection domain and using the mesh replication, translation and connection functions of ANSYS CFX software, the shell-side fluid domain with different tube row spacing (H=40–70 mm) is assembled and the effect of tube row spacing on the vibration response of each row of ETBs is investigated. For the calculation, the shell-side inlet velocity u_{in} = 0.4 m/s and the number of ETBs n=6.

Table 4.6 shows the vibration frequency (f_x) and amplitude (A_x) of the monitoring points on the two stainless steel connectors in the x-direction for the bottom row ETB (ETB 1) and the top row ETB (ETB 6) under different tube row spacing (H=40–70 mm) conditions.

From Table 4.6, it can be derived as follows.

(1) With the increase of the tube row spacing, the frequency of the monitoring points on the two ETBs gradually decreases, but the decrease is low, and the amplitude basically tends to increase first and then decrease.

(2) For the frequency, the decrease of monitoring point A_n on stainless steel connector Ⅲ (0.21%) is lower than that of monitoring point B_n on stainless steel connector Ⅳ (1.87%).

Table 4.6 Vibration frequency and amplitude of the monitoring points on the two stainless steel connectors in the x-direction for the bottom row ETB and the top row ETB under different tube row spacing conditions

Monitoring points	H/mm	Calculation results		Monitoring points	H/mm	Calculation results	
		f_x/Hz	A_x/mm			f_x/Hz	A_x/mm
A_1	40	23.37	0.0230	A_6	40	23.37	0.0206
	50	23.35	0.0295		50	23.35	0.0284
	60	23.33	0.0307		60	23.33	0.0306
	70	23.32	0.0310		70	23.32	0.0299
B_1	40	24.58	0.0170	B_6	40	24.58	0.0188
	50	24.32	0.0208		50	24.32	0.0246
	60	24.17	0.0220		60	24.17	0.0263
	70	24.12	0.0212		70	24.12	0.0240

(3) For the magnitude, the maximum variation of monitoring point A_n on stainless steel connector Ⅲ (A_1 is 34.78%, A_6 is 48.54%) is higher than the maximum variation of monitoring point B_n on stainless steel connector Ⅳ (B_1 is 29.41%, B_6 is 39.89%).

(4) The amplitude of the monitoring point on stainless steel connector Ⅲ is influenced by the tube row spacing, and the frequency of the monitoring point on stainless steel connector Ⅳ is influenced by the tube row spacing, which is consistent with the situation when the shell-side inlet velocity is affected.

Fig. 4.9 shows the variation of the amplitude A_x in the x-direction of the two monitoring points on each row of ETBs with the number of the ETBs for different tube row spacing (H=40–70 mm).

From Fig. 4.9, it can be seen as follows.

(1) For the stainless steel connector Ⅲ, the maximum relative error of the monitoring point amplitude is 44.47% when the tube row spacing H = 40 mm, and 11.28% when the tube row spacing H=70 mm. For the stainless steel connector Ⅳ, the maximum relative error of the monitoring point amplitude is 32.81% when the tube row spacing H = 40 mm, and 25.09% when the tube row spacing H = 70 mm. This indicates that when the tube row spacing is small, the amplitude changes more drastically with the number of the ETBs, and the interaction between each row of ETBs is stronger. In addition, the amplitude of the monitoring points on the stainless steel connector Ⅲ is more influenced by the tube row spacing.

(2) Under different tube row spacing conditions, the amplitude of the two monitoring points basically shows a trend of increasing first and then decreasing with the number of the ETBs. When the tube row spacing is small ($H = 40$ mm), the uniformity of vibration of stainless steel connector Ⅲ is poor. Differently, when the tube row spacing is large ($H=70$ mm), the uniformity of vibration of stainless steel connection Ⅳ is poor.

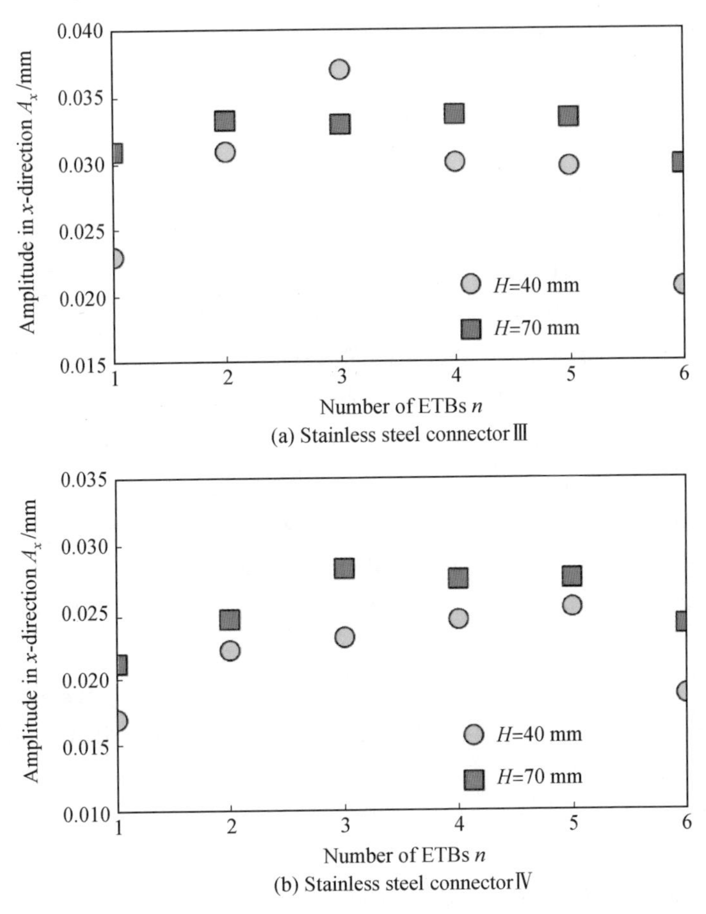

Fig. 4. 9 Variation of monitoring point amplitude with the number of the ETBs for different tube row spacing

4.2.4 Effect of Tube Row Number on Vibration

By changing the number of replicatesof the tube bundle domain and the middle connection domain, based on the mesh replication, translation and connection functions of ANSYS CFX software, a shell-side fluid domain with 12 tube rows ($n =$

12) is assembled, as shown in Fig. 4. 10, and the effect of tube row number on the vibration response of each row of ETBs is studied.

During the calculation, the shell-side inlet velocity $u_{in}=0.4$ m/s, and the tube row spacing $H=60$ mm. The ETB is numbered from bottom to top as 1, 2, ..., 12, and the corresponding monitoring points are A_n and B_n ($n=1, 2, \ldots, 12$).

It should be pointed out that in the first rough calculation process of the shell-side fluid domain containing 12 rows of ETBs, the calculation time is set to 600 s, and the time step is set to 0. 1 s. After calculation, those setting can ensure that the flow field is fully developed at the shell-side inlet velocity $u_{in}=0.4$ m/s.

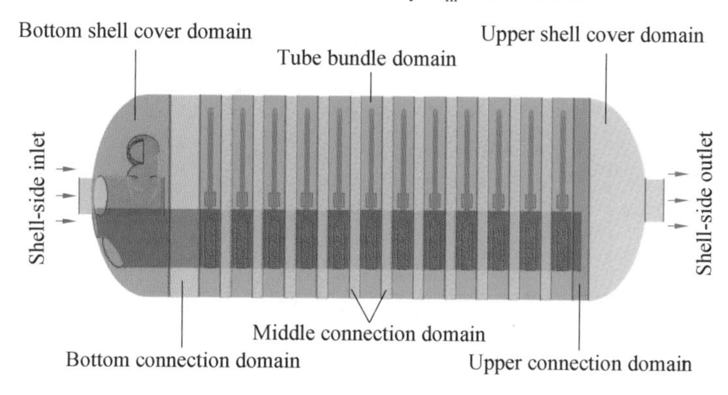

Fig. 4. 10 Model of shell-side fluid domain containing 12 rows of ETBs

Table 4. 7 shows the vibration frequencies and amplitudes in the x-direction of monitoring points on the stainless steel connectors of some ETBs.

Table 4. 7 Vibration frequencies and amplitudes in the x-direction of monitoring points on the stainless steel connectors of some ETBs

Monitoring points	Calculation results	Tube bundle number					
		2	4	6	8	10	12
A_n	f_x/Hz	23.33	23.33	23.33	23.33	23.33	23.33
	A/mm	0.0329	0.0339	0.0347	0.0349	0.0342	0.0300
B_n	f_x/Hz	24.17	24.17	24.17	24.17	24.17	24.17
	A_x/mm	0.0275	0.0308	0.0311	0.0306	0.0304	0.0278

From Table 4. 7, the following conclusions can be drawn.

(1) The vibration frequency of the monitoring points on the same stainless steel connector of each row of ETBs is consistent, and the vibration frequency of the monitoring points on the stainless steel connector Ⅳ is relatively high.

(2) The vibration amplitude of the monitoring points on the same stainless steel connector of the ETB increases with the increase of the number of ETBs and then decreases, and the vibration amplitude of the monitoring points on the stainless steel connector Ⅲ is higher.

Fig. 4.11 shows the variation of the amplitude A_x of the two monitoring points in the x-direction on each row of ETBs with different tube row numbers (6 rows, 12 rows) as a function of the number of ETBs.

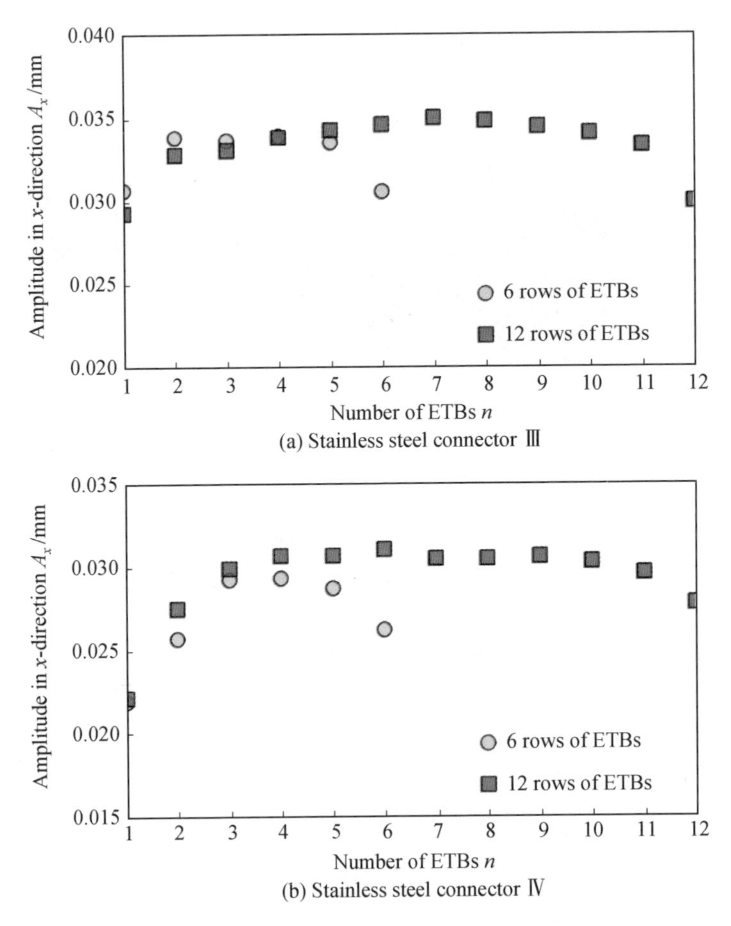

Fig. 4.11 Variation of monitoring point amplitude with the number of ETBs under different tube row numbers

From Fig. 4.11, it can be seen as follows.

(1) The vibration amplitude of the monitoring points on the same stainless steel connector of each row of ETBs increases first and then decreases with the increase of the number of ETBs, and this trend is not affected by the number of rows.

(2) Compared with the vibration amplitude of the monitoring points on the stainless steel connector Ⅳ, the vibration amplitude of the monitoring points on the stainless steel connector Ⅲ is higher, which is also not affected by the number of ETBs.

(3) For stainless steel connector Ⅲ, the maximum relative error of the amplitudes of all monitoring points for 12 rows of ETBs is 16.53%, and for 6 rows of ETBs is 10.00%. In addition, for stainless steel connector Ⅳ, the maximum relative error of the amplitudes of all monitoring points for 12 rows of ETBs is 28.70%, and for 6 rows of ETBs is 25.17%. This indicates that the non-uniformity of the vibration increases with the number of ETBs increase.

4.2.5 Heat Transfer Performance Analysis

Fig. 4.12 shows the heat transfer coefficient (h) of each row of ETBs in the six-row ETB heat exchanger when the shell-side inlet velocity $u_{in}=0.1$ m/s and 0.4 m/s, where the tube row spacing $H=60$ mm, the shell-side inlet fluid temperature $T_{in}=293.15$ K, and the ETB wall temperature $T_w=353.15$ K.

In Fig. 4.12, "no vibration" refers to the heat transfer coefficient of each row of ETBs without flow-induced vibration when only ANSYS CFX software is used to calculate the shell-side fluid domain. And also, "vibration" refers to the heat transfer coefficient of each row of ETBs when the ETBs vibrate under the guidance of actual shell-side fluid after FSI calculation of shell-side fluid and structural domains.

From Fig. 4.12, it can be seen as follows.

(1) The heat transfer coefficient of each row of ETBs increases with the increase of shell-side inlet velocity. When the shell-side inlet velocity increases from 0.1 m/s to 0.4 m/s, the heat transfer coefficient of the ETB without vibration increases by 2.15 times on average, and the heat transfer coefficient of the ETB with vibration increases by 1.68 times on average.

(2) When the shell-side inlet velocity is low, the heat transfer coefficients of each row of ETBs under flow-induced vibration conditions are significantly increased. When $u_{in}=0.1$ m/s, the heat transfer coefficient of each row of ETBs is increased by a maximum of 1.31 times.

(3) When the shell-side inlet velocity is high, the heat transfer coefficient of each row of ETBs under flow-induced vibration condition does not change significantly, and the average heat transfer coefficient of each row of ETBs under flow-induced is only increased by about 0.73%. When $u_{in}=0.4$ m/s, the heat transfer

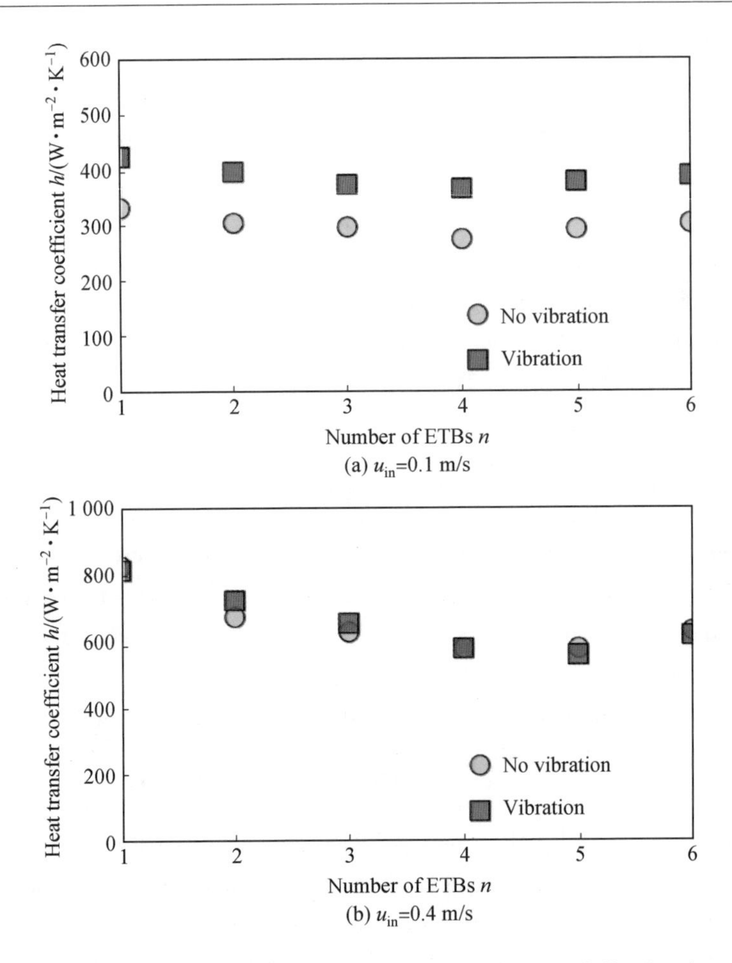

Fig. 4. 12 Heat transfer coefficient of each row of ETBs at different shell-side inlet velocities

coefficients of the second, third and fourth rows of ETBs increases slightly, but the heat transfer coefficient of the first, fifth and sixth rows of ETBs decreases to a certain extent.

In summary, the actual shell-side fluid induces the vibration of the ETBs in the heat exchanger, which can significantly improve the heat transfer coefficient of each row of ETBs at low shell-side inlet velocity (or low *Re*), but when the *f* shell-side inlet velocity is high, the effect of heat transfer enhancement is not obvious.

4.3 IETB and Effect of Flow Direction

4.3.1 IETB and Hexahedron Shell-side Fluid Domain

The schematic diagram of the structure of an IETB is shown in Fig. 4. 13. Compared with TETB, the IETB consists of three stainless steel connectors (Ⅲ, Ⅳ, Ⅴ) of the same size (40 mm×20 mm×20 mm), which can effectively reduce the inherent characteristics of the ETB and make it more prone to vibration under the impact of slower flowing fluids[7].

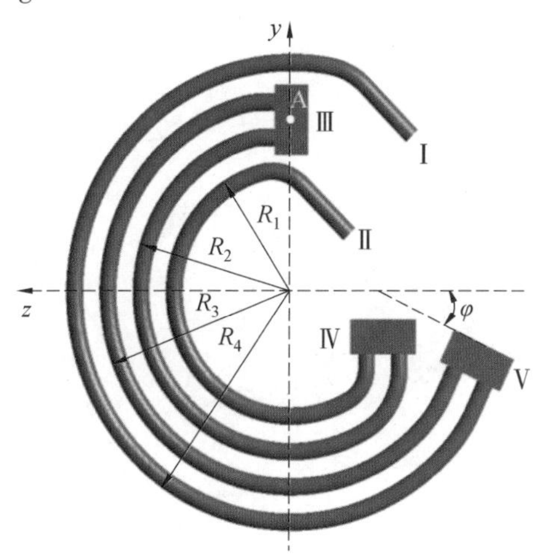

Fig. 4. 13 Schematic diagram of the structure of an IETB

The IETB is also assembled from four copper bent tubes, with Ⅰ and Ⅱ as fixed ends. In order to facilitate the analysis of the vibration characteristics of the IETB under the impacts of different flow directions of the shell-side fluid, a monitoring point (marked as A in the diagram) is set at the top center of the stainless steel connector Ⅲ. The main structural parameters of the IETB are shown in Table 3. 1, and the specific material parameters are shown in Table 3. 2.

In order to study the vibration and heat transfer characteristics of the IETB under the impact of shell-side fluid with different flow directions, a hexahedron shaped shell-side fluid domain is proposed, as shown in Fig. 4. 14, where the length, width and height of the fluid domain are $L=320$ mm, $W=320$ mm and $H=30$ mm, respectively. The six faces are labeled as F_1–F_6.

In order to study the influence of the shell-side fluid on the vibration and heat transfer characteristics of the IETB under different flow directions, three sets of shell-side fluid with different flow directions are set up. The specific flow direction settings of the shell-side fluid are shown in Table 4.8.

Fig. 4.15 shows the mesh division of the structural domain and the shell-side fluid domain of IETB. Four copper bent tubes are divided into hexahedron meshes, three stainless steel connectors are divided into tetrahedron meshes. The shell-side fluid domain is divided into tetrahedron meshes, and six layers of boundary layer meshes are set near the inner wall surfaces. The number of elements in the structural domain is 5,014, and the number of nodes is 23,053. And also, the number of elements in the shell-side fluid domain is 963,416, and the number of nodes is 176,304.

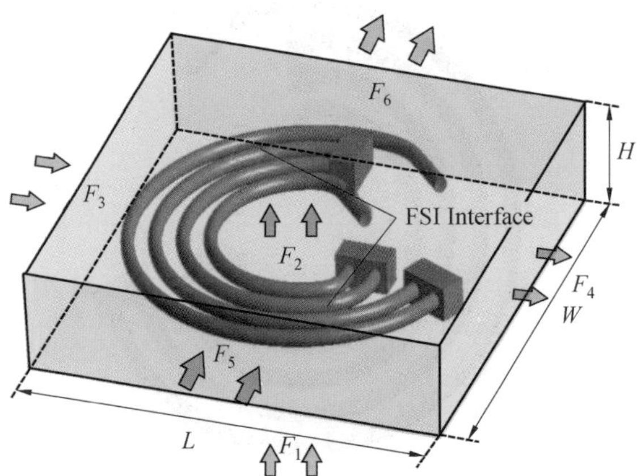

Fig. 4.14 Schematic of the hexahedron shaped shell-side fluid domain

Table 4.8 Three sets of shell-side fluid with different flow directions

Mark	Shell-sideinlet face	Shell-side outlet face
D_{1-2}	F_1	F_2
D_{3-4}	F_3	F_4
D_{5-6}	F_5	F_6

Through trial calculation, further increasing the number of meshes in the structural and shell-side fluid domains, and the number of boundary layers have no effect on the numerical analysis results.

The boundary conditions of the structural domain are set as follows. The outer surfaces of the copper bent tubes and three stainless steel connectors are set as "Fluid

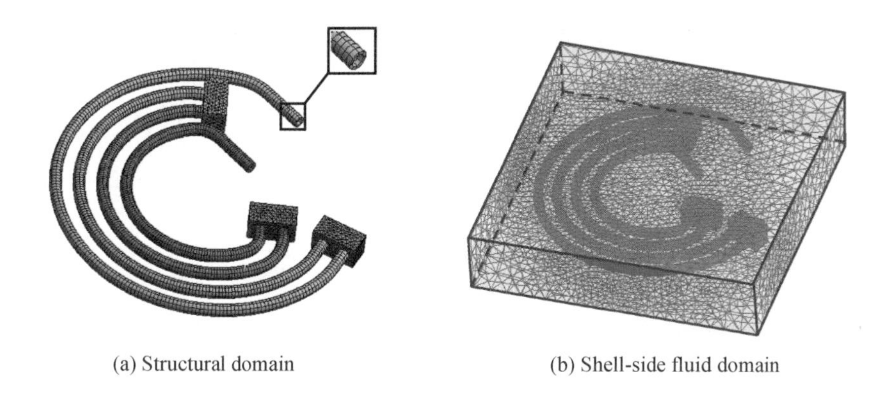

(a) Structural domain (b) Shell-side fluid domain

Fig. 4. 15 Mesh division of the structural domain and the shell-side fluid domain of IETB

Solid Interface". The two end faces of Ⅰ and Ⅱ are set as "Fixed Support". The direction of gravitational acceleration (Standard Earth Gravity) is set as the x-direction, and its value is 9.8066 m/s^2.

The boundary conditions of the fluid domain are set as follows. The inlet adopts the "Inlet" type, with a given inlet velocity (u_{in} = 0.1 m/s, 0.3 m/s, 0.5 m/s, 0.7 m/s, 0.9 m/s, 1.1 m/s), and an inlet temperature (T_{in} = 293.15 K). The outlet adopts the "Outlet" type and the outlet pressure is set to 0 Pa. The inner surface of the shell-side fluid domain is set as "Fluid Solid Interface", and the temperature is set as T_{wall} = 333.15 K. The other wall surface adopts a non-slip adiabatic "Wall".

When calculating, the structuraland shell-side fluid domains use the same total calculation time (0.96 s) and time step (0.003 s).

4.3.2 Vibration and Heat Transfer Analyses of IETB

Fig. 4.16 shows the variation amplitude (A) of the monitoring point A in all directions with inlet velocity (u_{in}) under different shell-side flow directions.

From Fig. 4.16, it can be seen as follows.

(1) When the shell-side fluid impacts the IETB from the D_{1-2} direction, the variation amplitude fluctuation of monitoring point A in each direction is relatively small within the range of 0.1 – 0.5 m/s, and the amplitude in the x-direction is relatively large, manifested as an out-plane vibration. The variation amplitude is relatively significant within the u_{in} range of 0.5 – 1.1 m/s. The amplitudes in the y- and z-directions increase with the increase of inlet velocity, while the amplitudes in the x-direction decrease with the increase of inlet velocity. And also, the vibration

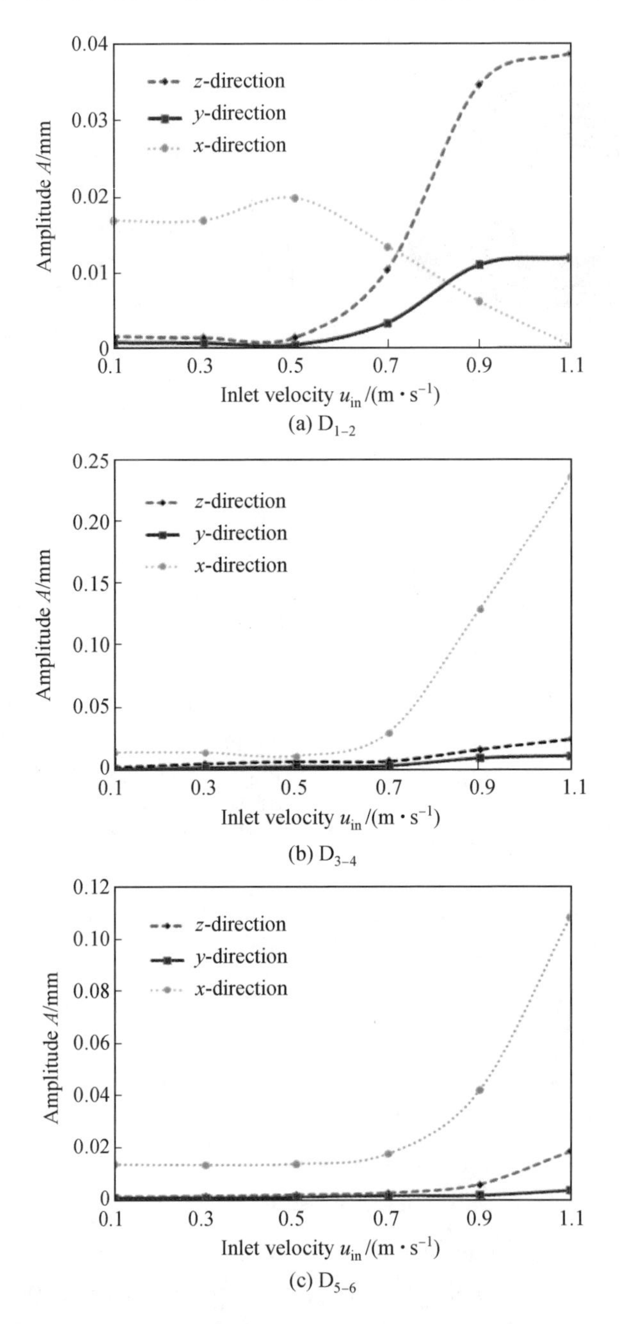

Fig. 4. 16 Variation amplitude of the monitoring point A in all directions with inlet velocity under different shell-side flow directions

mode gradually changes to in-plane vibration.

(2) When the shell-side fluid impacts the IETB from the D_{3-4} and D_{5-6} directions, the variation amplitude of monitoring point A in all directions with velocity is basically the same, and the amplitude fluctuation of monitoring point A in all directions is relatively small within the u_{in} range of 0.1 – 0.7 m/s. The variation amplitude changes significantly within the u_{in} range of 0.7 – 1.1 m/s. The variation amplitude continues to increase, and the variation amplitude in the x-direction remains relatively large. In addition, the vibration mainly manifests as an out-plane vibration perpendicular to the plane of the IETB.

(3) At low inlet velocities (u_{in} = 0.1–0.7 m/s), there is little difference in the vibration intensity of the shell-side fluid impacting the IETB from all directions, and the maximum vibration amplitude in all directions of the monitoring point is less than 0.03 mm. At high inlet velocities (u_{in} = 0.7–1.1 m/s), the vibration intensity of the shell-side fluid impacting the IETB from the D_{3-4} and D_{5-6} directions is relatively high, indicating that the vibration of the IETB is more intense when the fluid is impacted from the side.

Table 4.9 shows the heat transfer coefficient of the IETBs with different inlet velocities under different shell-side flow directions. During the calculation, the inlet fluid temperature is set to T_{in} = 293.15 K, and the IETB wall temperature is set to T_{wall} = 333.15 K.

Table 4.9 Heat transfer coefficient of the IETB with different inlet velocities under different shell-side flow directions

$u_{in}/(m \cdot s^{-1})$	Heat transfer coefficient $h/(W \cdot m^{-2} \cdot K^{-1})$		
	D_{1-2}	D_{3-4}	D_{5-6}
0.1	1,027.56	757.437	739.49
0.3	2,346.86	1,612.96	1,556.58
0.5	3,586.67	2,396.79	2,336.49
0.7	4,785.58	3,237.72	3,165.94
0.9	5,898.56	4,100.22	4,109.72
1.1	6,925.36	4,953.73	5,042.25

From Table 4.9, it can be concluded as follows.

(1) When the shell-side fluid impacts the IETB from different directions, the flow rate of the shell side fluid increases, the heat transfer coefficient of the IETB

increases with the increase of the inlet velocity, which is consistent with the conclusion in the literatures [2-3,8-9].

(2) When the shell-side fluid impacts the IETB from the D_{1-2} direction, the heat transfer coefficient of the IETB is greater than that of the other two directions. This indicates that the heat transfer performance of IETB is the best when the shell-side fluid impacts the IETB from the D_{1-2} direction.

Based on the previous vibration analysis of the IETB, it can be seen that the vibration intensity is not the only factor causing enhanced heat transfer. For the IETB, the direction of shell-side fluid impact is an important factor affecting its heat transfer characteristics. Therefore, when designing the ETB heat exchangers, it is not necessary to blindly pursue high intensity vibration.

4.3.3 Effect of Installation Angle on IETB Heat Exchanger

1. IETB Heat Exchanger and Its Fluid Domain

The above analysis of a single row IETB shows that when the fluid impacts the plane of the IETB in a forward direction, it can cause the minimum vibration and the maximum heat transfer coefficient. However, due to the structural characteristics of the shell-side inlet tube and the arrangement of the IETB and the tube-side inlet and outlet tubes inside the heat exchanger, the shell-side fluid does not directly impact the IETB. Therefore, based on the heat exchanger with six-row ETBs and its shell-side fluid domain, the vibration response and heat transfer performance of the six-row ETBs in heat exchanger with different installation angles are studied by using the sequential solution method for bi-directional FSI.

Fig. 4.17 shows the IETB heat exchanger and its shell-side fluid domain.

Consistent with the TETB heat exchangers, as shown in Fig. 4.1, the shell-side fluid flows into the heat exchanger from the shell-side inlet tube at the bottom of the heat exchanger. The opening of the shell-side inlet tube is tilted downwards to impact the inner wall of the bottom shell cover of the heat exchanger, with the aim of achieving an approximate spiral upward flow of the shell-side fluid. In Fig. 4.17, α is the installation angle between the IETB and the horizontal plane.

In the calculation, the cylinder of the shell-side fluid domain $D = 300$ mm, the tube row spacing $H = 40$ mm, the diameters of the shell-side inlet and outlet tubes of the fluid domain are 65 mm and 54 mm, respectively.

The mesh generation strategy of the IETB and the shell-side fluid domain are

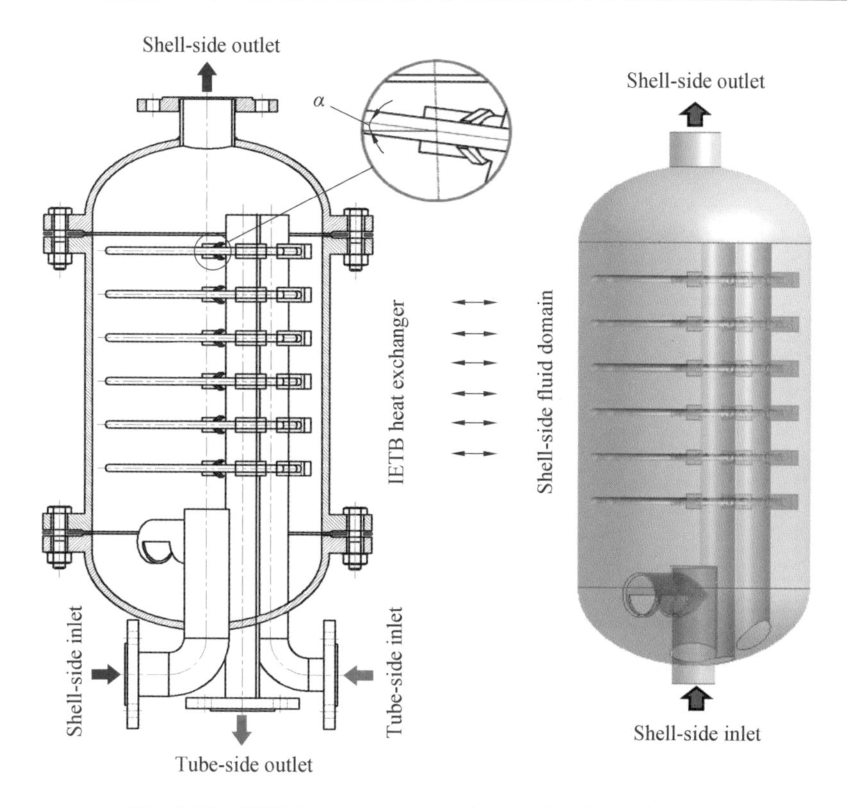

Fig. 4. 17 IETB heat exchanger and its shell-side fluid domain

consistent with the above TETB heat exchanger. Multiple trials are conducted using different meshes to ensure that the mesh meets its independence requirements. It is finally determined that the total number of nodes in the all computational domain is 1,203,050 and the number of elements is 1,467,214.

2. Streamline Distribution in the Shell Side Fluid Domain

The distribution of streamline in the shell-side fluid domain is shown in Fig. 4. 18, and during the calculation process, $u_{in}=0.5$ m/s.

From Fig. 4. 18, it can be seen that the shell-side fluid flows approximately spirally upwards after impacting the bottom shell cover of the IETB heat exchanger. The fluid mainly flows through the area where the stainless steel connector V is located, and then flows approximately spirally downwards after passing through the upper shell cover of the heat exchanger. Then, it impacts the IETB from various directions in the middle area of the heat exchanger, and finally flows out from the shell-side outlet. This demonstrates the chaotic nature of shell-side fluid.

In summary, further determination is needed to determine whether the horizontal

Fig. 4. 18 Distribution of streamline in the shell-side fluid domain

arrangement of the ETBs inside the TETB heat exchanger is reasonable. For this purpose, different installation angles of the IETB are proposed ($\alpha=-30°$, $-20°$, $-10°$, $0°$, $10°$ and $20°$). Then, based on different inlet velocities of the shell-side fluid, vibration response and heat transfer performance of the IETB are studied in this chapter.

3. Effect of Installation Angle on Vibration

Table 4. 10 shows the amplitude of monitoring point A under different installation angles ($\alpha=-30°, -20°, -10°, 0°, 10°, 20°$) and inlet velocity ($u_{in}=0.1$ m/s, 0.5 m/s) conditions.

Table 4. 10 Amplitude of the monitoring point A under different installation angles and inlet velocity conditions

u_{in}	Number of	Amplitude $A_x/(\times10^{-2}$ mm)					
/(m · s^{-1})	ETBs	−30°	−20°	−10°	0°	10°	20°
0.1	1	1.41	1.73	1.94	2.08	2.17	1.91
	2	1.40	1.74	2.01	2.22	2.29	2.16
	3	1.44	1.82	2.08	2.28	2.24	1.88
	4	1.49	1.85	2.14	2.29	2.25	1.95
	5	1.50	1.88	2.12	2.27	2.36	1.91
	6	1.51	1.80	1.96	2.16	2.18	1.95

Table 4.10 (continued)

u_{in} /(m·s^{-1})	Number of ETBs	Amplitude A_x/(×10^{-2} mm)					
		-30°	-20°	-10°	0°	10°	20°
0.5	1	12.07	12.39	12.99	12.16	14.09	13.16
	2	11.57	13.26	13.84	14.25	15.13	14.97
	3	11.46	13.35	14.96	15.11	15.19	13.53
	4	11.05	12.92	14.77	15.39	15.29	13.97
	5	11.29	13.01	14.93	15.09	16.80	13.79
	6	11.68	12.66	15.37	14.92	16.48	14.82

The calculation shows that the vibration of monitoring point A is the most severe in the x-direction. Based on the previous analysis results shown in Fig. 4.16 (a), Table 4.10 only lists the amplitudes of monitoring point A in the x-direction (A_x). For the convenience of analysis, the IETB is numbered from bottom to top as 1–6.

From Table 4.10, it can be seen that the shell-side inlet velocity has a significant impact on the vibration response of the monitoring point. The results indicate that the higher the inlet velocity, the greater the amplitude of monitoring point A in the x-direction. In addition, the installation angle of the IETB has a significant impact on the vibration response. The calculation results at different flow rates indicate that the maximum amplitude occurs when the installation angle is 10°.

4. Effect of Installation Angle on Heat Transfer

Table 4.11 shows the average heat transfer coefficient (h_a) of the IETBs under different installation angles. During the calculation, the inlet velocity u_{in} = 0.1 m/s and 0.5 m/s, the inlet fluid temperature is set to T_{in} = 293.15 K, and the IETB wall temperature is set to T_{wall} = 333.15 K.

Table 4.11 Average heat transfer coefficient of the IETBs under different installation angles

u_{in}/(m·s^{-1})	h_a/(W·m^{-2}·K^{-1})					
	-30°	-20°	-10°	0°	10°	20°
0.1	289.28	294.61	296.66	293.09	291.98	290.70
0.5	686.31	740.42	748.70	715.05	680.61	699.81

It can be seen from Table 4.11 that the installation angle of the IETB has an impact on the average heat transfer coefficient of the heat exchanger under different inlet velocities. When $\alpha=-10°$, the average heat transfer coefficient of the IETB is the largest. This indicates that within the parameter range calculated in this book, $-10°$ is the optimal installation angle.

Combined with the analysis of the vibration responses of the IETB, the installation angle corresponding to the maximum average heat transfer coefficient is not consistent with the installation angle corresponding to the maximum vibration amplitude, which is consistent with the analysis conclusion of the single-row IETB. This further indicates that vibration intensity is not the only factor causing enhanced heat transfer of the ETB. In addition, based on the previous research on the heat transfer performance of the single-row IETB under the impact of different directions of incoming flow, the average heat transfer coefficient of the IETB is the highest when the fluid impacts the plane of the ETB in the positive direction. The analysis of different installation angles shows that when the installation angle of the IETB is $-10°$, the shell-side fluid impact of the IETB tends to be more frontal impact.

References

[1] JI J D, GE P Q, BI W B. Numerical analysis on shell-side flow-induced vibration and heat transfer characteristics of elastic tube bundle in heat exchanger [J]. Applied thermal engineering, 2016, 107: 544-551.

[2] DUAN D R, GE P Q, BI W B, et al. An empirical correlation for the heat transfer enhancement of planar elastic tube bundle by flow-induced vibration [J]. International journal of thermal sciences, 2020, 155: 106405.

[3] JI J D, GE P Q, LIU P, et al. Design and application of a new distributed pulsating flow generator in elastic tube bundle heat exchanger [J]. International journal of thermal sciences, 2018, 130: 216-226.

[4] JI J D, ZHANG J W, GAO R M, et al. Tests for pulsating flow generator-induced vibration of elastic tube bundle [J]. Journal of vibration and shock, 2021, 40 (3): 291-296.

[5] JI J D, NI X W, SHI B J, et al. Influence of deflector direction on heat transfer capacity of spiral elastic tube heat exchanger [J]. Applied thermal engineering, 2024, 236: 121754.

[6] SU Y C. A study on the characteristics of the flow-induced vibration and heat transfer of elastic tube bundle [D]. Jinan: Shandong University, 2012.

[7] JI J D, CHEN W Q, ZHOU R, et al. Influence of flow direction on vibration and

heat transfer characteristics of elastic tube bundle [J]. Journal of vibration and shock, 2022, 41(18): 252-257.
[8]JI J D, ZHOU R, GAO R M, et al. Analysis on heat transfer characteristic of spiral elastic tube bundle heat exchangers with different cross section shapes [J]. Journal of vibration and shock, 2022, 41(22): 219-225.
[9]JI J D, LU Y, SHI B J, et al. Numerical research on vibration and heat transfer performance of a conical spiral elastic bundle heat exchanger with baffles [J]. Applied thermal engineering, 2023, 232: 121036.

Chapter 5 Design of a Shell-side Distributed Pulsating Flow Generator

It can be seen from the above statement that the ETB in the heat exchanger has the problem of uneven vibration of each row of ETBs. In this way, parts of the ETBs are prone to fatigue damage, and parts of the ETBs heat transfer effect are poor, affecting the service life and heat transfer efficiency of the heat exchanger. Therefore, the research of pulsating flow generators which can reasonably induce and control the vibration of each ETB in heat exchanger has gradually attracted the attention of academic community[1-8].

When a fluid flows around avortex element with a certain shape and size, it will form a pulsating flow with the velocity and direction changing with time in its wake. By using pulsating flow to impact the appropriate position of the ETB, the ETB can be stimulated to vibrate with a certain strength and according to the frequency of the pulsating flow, and then the vibration of the ETB can be properly adjusted, and controlled, and finally, the vibration of the ETB can be controlled. Based on the studies of Liu et al. [8] and Tian et al. [9], by installing a certain shape of vortex element in the tubular flow passage, it was proved that the feasibility of using the pulsating flow formed by the fluid flowing around the vortex element to impact the ETB and thus stimulate the vibration.

In this chapter, through the study of the two-dimensional fluid flow around vortex elements with different shapes, the appropriate vortex element was selected. Based on the study of two-dimensional fluid flow around the vortex element in the branch tube, the flow passage structure was designed, and the influence of inlet velocity, inlet fluid flow direction, vortex element structural parameters and external fluid domain on the pulsating flow formed by the vortex element was studied. Combined with the selected vortex element and the designed flow channel for the branch tube, a distributed pulsating flow generator and its installation were designed. The effects of the structural parameters of the distributed pulsating flow generator on the fluid flow at the outlet of each branch tube and the frequency and intensity of the pulsating flow were studied. The suitable structural parameters of distributed pulsating flow generator in the ETB heat exchanger were selected.

5.1 Selection of Vortex Element

5.1.1 Computation Model and Verification

The frequency, intensity and stability of pulsating flow formed in the wake of fluid flowing around different shapes of vortex elements are different. To select a suitable vortex element for pulsating flow generator, the pulsating flow formed by uniform flow around the vortex element is analyzed based on three common simple vortex elements: circular cylinder, square cylinder and right triangular cylinder.

Taking the flow around a two-dimensional circular cylinder as an example, the calculation domain is shown in Fig. 5.1. Where, L is the length of the fluid domain; W is the width of the fluid domain; S is the distance between the circular cylinder oncoming surface and the fluid outlet; d is the diameter of the circular cylinder is used, d is the width of the square cylinder is used, d is the hypotenuse length (or longitudinal dimension) of the right triangular cylinder is used. In the calculation process, unless otherwise specified, $L=25$ mm, $W=10$ mm, $S=15$ mm, $d=2$ mm.

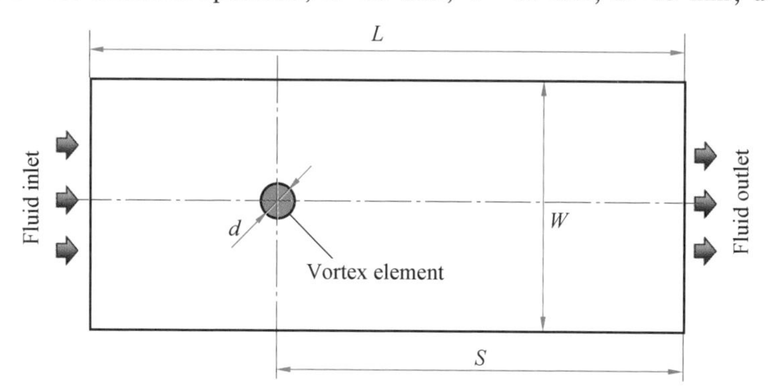

Fig. 5.1 Schematic diagram of the computational domain for flow around a two-dimensional circular cylinder

The simulationis carried out by using ANSYS FLUENT software, the second-order implicit algorithm is solved by the uncoupled two-dimension single-precision solver. The pressure-velocity coupling is performed in SIMPLEC mode. For the fluid inlet, the boundary condition is set to velocity inlet, and the inlet velocity is given. The fluid outlet is set to pressure-outlet, and the outlet pressure is 0 Pa. The surfaces of the vortex element and the upper and lower wall surfaces are both arranged as non-slip surfaces. In addition, the total computation time is 100 s, and the time step is 0.001 s.

To verify the correctness of the numerical simulation method, based on the research of Norberg[10] and Zhang et al.[11], a corresponding geometric model (L = 280 mm, W = 160 mm, S = 200 mm, d = 10 mm) is established to numerically simulate the flow around a two-dimensional circular cylinder. In the calculation process, the fluid medium is water, and the inlet velocity is u_{in} = 0.02 m/s (Re = 200). The meshing is done by ICEM, which is not discussed here.

Fig. 5.2 shows the variation of the lift coefficient (C_l) and drag coefficient (C_d) with the calculation time (t).

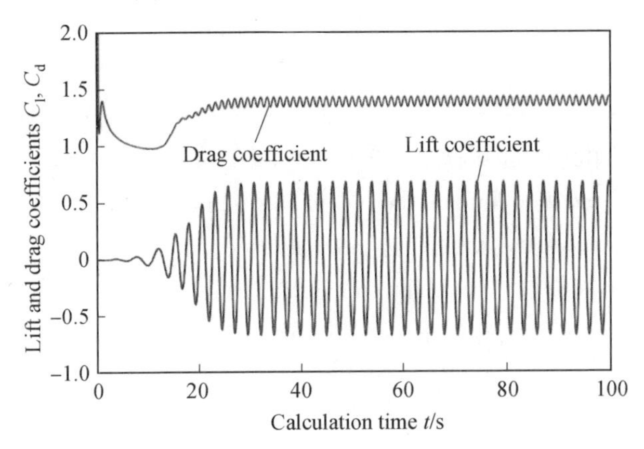

Fig. 5.2 Variation of the lift coefficient and drag coefficient with the calculation time

As can be seen from Fig. 5.2, when t = 35 s, the lift coefficient and drag coefficient are stable. Through the FFT of the lift coefficient and drag coefficient, it is obtained that the frequency of the lift coefficient is 0.41 Hz, the frequency of the drag coefficient is 0.8 Hz, and the frequency of the drag coefficient is about twice the frequency of the lift coefficient.

To verify the accuracy of the simulation, the calculated results in this book are compared with those of Norberg[10] and Zhang et al.[11], as shown in Table 5.1.

Table 5.1 Verification of numerical simulation results

	St	C_l	C_d
Calculation results of this book	0.205	0.483	1.380
Experimental results[10]	0.190–0.210	0.320–0.500	1.300–1.350
Literature calculation results[11]	0.200	0.485	1.355

As can be seen from Table 5. 1, the calculated results are consistent with the experimental results of Norberg and the simulation results of Zhang et al.

5.1.2 Characterization of Pulsating Flow Intensity

Pulsating flow is a fluid whose velocity and direction change periodically with time. To characterize the ability of pulsating flow to induce the vibration strength of the ETBs, a monitoring point is set in the wake of the vortex element to monitor the velocity change with time. The root-mean-square (RMS) of the fluid velocity at the monitoring point over time represents the work capacity of the fluid. The larger the RMS of the fluid velocity is, the more intense it induces the vibration of the ETB.

In this book, the RMS of the fluid velocity at the monitoring point is used to characterize the intensity of pulsating flow. The expression is

$$\zeta = \left(\frac{1}{n}\sum_{i=1}^{n} u_i^2\right)^{1/2} \tag{5.1}$$

where ζ is the RMS of the fluid velocity at the monitoring point; u_i is the fluid velocity value of the monitoring point after the velocity changes steadily; n is the number of fluid velocity values, $n=300$.

5.1.3 Analysis of Fluid Flow Around Vortex Elements

Fig. 5. 3 shows the calculation domain and mesh division when flowing around the circular cylinder. Set monitoring point A in the wake of the vortex element (away from the outlet—3 mm, away from the lower wall—2. 5 mm). The mesh is divided by ICEM software, all of which are quadrilateral meshes. The minimum mesh quality is 0. 94, the number of elements is 89,846, and the number of nodes is 90,584. The effect of mesh encryption (increase by approximately 2 times) on the lift coefficient and its RMS is less than 3. 0%.

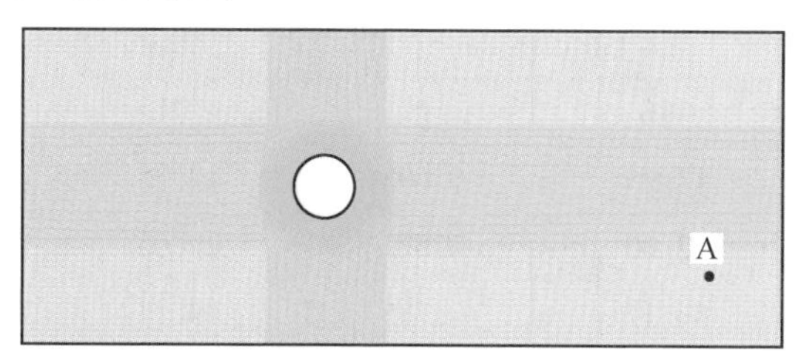

Fig. 5. 3 Calculation domain and mesh division when flowing around the circular cylinder

Fig. 5. 4 shows the velocity cloud diagram of flow around the circular cylinder at different inlet velocities ($u_{in}=0.1$ m/s and 0. 4 m/s). In the calculation process, the fluid medium is water. As can be seen from Fig. 5.4, pulsating flows can be generated in the wake of the circular cylinder under different inlet velocity conditions, and the higher the inlet velocity, the higher the maximum velocity of pulsating flow.

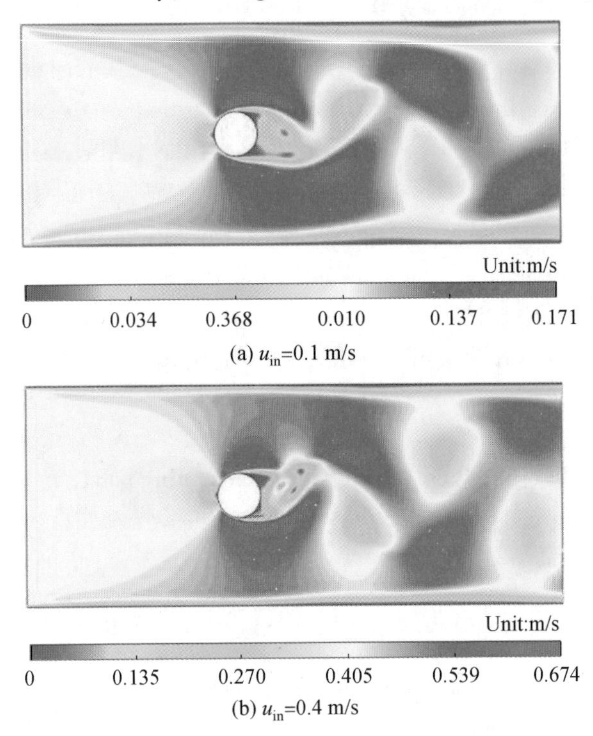

Fig. 5.4 Velocity cloud diagram of flow around the circular cylinder at different inlet velocities

Based on different inlet velocities, the frequency and RMS of the lift coefficient and the frequency and RMS of the fluid velocity at the monitoring point are calculated, as shown in Table 5.2.

Table 5.2 Calculation results of flow around the circular cylinder

$u_{in}/(m \cdot s^{-1})$	Lift coefficient		Monitoring point velocity	
	f/Hz	ζ	f/Hz	ζ
0.1	12.25	0.608	12.25	0.122
0.2	22.00	0.622	22.00	0.231
0.3	34.17	0.785	34.17	0.337
0.4	45.43	0.901	45.43	0.445

From Table 5.2, it can be seen as follows.

(1) The variation frequency of the lift coefficient is consistent with that of the fluid velocity at the monitoring point (or pulsating flow), which is consistent with the conclusions in the literature [12].

(2) With the increase of inlet velocity, the frequency of lift coefficient and the frequency of monitoring point velocity increases, while the RMS of lift coefficient and the root mean square of monitoring point velocity increase too.

(3) The larger the RMS of the lift coefficient is, the larger the RMS of the monitoring point velocity is, and the higher the intensity of the pulsating flow is characterized.

In summary, the frequency of the lift coefficient is consistent with that of pulsating flow. The larger the RMS of the lift coefficient is, the larger the RMS of the monitoring point velocity is, and the higher the intensity of pulsating flow is formed.

Fig. 5.5 shows the calculation domains and mesh divisions when flowing around the square cylinder and the right triangular cylinder. The meshes are divided by ICEM software and are quadrilateral meshes.

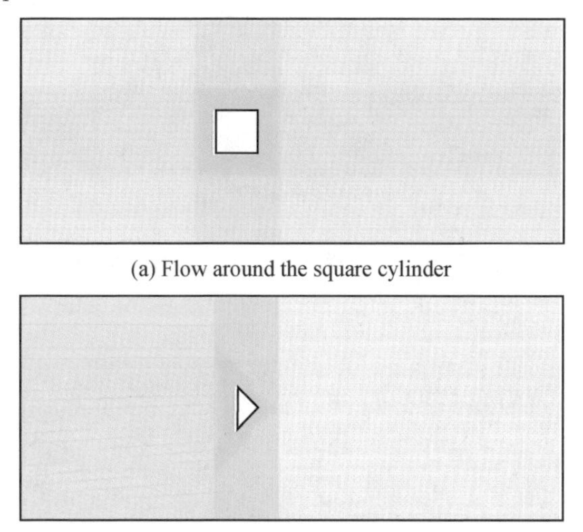

(a) Flow around the square cylinder

(b) Flow around the right triangular cylinder

Fig. 5.5 Calculation domains and mesh divisions when flowing around the square cylinder and the right triangular cylinder

For the fluid domain of flow around the square cylinder, the minimum mesh quality is 0.94, the number of elements is 89,846, and the number of nodes is 90,584. For the fluid domain of flow around the right triangular cylinder, the minimum mesh quality is 0.92, the number of elements is 69,236, and the number of nodes is 69,923. The effect of mesh encryption (increase by approximately 2 times) on the lift coefficient and its RMS is less than 3.0%.

Fig. 5.6 and Fig. 5.7 show velocity cloud diagrams of flow around the square cylinder and right triangular cylinder under different inlet velocities (u_{in} = 0.1 m/s and 0.4 m/s). In the calculation process, the fluid medium is water.

In combination with the velocity cloud diagram of flow around the circular cylinder shown in Fig. 5.4, the following conclusions can be drawn.

(1) When the inlet velocity is low (u_{in} = 0.1 m/s), pulsating flows are formed in the wake of the three common simple vortex elements: circular cylinder, square cylinder and right triangular cylinder.

(2) When the inlet velocity is high (u_{in} = 0.4 m/s), pulsating flows are formed in the wake of the circular cylinder and the right triangular cylinder. However, pulsating flow cannot be formed in the wake of the square cylinder, indicating that the limitation of using the square cylinder is large.

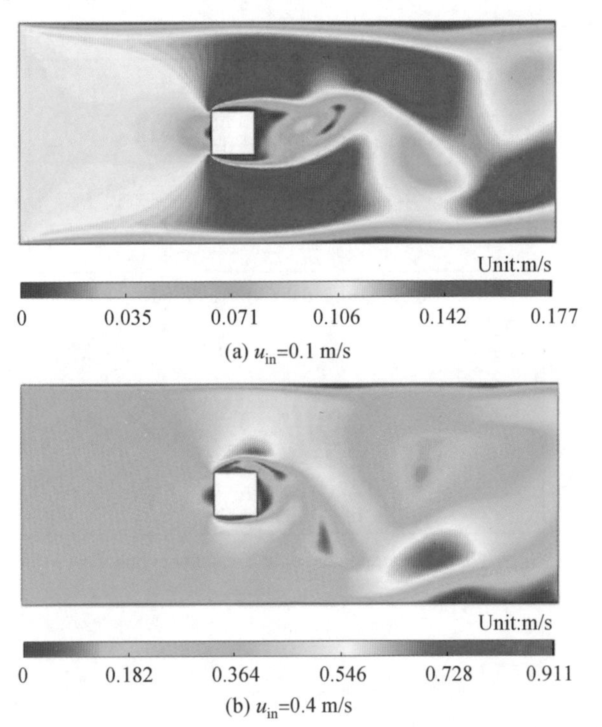

Fig. 5.6 Velocity cloud diagrams of flow around the square cylinder under different inlet velocities

Based on the limitations of the above analysis of the flow aroundthe square cylinder, the following analysis will only be carried out on the lift and drag coefficients

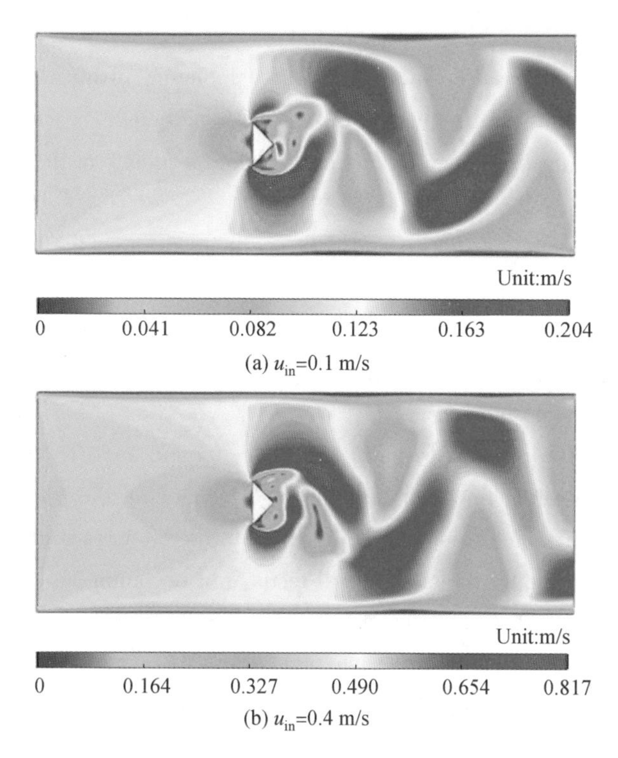

Fig. 5.7 Velocity cloud diagram of flow around the right triangular cylinder under different inlet velocities

of the flow around the circular cylinder and the right triangular cylinder.

Table 5.3 shows the variation of the frequency and RMS of the lift coefficient and the frequency and average value of the drag coefficient of the flow around the circular and right triangular cylinders at different inlet velocities.

Table 5.3 Lift and drag coefficients for two kinds of vortex elements at different inlet velocities

Vortex element	u_{in} /(m · s^{-1})	C_l		C_d	
		f/Hz	RMS	f/Hz	Average value
Circular cylinder	0.1	12.25	0.608	24.0	1.83
	0.4	45.43	0.901	90.75	1.68
Right triangular cylinder	0.1	11.75	1.244	23.5	4.10
	0.4	43.5	1.344	86.75	4.21

From Table 5.3, the followings can be seen.

(1) The frequency of the drag coefficient is about twice that of the lift

coefficient, which is consistent with the previous conclusion.

(2) With the increase of flow velocity, the frequency of lift and drag coefficients both increase, while the frequency of the lift coefficient is consistent with the frequency of the pulsating flow, indicating that the frequency of the forming pulsating flow increases.

(3) The frequency of the lift and drag coefficients under the two kinds of vortex elements at different inlet velocities have little difference. However, the flow around the right triangular cylinder, the EMS of the lift coefficient, and the average value of the drag coefficient increase, indicating that the intensity of the formed pulsating flow and the flow resistance increase.

On the premise of not changing the characteristic length, the frequency and intensity of the pulsating flow can be adjusted by changing the transverse size of the vortex element. This conclusion can be obtained in the later research. For the circular cylinder, changing the transverse size will result in it becoming an ellipse cylinder. In addition, for the right triangular cylinder, changing the transverse size will result in a right angle becoming an acute or obtuse angle.

Considering the difficulty of processing, the processing of a triangular cylinder is easier than that of the an elliptical cylinder. Therefore, the use of the triangular cylinder is more flexible in adjusting the size of the spoiler and processing.

5.2 Design of Branch Tube Flow Channel and Fluid Flow Analysis

5.2.1 Flow Around Vortex Element in Branch Tube

For the pulsating flow generator with multiple branch tubes (Fig. 1.14), the fluid domain structure and the inlet flow direction are different from the computational domain shown in Fig. 5.1 for the flow around a vortex element.

Fig. 5.8 shows the schematic diagram of the branch domain with the right triangular cylinder as the vortex element.

In Fig. 5.8, w is the longitudinal dimension of the right triangular cylinder; l is the transverse dimension of the right triangular cylinder; α is the included angle between the branch tube and the riser tube; γ is the direction angle between the flow direction of the inlet fluid and the axis of the branch tube. In the calculation process, $L=25$ mm, $W=10$ mm, $S=15$ mm, $w=2$ mm, $l=1$ mm, $\alpha=60°$.

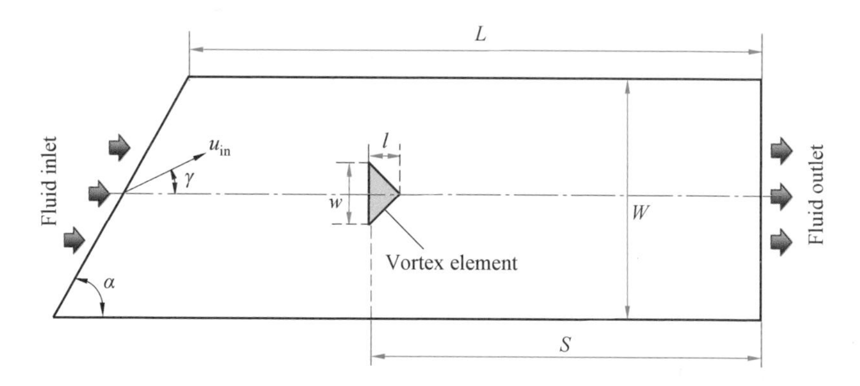

Fig. 5.8 Schematic diagram of the branch domain with the right triangular cylinder as the vortex element

The flow aroundthe vortex element (right triangular cylinder) in the branch domain is numerically calculated based on different inlet flow directions ($\gamma = 15°$, $30°$, $45°$ and $60°$). Fig. 5.9 shows the velocity cloud diagram of the flow around the right triangular cylinder in the branch domain under different inlet fluid flow direction conditions, where the inlet velocity $u_{in} = 0.1$ m/s, and the fluid medium is water.

From Fig. 5.9, it can be seen as follows.

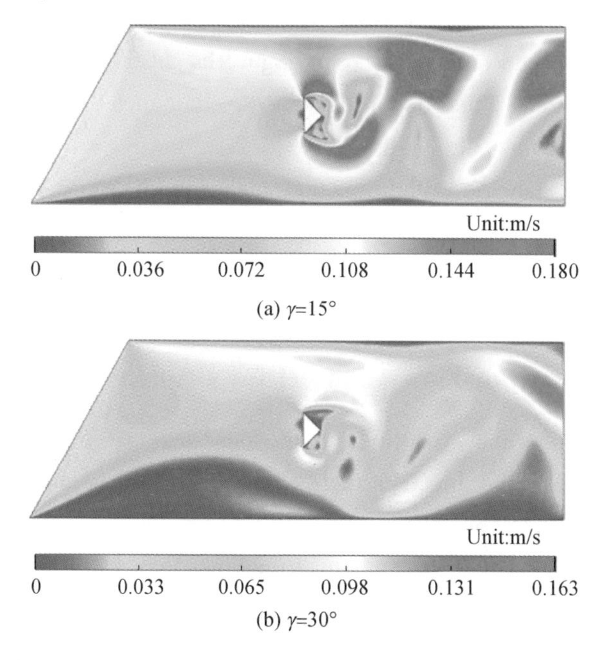

Fig. 5.9 Velocity cloud diagram of the flow around the right triangular cylinder in the branch domain under different inlet fluid flow direction conditions

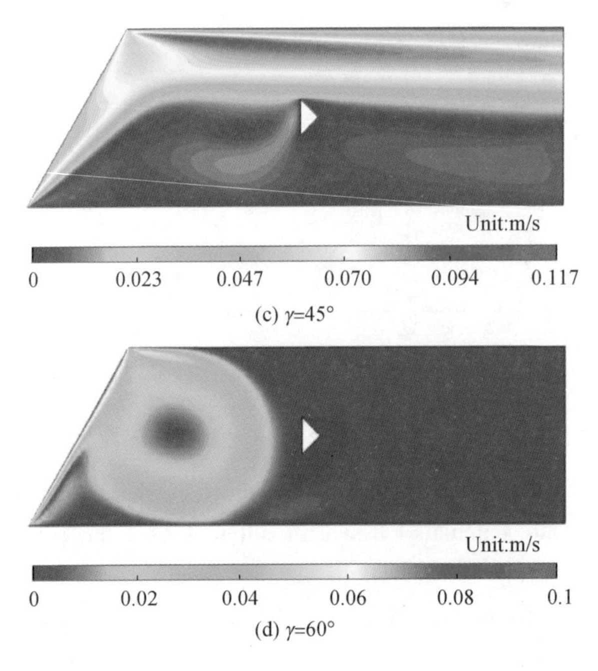

Fig. 5. 9(continued)

(1) When the direction angle is small ($\gamma = 15°$, $30°$), pulsating flow can be formed in the upper part of the wake of the right triangular cylinder, but the regularity of pulsating flow is poor, and the greater the direction angle, the lower the maximum velocity of pulsating flow.

(2) With the increase of the direction angle, the pulsating flow in the wake of the right triangular cylinder gradually disappeared. When $\gamma = \alpha = 60°$, the fluid could not flow out from the fluid outlet of the branch domain, and only a fluid vortex is formed in front of the oncoming surface of the right triangular cylinder.

5.2.2 Design of Flow Channel for Branch Tube

According to the above research on the flow around the vortex element in the branch domain, when the direction angle of the inlet fluid is large, the wake of the vortex element cannot form pulsating flow. To form pulsating flow in each branch domain of the pulsating flow generator with multiple branch tubes, it is necessary to improve the flow channel structure of the branch tube.

Fig. 5. 10 shows a schematic diagram of the branch domain after improving the channel structure. By setting the diversion wall, the flow direction of the fluid is changed to guide the fluid to impact the vortex element arranged in the branch domain, and the influence of the flow direction angle of the inlet fluid on the pulsating flow in the wake of the vortex element is weakened.

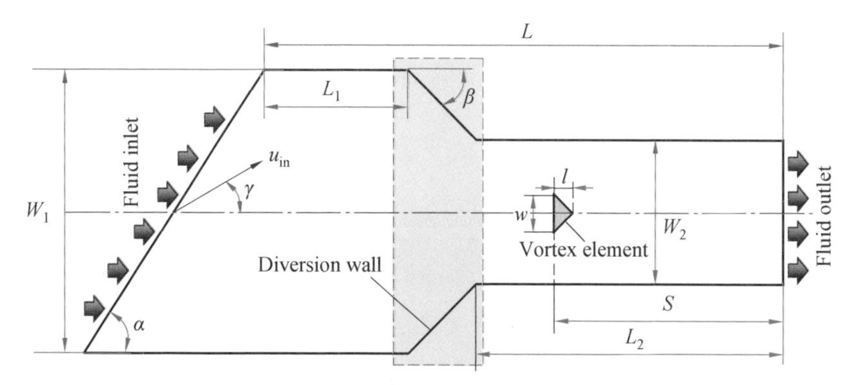

Fig. 5. 10 Schematic diagram of the branch domain after improving the channel structure

In Fig. 5. 10, L is the length of the top edge of the entire fluid domain; L_1 is the length of the top edge of the inlet domain; L_2 is the length of the outlet domain; W_1 is the width of the inlet domain; W_2 is the width of the outlet domain; β is the transition angle between the inlet domain and the outlet domain. Unless otherwise specified in the calculation process, $L = 30$ mm, $L_1 = 5$ mm, $L_2 = 20$ mm, $W_1 = 20$ mm, $W_2 = 10$ mm, $S = 15$ mm, $w = 2$ mm, $l = 1$ mm, $\alpha = 60°$, $\beta = 45°$, $\gamma = 30°$.

5.2.3 Fluid Flow Analysis of the Improved Branch Domain

Based on the branch domain of the improved flow channel, the effects of direction angle, inlet velocity, vortex element size, wake length and external fluid domain on the formation of pulsating flow are studied. During the calculation process, the fluid medium is water.

1. Effect of Direction Angle

Based on the different inlet flow direction angles ($\gamma = 15°$, $30°$ and $45°$), the flow around the right triangular cylinder in the branch domain of the improved flow channel is numerically calculated.

Fig. 5. 11 shows the velocity cloud diagram of the branch domain under the condition of different direction angles, where the inlet velocity $u_{in} = 0.1$ m/s. It can be seen from Fig. 5. 11 as follows.

(1) Under the condition ofdifferent direction angles, the wake of the right triangular cylinder can form pulsating flow. The larger the direction angle, the lower the maximum velocity of the pulsating flow, which is consistent with the conclusion

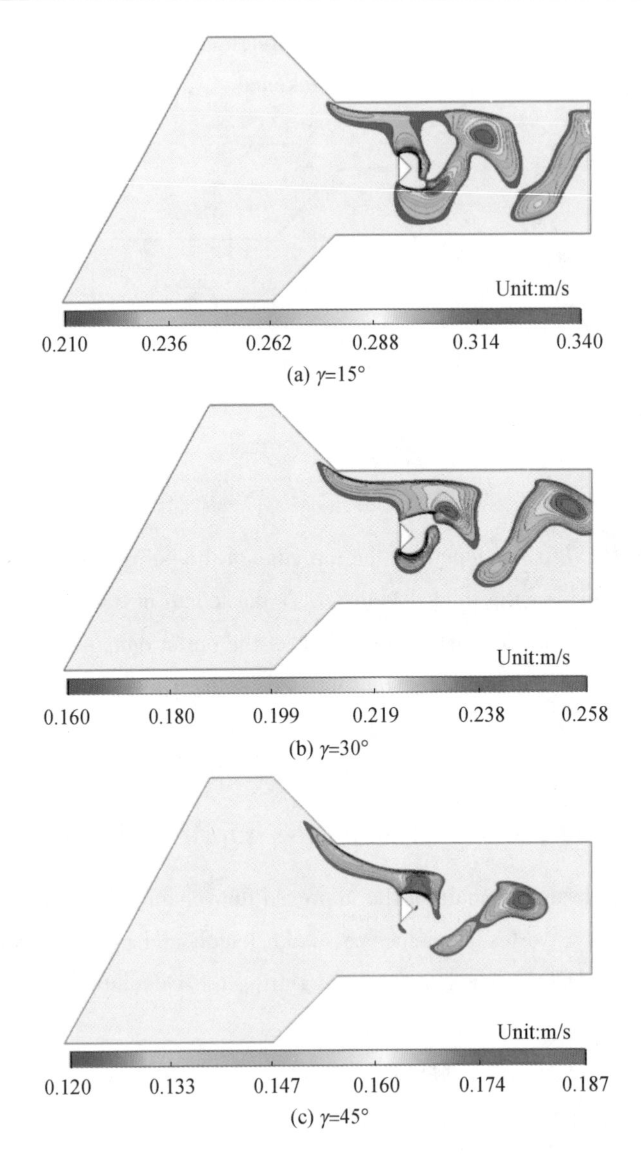

Fig. 5. 11 Velocity cloud diagram of the branch domain under the condition of different direction angles

drawn in Fig. 5. 9.

(2) When the direction angle is large, obvious pulsating flow can also be observed in the wake of the vortex element, indicating that the parameter range of the direction angle forming the pulsating flow is widened through the diversion effect of the diversion wall.

Table 5.4 shows the variation of the frequency and RMS of the lift coefficient and the frequency and average value of the drag coefficient under the condition of different direction angles.

Table 5.4 Variation of the frequency and RMS of the lift coefficient and the frequency and average value of the drag coeffident under the condition of different direction angles

γ	C_l		C_d	
	f/Hz	RMS	f/Hz	Average value
15°	18.67	3.487	36.67	11.4
30°	14.00	1.823	27.33	5.83
45°	9.33	0.790	18.67	2.59

The following conclusions can be drawn from Table 5.4.

(1) The frequency of the drag coefficient is about twice that of the lift coefficient, which is consistent with the previous conclusion.

(2) With the increase of the direction angle, the frequency of the lift coefficient and drag coefficient decreases, indicating that the frequency of the pulsating flow is reduced.

(3) With the increase of the direction angle, the RMS of the lift coefficient and the average value of the drag coefficient both decrease, indicating that the intensity of the pulsating flow and the flow resistance both decrease.

2. Effect of inlet velocity

Based on the different inlet velocities (u_{in} = 0.2 m/s, 0.3 m/s and 0.4 m/s), the numerical calculation of the flow around the right triangular cylinder in the branch domain of the improved flow channel is carried out.

Fig. 5.12 shows the velocity cloud diagram of the branch domain under the condition of different inlet velocities.

Combined with the velocity clouddiagram of $\gamma = 30°$ (u_{in} = 0.1 m/s) shown in Fig. 5.11(b), it can be seen from Fig. 5.12 that the wake of the vortex element can form pulsating flow under different inlet velocity conditions, and the higher the inlet velocity, the higher the maximum velocity of the pulsating flow.

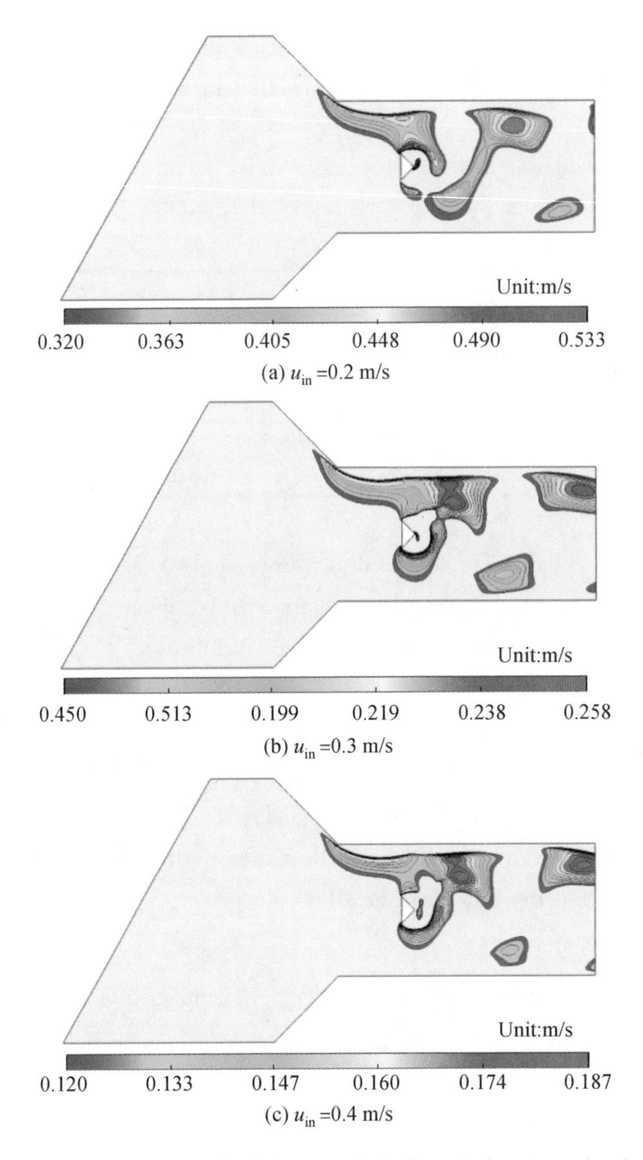

(a) u_{in} =0.2 m/s

(b) u_{in} =0.3 m/s

(c) u_{in} =0.4 m/s

Fig. 5. 12 Velocity cloud diagram of the branch domain under the condition of different inlet velocities

Table 5.5 shows the variation of the frequency and RMS of the lift coefficient and the frequency and average value of the drag coefficient under the condition of different inlet velocities.

Table 5.5 Variation of the frequency and RMS of the lift coefficient and the frequency and average value of the drag coefficient under the condition of different inlet velocities

$u_{in}/(m \cdot s^{-1})$	C_l		C_d	
	f/Hz	RMS	f/Hz	Average value
0.2	23.33	1.810	46.67	5.94
0.3	30.00	1.685	60.00	5.82
0.4	46.22	1.870	92.00	6.15

Combined with the lift and drag coefficients when $\gamma=30°$ ($u_{in}=0.1$ m/s) shown in Table 5.4, the following conclusions can be drawn from Table 5.5.

(1) With the increase of the inlet velocity, the frequency of the lift coefficient and drag coefficient increases, indicating that the frequency of pulsating flow is increased.

(2) With the increase ofthe inlet velocity, the RMS of the lift coefficient and the average value of the drag coefficient both decrease first and then increase, and the changes are not large. It shows that the inlet velocity has little effect on the lift and drag coefficients after increasing the diversion wall.

3. Effect of Vortex Element Size

The dimensions of the vortex element include longitudinal dimension w and transverse dimension l. To study the influence of the vortex element size on the formation of the pulsating flow, the numerical calculation of the flow around the right triangular cylinder in the branch domain of the improved flow channel is first carried out based on different longitudinal dimensions ($w=1.0$ mm and 1.5 mm).

Fig. 5.13 shows the velocity cloud diagram of the branch domain under the condition of different longitudinal dimensions, where $l=1.0$ mm, $u_{in}=0.1$ m/s.

Combined with the velocity cloud diagram of $\gamma=30°$ ($w=2.0$ mm) shown in Fig. 5.11(b), it can be seen from Fig. 5.13 that the wake of the vortex element can form pulsating flow under different longitudinal dimensions, and the larger the longitudinal dimension, the higher the maximum velocity of the pulsating flow.

Table 5.6 shows the variation of the frequency and RMS of the lift coefficient and

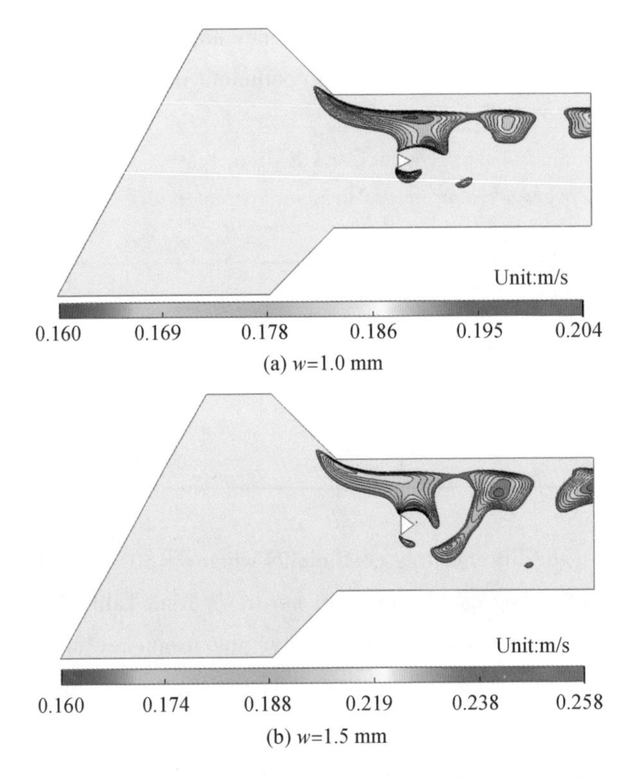

Fig. 5. 13 Velocity cloud diagram of the branch domain under the condition of different longitudinal dimensions

the frequency and average value of the drag coefficient under the condition of different longitudinal dimensions.

Combined with the lift and drag coefficients when $\gamma = 30°$ ($w = 2.0$ mm) shown in Table 5.4, the following conclusions can be drawn from Table 5.6.

(1) With the increase of the longitudinal dimension, the frequency of the lift coefficient and drag coefficient decreases, indicating that the frequency of pulsating flow is decreased.

(2) With the increase of the longitudinal dimension, the RMS of the lift coefficient and the average value of the drag coefficient both increase, indicating that the intensity of the pulsating flow and the flow resistance both increase.

Table 5.6 Variation of the frequency and RMS of the lift coefficient and the frequency and average value of the drag coefficient under the condition of different longitudinal dimensions

w/mm	C_l		C_d	
	f/Hz	RMS	f/Hz	Average value
1.0	18.67	1.495	37.33	3.90
1.5	16.00	1.730	31.67	4.90

Similarly, based on different transverse dimensions (l=0.5 mm and 1.5 mm), the numerical calculation of the flow around the right triangular cylinder in the branch domain of the improved flow channel is carried out.

Fig. 5.14 shows the velocity cloud diagram of the branch domain under the condition of different transverse dimensions, where w=2.0 mm, u_{in}=0.1 m/s.

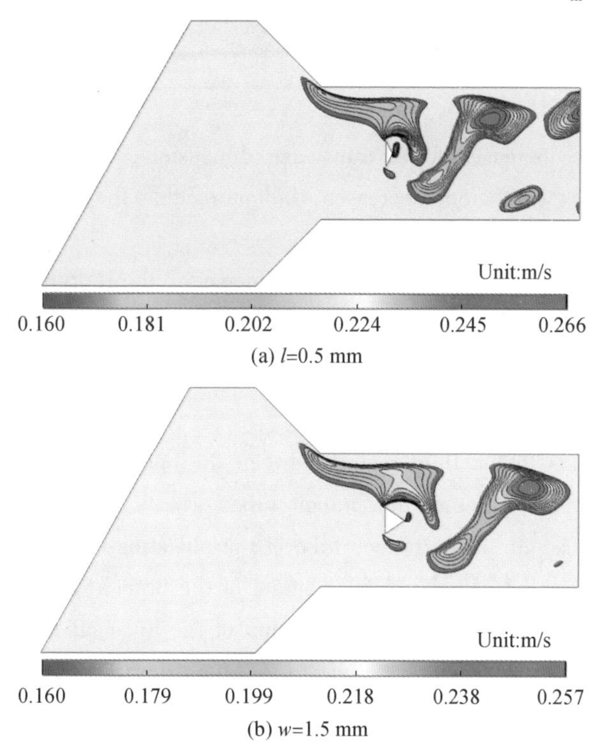

Fig. 5.14 Velocity cloud diagram of the branch domain under the condition of different transverse dimensions

Combined with the velocity clouddiagram of γ=30° (l=1.0 mm) shown in Fig.

5.11(b), it can be seen from Fig. 5.14 that the wake of the vortex element can form pulsating flow under different transverse dimensions, and the larger the transverse dimension, the lower the maximum velocity of the pulsating flow.

Table 5.7 shows the variation of the frequency and RMS of the lift coefficient and the frequency and average value of the drag coefficient under the condition of different transverse dimensions.

Combined with the lift and drag coefficients when $\gamma=30°$ ($l=1.0$ mm) shown in Table 5.4, the following conclusions can be drawn from Table 5.7.

Table 5.7 Variation of the frequency and RMS of the lift coefficient and the frequenay and average value of the drag coefficient under the condition of different transverse dimensions

l/mm	C_l		C_d	
	f/Hz	RMS	f/Hz	Average value
0.5	14.00	1.206	28.00	6.90
1.5	13.33	2.262	25.33	5.21

(1) With the increase of the transverse dimension, the frequency of the lift coefficient and drag coefficient decreases, indicating that the frequency of pulsating flow is decreased.

(2) With the increase of the transverse dimension, the RMS of the lift coefficient increases, and the average value of the drag coefficient decreases, indicating that the intensity of pulsating flow increases and the flow resistance decreases.

4. Effect of Wake Length

The wake refers to the fluid region between the oncoming surface of the vortex element and the fluid outlet of the branch tube. The length of the wake (S) is determined by the length of the branch tube of the pulsating flow generator.

Fig. 5.15 shows the velocity cloud diagram of the branch domain under different wake lengths, and Fig. 5.16 shows the variation of the lift coefficient (C_l) and drag coefficient (C_d) with the calculation time (t), where $u_{in}=0.1$ m/s.

Combined with the velocity clouddiagram of $\gamma=30°$ ($S=15$ mm) shown in Fig. 5.11(b), it can be seen from Fig. 5.15 and Fig. 5.16 that when the wake length is small ($S=5$ mm), the wake of the vortex element cannot form pulsating flow, indicating that the wake length has an important influence on the formation of pulsating flow.

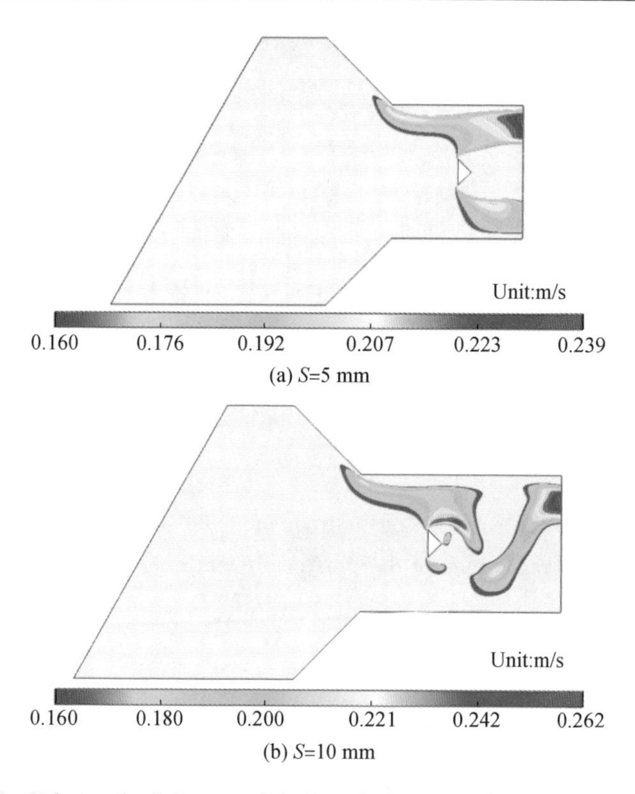

Fig. 5. 15 Velocity cloud diagram of the branch domain under different wake lengths

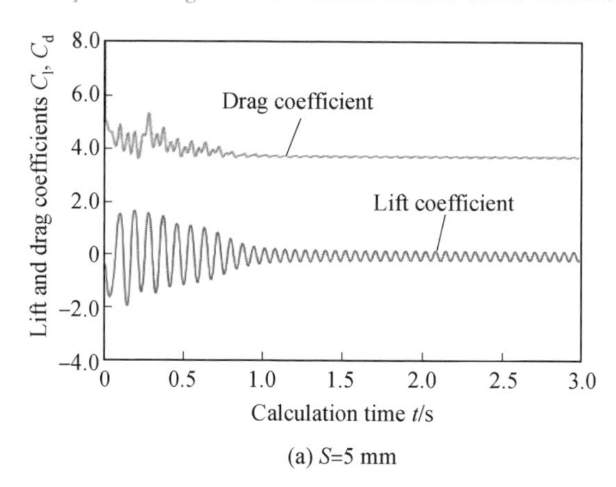

Fig. 5. 16 Variation of the lift coefficient and drag coefficient with the calculation time

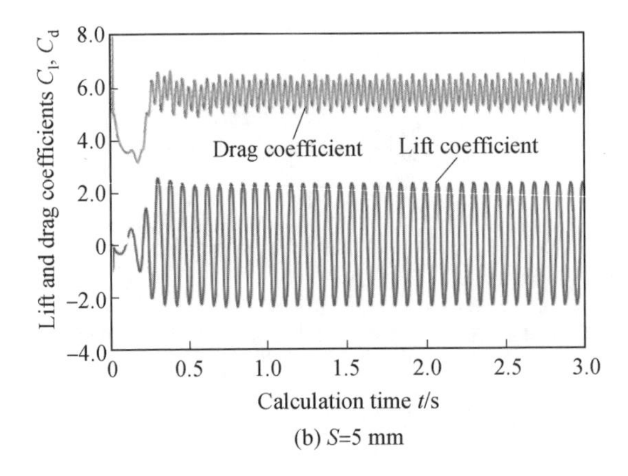

(b) S=5 mm

Fig. 5. 16(continued)

Table 5. 8 shows the variation of the frequency and RMS of the lift coefficient and the frequency and average value of the drag coefficient under the condition of different wake lengths.

Table 5. 8 Variation of the frequency and RMS of the lift coefficient and the frequency and average value of the drag coefficient under the condition of different wake lengths

S/mm	C_l		C_d	
	f/Hz	RMS	f/Hz	Average value
5	16. 67	0. 128	16. 67	3. 75
10	14. 00	1. 804	26. 67	5. 76

Combined with the lift and drag coefficients when $\gamma=30°$ ($S=15$ mm) shown in Table 5.4, the following conclusions can be drawn from Table 5. 8.

(1) When the wake length is small ($S=5$ mm), the lift coefficient is the same as the drag coefficient, and the RMS of the lift coefficient and the mean value of the drag coefficient are small. In this case, the wake of the vortex element cannot form pulsating flow.

(2) When the wake length is large ($S=10$ mm, 15 mm), the frequency of the drag coefficient is about twice that of the lift coefficient. The frequency of the lift coefficient and drag coefficient, the RMS of the lift coefficient, and the average value of the drag coefficient are consistent under different wake lengths. This indicates that when the wake length is sufficient to ensure the formation of pulsating flow, it has little influence on the frequency and intensity of the pulsating flow.

5. Effect of the External Fluid Domain

In the actual working process of the pulsating flow generator, there exists shell-side fluid with a certain velocity at the fluid outlet of the branch tube, which has an important effect on the pulsating flow out of the branch tube.

To explore the influence of external shell-side fluid on the formation of pulsating flow, an external fluid domain is added based on the fluid domain shown in Fig. 5. 10, as shown in Fig. 5. 17.

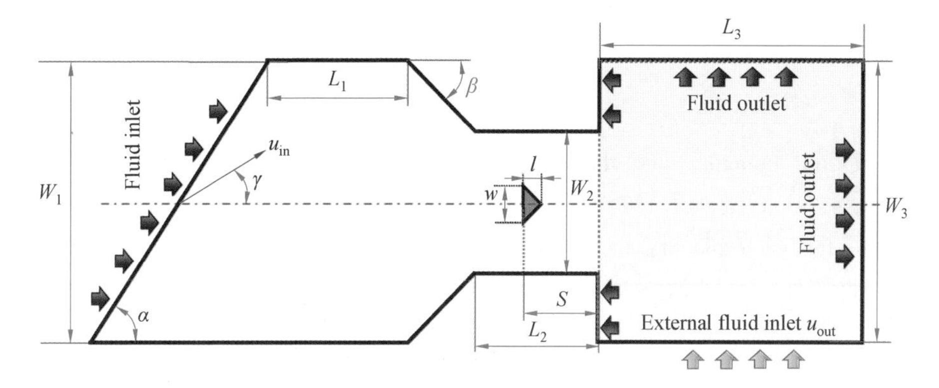

Fig. 5. 17 Schematic diagram of the branch domain with external fluid domain

In Fig. 5. 17, the bottom boundary of the external fluid domain is the external fluid inlet, and the other boundaries are the external fluid outlet. L_3 is the length of the external fluid domain, and W_3 is the width of the external fluid domain. In this chapter, unless otherwise specified during the calculation, $L_2 = 10$ mm, $L_3 = 20$ mm, $W_3 = 20$ mm, $S = 5$ mm. And also, other dimensions are consistent with those shown in Fig. 5. 10.

Fig. 5. 18 shows the velocity cloud diagram of the branch domain with the external fluid domain when the external fluid inlet velocity $u_{out} = 0$ m/s, where $u_{in} =$ 0. 1 m/s.

As can be seen from Fig. 5. 18, when the external fluid domain is added at the outlet of the branch tube, pulsating flow is formed in the wake of the vortex element. This shows that if the fluid domain of the pulsating flow generator is analyzed separately, the length of each branch tube should be kept sufficient. However, when analyzing the fluid domain composed of shell-side fluid and pulsating fluid inside the heat exchanger, the length of the branch tube can be reduced appropriately according to the needs.

Fig. 5. 19 shows the velocity cloud diagram of the branch domain with the

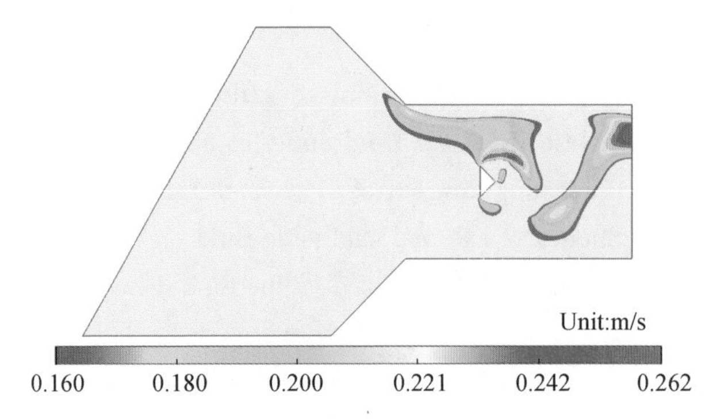

Fig. 5. 18 Velocity cloud diagram of the branch domain with the external fluid domain when the external fluid inlet velocity $u_{out}=0$ m/s

external fluid domain when the external fluid inlet velocity u_{out} = 0. 05 m/s and 0. 10 m/s, where u_{in} = 0. 1 m/s.

The following conclusions can be drawn from Fig. 5. 19.

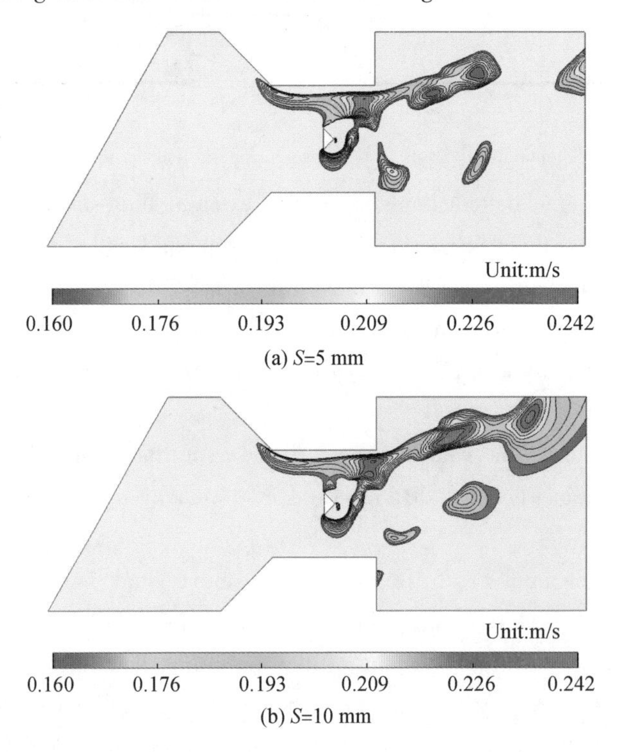

Fig. 5. 19 Velocity cloud diagram of the branch domain with the external fluid domain when the external fluid inlet velocity u_{out} = 0. 05 m/s and 0. 10 m/s

(1) The external fluid inlet velocity in the external fluid domain has a significant influence on the pulsating flow, and the pulsating flow formed by the vortex element wake deviates along the inlet velocity direction of the external fluid domain.

(2) The greater the external fluid inlet velocity of the external fluid domain, the greater the offset of the outlet pulsating flow along this velocity direction.

Table 5.9 shows the variation of the frequency and RMS of the lift coefficient and the frequency and average value of the drag coefficient under the condition of different external fluid inlet velocities.

Table 5.9 Variation of the frequency and RMS of the lift coefficient and the frequency and average value of the drag coefficient under the condition of different external fluid inlet velocities

$u_{out}/(m \cdot s^{-1})$	C_l		C_d	
	f/Hz	RMS	f/Hz	Average value
0	13.33	1.785	26.00	5.67
0.05	12.67	1.743	24.67	5.51
0.10	12.00	1.736	24.00	5.44

The following conclusions can be drawn from Table 5.9.

(1) The external fluid inlet velocity of the external fluid domain has a certain influence on the frequency ofthe pulsating flow. The higher the external fluid inlet velocity, the lower the frequency of the lift drag coefficient, but the change range is not large.

(2) The greater the external fluid inlet velocity, the RMS of the lift coefficient and the average value of the drag coefficient both decrease more, indicating that the external fluid inlet velocity of the external fluid domain has a certain influence on the strength of the pulsating flow and the flow resistance.

5.3 Shell-side Distributed Pulsating Flow Generator

5.3.1 Structure Design

Based on the above selected vortex element and the fluid flow analysis of the improved branch domain, combined with the structural characteristics and shortcomings of the existing independent pulsating flow generator (Fig. 1.14), the shell-side distributed pulsating flow generator used in the ETB heat exchanger is designed, as shown in Fig. 5.20.

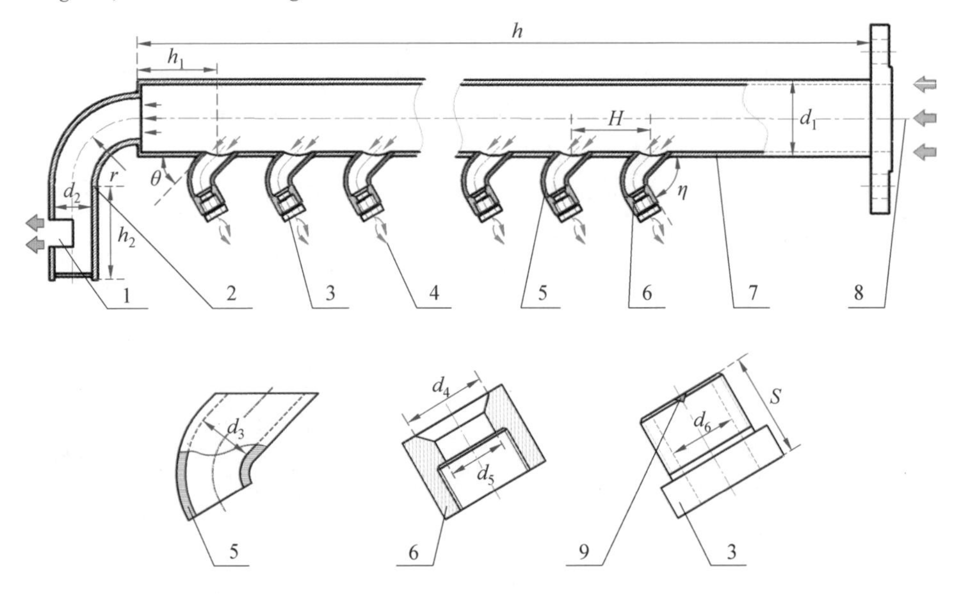

1—Opening; 2—Horizontal tube; 3—Pulsating flow tube; 4—Pulsating flow outlet; 5—Branch elbow; 6—Diversion tube; 7—Vertical tube; 8—Fluid inlet; 9—Vortex element

Fig. 5.20 Schematic diagram of the shell-side distributed pulsating flow generator

The shell-side distributed pulsating flow generator includes the vertical tube, horizontal tube, branch elbow, diversion tube, pulsating flow tube and vortex element. The branch elbow, diversion tube and pulsating flow tube together constitute the branch tube. All branch elbows are evenly welded to the vertical tube, and the diversion tubes are welded to the branch elbows. The diversion tubes and the pulsating flow tubes are connected by threads. Triangular slots are provided at one end of the pulsating flow tubes for the placing and fixing of the vortex elements. The horizontal tube is welded to the lower end of the vertical tube, and its incision acts as the shell-side fluid inlet.

In Fig. 5.20, H is the distance between branch tubes; h is the height of the vertical tube; h_1 is the distance between the bottom of the vertical tube and the first branch tube; h_2 is the straight tube length of the horizontal tube; r is the bending radius of the horizontal tube; θ is the included angle between the branch elbow and the vertical tube; η is the angle between the diversion tube (or pulsating flow tube) and the vertical tube; d_1 is the inner diameter of the vertical tube; d_2 is the inner diameter of the horizontal tube; d_3 is the inside diameter of the branch elbow; d_4 and

d_5 are the two inner diameters of the diversion tube; d_6 is the inner diameter of the pulsating flow tube; S is the distance between the oncoming surface of the vortex element and the outlet of the pulsating flow tube. In addition, the vortex element is a right triangular cylinder.

In the calculation process of this chapter, unless otherwise specified: H = 60 mm, h=530 mm, h_1=60 mm, h_2=70 mm, r=50 mm, θ=45°, η=60°, d_3 = d_4=15 mm, d_5=d_6=10 mm, S=15 mm, the length of the hypotenuse of the vortex element is 2 mm, and the width of the opening on the horizontal tube is 20 mm.

To facilitate manufacturing, the dimensions of the vertical tube and the horizontal tube of the pulsating flow generator are selected from China's commonly used seamless steel tube size specification table. Take the heat exchanger with six rows of ETBs as an example, as shown in Fig. 4. 1, three specifications of the vertical tube and three specifications of the horizontal tube are selected based on the structural dimensions of the heat exchanger, as shown in Table 5. 10.

Table 5. 10 Selected steel tube sizes

Number	Vertical tube		Horizontal tube	
	Outer diameter /mm	Wall thickness /mm	Outer diameter /mm	Wall thickness /mm
1	45	2.5	32	3.5
2	50	2.5	38	4.0
3	60	3.0	42	3.5

In this way, the inner diameters of the vertical tube (d_1) are 40 mm, 45 mm and 54 mm, and the inner diameters of the horizontal tube (d_1) are 25 mm, 30 mm and 35 mm. Then, six tube inner diameter combinations can be formed based on Table 5. 10, as shown in Table 5. 11.

Table 5. 11 Six tube inner diameter combinations

Combination	d_1/mm	d_2/mm	Combination	d_1/mm	d_2/mm
Ⅰ	40	25	Ⅳ	45	30
Ⅱ	40	30	Ⅴ	54	30
Ⅲ	45	25	Ⅵ	54	35

Compared with the branch structure shown in Fig. 5. 10, based on the practical application of the distributed pulsating flow generator, the inlet domain tube (large diameter tube) of the branch tube is changed from a straight tube to a curved tube (i. e., the branch elbow shown in Fig. 5. 20). This structural improvement has the following characteristics.

(1) The fluid inlet of the branch elbow is inclined downward so that the fluid in the vertical tube is easy to flow into the branch elbow.

(2) The bending structure of the branch elbow has a certain diversion effect. Through the joint guidance of the branch elbow and the diversion tube, the fluid impacts the vortex element fixed on the triangular slot.

(3) The bending structure of the branch elbow uniformly changes the flow direction of the fluid, making the pulsating flow formed in the pulsating flow tube tilt upward and impact the appropriate position of the ETB, weakening the influence of shell-side fluid on the pulsating flow.

5.3.2 Installation of Distributed Pulsating Flow Generator

Fig. 5.21 shows the installation of distributed pulsating flow generator in the ETB heat exchanger, where taking the heat exchanger with six rows of ETBs as an example.

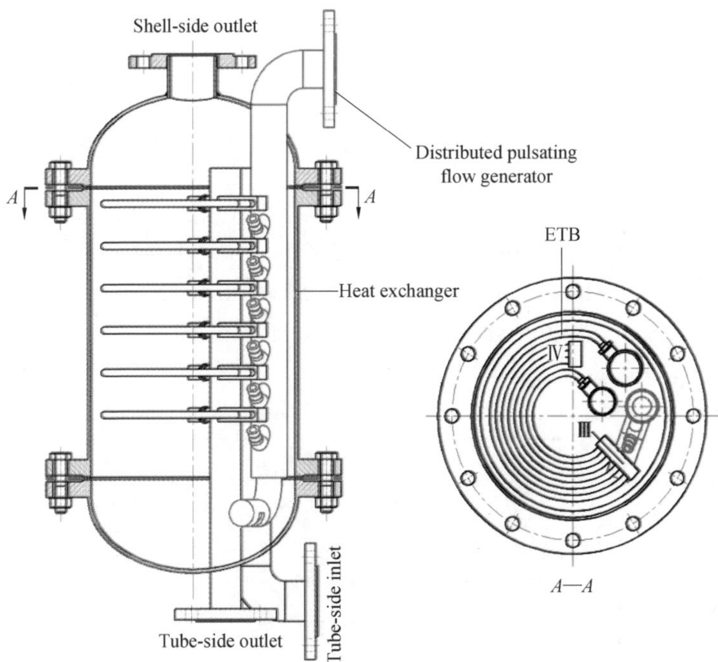

Fig. 5.21 Installation of distributed pulsating flow generator in the ETB heat exchanger

As shown in Fig. 5. 21, one end of the distributed pulsating flow generator is mounted on the upper shell cover of the heat exchanger, and the other end is suspended at the bottom of the heat exchanger. There is a one-to-one correspondence between the pulsating flow outlets of the distributed pulsating flow generator and the stainless steel connector Ⅲ of the ETB. When the fluid flows around the vortex elements, pulsating flows with a certain frequency and intensity can be formed. In this way, the ETBs are vibrated by the coupling of the pulsating fluid and the shell-side fluid. In addition, the horizontal tube is welded to the lower end of the vertical tube, which incision impacts the inner wall of the bottom shell cover as the shell-side fluid inlet. This arrangement allows the shell-side fluid to flow upward approximately spirally in the heat exchanger. In this way, the heat transfer performance of the ETBs can be sufficiently improved.

The ETB heat exchanger enhances heat transfer by inducing flow-induced vibration of the internal ETBs, but this flow-induced vibration can cause fatigue damage to the ETBs. To achieve the ETB heat exchanger that not only enhances heat transfer but also considers the fatigue life of the internal ETBs, reasonable induction and control of ETB vibration under actual operating conditions has become the key to the design of the ETB heat exchangers. The design of a shell-side distributed pulsating flow generator makes it possible to reasonably induce and control this vibration.

5.3.3 Meshing of the Fluid Domain

Fig. 5. 22 shows a schematic diagram of the fluid domain of a six-branch distributed pulsating flow generator. According to the geometric characteristics and computational requirements of the fluid domain, it can be divided into the vertical tube domain, curved tube domain, pulsating flow domain and horizontal tube domain.

To generate a reasonable mesh, the following points should be noted.

(1) To generate as few elements as possible, the mesh size in the vertical tube domain should be as large as possible without affecting the calculation accuracy.

(2) The element size of the vortex element in the pulsating flow domain is small, so sufficient mesh density is needed to capture the clear pulsating flow in the wake of the vortex element.

(3) The mesh transition should be reasonable at the connections between the vertical tube domain and the horizontal tube domain, the curved tube domain and the pulsating flow domain, as well as between the vertical tube domain and the curved tube domain.

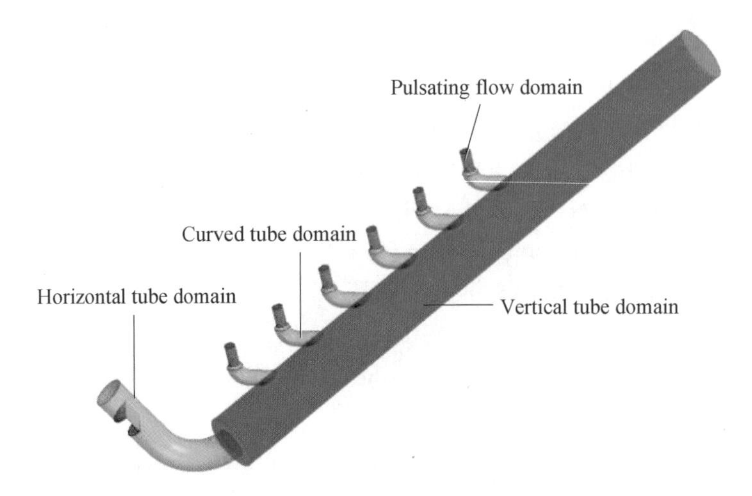

Fig. 5. 22 Schematic diagram of the fluid domain of a six-branch distributed pulsating flow generator

Therefore, the fluid domain should be properly divided before meshing, as shown in Fig. 5. 23.

In Fig. 5. 23, the projection surface is the projection of the end face of the curved tube domain on the vertical tube domain. The purpose of this method is to facilitate the reasonable transition of meshes between fluid domains with different mesh densities. During calculation, interfaces need to be established between the vertical tube domain and the curved tube domain, the curved tube domain and the pulsating flow domain, and the vertical tube domain and the horizontal tube domain respectively, that is, interface Ⅰ, interface Ⅱ and interface Ⅲ, as shown in Fig. 5. 23. In addition, to facilitate the analysis of the calculation results, the branch tubes are numbered successively from bottom to top as $n=1, 2, \ldots, 6$.

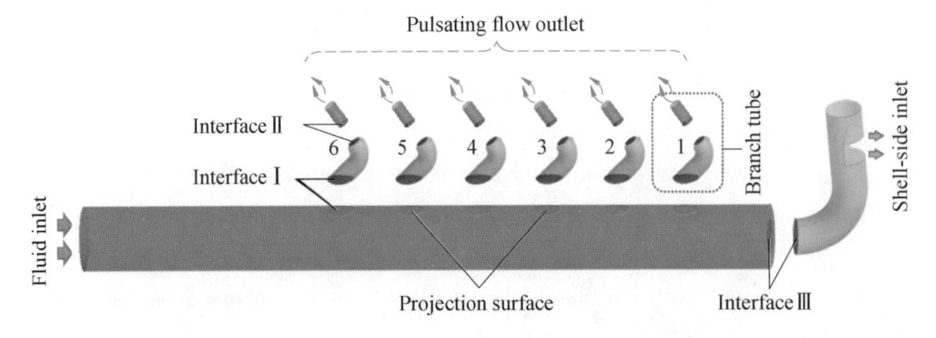

Fig. 5. 23 Schematic diagram of the fluid domain of a six-branch distributed pulsating flow generator

Fig. 5.24 shows the mesh division of the fluid domain of the distributed pulsating flow generator. The mesh division is performed by the Workbench platform's mesh division module, and tetrahedral meshes are used in all domains, with 872,099 elements and 189,305 nodes in total. Based on the calculation requirements, six boundary layer meshes with a total thickness of 0.1 mm are applied near the wall of the vortex element. As can be seen from Fig. 5.24, the mesh transition between each segmentation domain is reasonable.

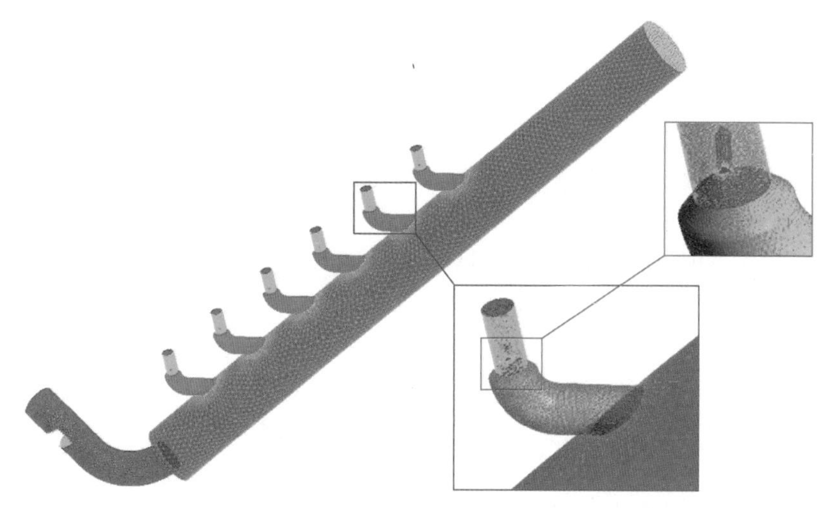

Fig. 5.24 Mesh division of the fluid domain of the distributed pulsating flow generator

To facilitate the analysis of the pulsating flow formed in each branch tube domain, monitoring points are placed at appropriate positions in the pulsating flow domain to monitor the change of the velocity with the calculation time. The locations of monitoring points are shown in Fig. 5.25. The monitoring point is located at the lower side of the pulsating flow domain ($0.1l$, $0.2d$), where l and d are the length and diameter of the pulsating flow domain. The numbers of monitoring points are the same as those of the branch tubes, that is, they are numbered successively from bottom to top as $n=1, 2, \ldots, 6$.

The simulationis carried out by using ANSYS FLUENT software, the second-order implicit algorithm is solved by the uncoupled three-dimension single-precision solver, and the pressure-velocity coupling is performed by SIMPLEC mode.

The fluid inlet is set to velocity inlet, and the inlet velocity is given. The opening and the pulsating flow outlet are set to pressure-outlet, and the outlet pressure is 0 Pa. The three pairs of interfaces as shown in Fig. 5.23 are set to "Interface", and other

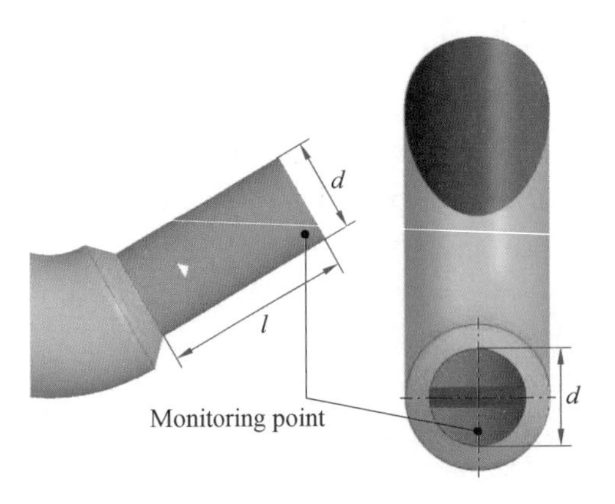

Fig. 5.25 Locations of monitoring points

surfaces are set as non-sliding walls. In addition, the mesh connections between the above segmented fluid domains are made by the GGI method.

The calculation adopts the step-by-step calculation strategy of rough calculation plus actuarial calculation. For the rough calculation, the time step is 0.1 s, and the calculation steps are 100. And also, for the actuarial calculation, the time step is 0.001 s, and the calculation steps are 2,000.

5.3.4 Mesh Independence Analysis

To analyze the independence of the mesh, based on different mesh division schemes, a trial calculation is carried out on the six-branch distributed pulsating flow generator with an internal diameter of 40 mm in the vertical tube and an internal diameter of 25 mm in the horizontal tube. The frequency (f_{pul}) and RMS of the velocity change over time at the monitoring points in the first and sixth branch tubes are obtained, as shown in Table 5.12. In the calculation process, the inlet fluid medium is water, and the inlet velocity is $u_{in}=0.4$ m/s.

Table 5.12 Fluid domain grid independence analysis of distributed pulsating flow generator

Case	Branch tube 1		Relative error /%		Branch tube 6		Relative error /%		Calculation time/h
	f_{pul}/Hz	R_{pul}/(m·s^{-1})	f_{pul}	R_{pul}	f_{pul}/Hz	R_{pul}/(m·s^{-1})	f_{pul}	R_{pul}	
1	53.3	0.610	5.04	9.63	54.6	0.602	7.06	8.51	18
2	49.0	0.675	—	—	51.0	0.658	—	—	21
3	48.0	0.693	2.04	2.67	49.6	0.679	2.75	3.19	34

Note: The relative error is calculated based on the calculation result of case 2.

In Table 5.12, case 2 is the mesh partitioning strategy mentioned above, that is, the number of nodes is 189,305. Cases 1 and 3 are obtained by reducing or increasing the number of elements and the local mesh density based on case 2. Case 1 has a total of 127,648 nodes, and case 3 has a total of 314,529 nodes.

As can be seen from Table 5.12, when the number of elements is increased (case 3), it has little impact on the calculation results. The maximum relative error is only 3.19%, but the calculation time is about 1.62 times that of case 2. When the number of elements is reduced (case 1), the maximum relative error can reach 9.63%, and the calculation time is about 0.86 times that of case 2. During the calculation, considering the calculation accuracy and efficiency, the mesh partitioning strategy mentioned above (case 2) is selected.

5.4 Analysis of the Pulsating Flow

5.4.1 Export Flow Analysis

To ensure that the frequency and intensity of pulsating flow in each branch tube of the distributed pulsating flow generator are consistent, the key is to ensure that the flow rate (M) of the pulsating flow outlet of each branch tube with the same cross-sectional is consistent.

Fig. 5.26 shows the flow rate of the pulsating flow outlet of each branch tube at different calculation times (t) under the conditions of $d_1 = 40$ mm, $d_2 = 25$ mm (the combination Ⅰ as shown in Table 5.11), and the fluid inlet velocity $u_{in} = 0.1$ m/s.

Table 5.13 shows the flow rate of the pulsating flow outlet of each branch tube under different inlet velocity conditions for the six combinations (as shown in Table 5.11) at the end of the calculation ($t = 12.0$ s).

From Fig. 5.26 and Table 5.13, it can be concluded as follows.

(1) When the calculation time is short, the flow rate of the pulsating flow outlet of the branch tube changes sharply with the number of branch tubes. As the calculation time increases, the trend gradually stabilizes.

(2) The flow rate of the pulsating flow outlet of the branch tube decreases with the increase of the branch tube number, but the change amplitude is relatively low.

(3) After calculation, the flow rate ofthe pulsating flow outlet of the branch tube is basically consistent under different conditions, and the maximum relative error is only 7.30%.

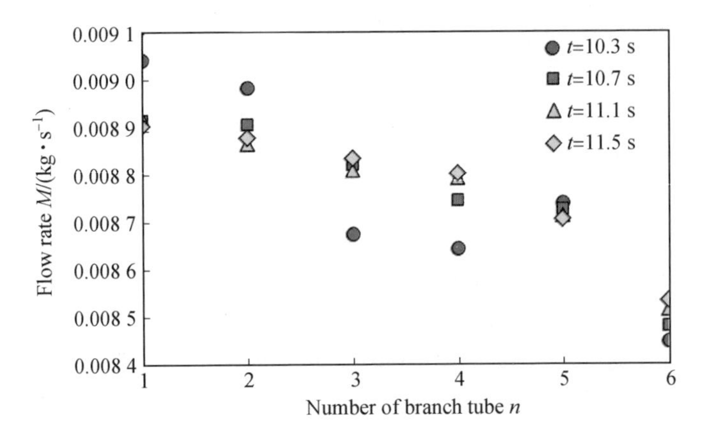

Fig. 5.26 Flow rate of the pulsating flow outlet of each branch tube at different calculation times generator

Table 5.13 Outlet flows of each branch at different velocities

Combinations	u_{in} /(m · s^{-1})	Flow rate of the pulsating flow outlet M/(kg · s^{-1})						Maximum relative error /%
		1	2	3	4	5	6	
I	0.1	0.0089	0.0089	0.0088	0.0088	0.0087	0.0085	4.49
	0.2	0.0176	0.0181	0.0178	0.0179	0.0174	0.0170	6.08
	0.3	0.0270	0.0272	0.0269	0.0265	0.0260	0.0258	5.15
	0.4	0.0359	0.0357	0.0355	0.0360	0.0352	0.0343	4.72
II	0.1	0.0067	0.0067	0.0066	0.0066	0.0065	0.0063	5.97
	0.2	0.0137	0.0135	0.0134	0.0133	0.0131	0.0127	7.30
	0.3	0.0206	0.0203	0.0207	0.0201	0.0197	0.0191	7.28
	0.4	0.0275	0.0273	0.0274	0.0266	0.0265	0.0260	5.45
III	0.1	0.0114	0.0114	0.0114	0.0114	0.0113	0.0111	2.63
	0.2	0.0231	0.0231	0.0228	0.0227	0.0229	0.0221	4.33
	0.3	0.0347	0.0345	0.0342	0.0342	0.0345	0.0339	2.31
	0.4	0.0450	0.0466	0.0466	0.0464	0.0461	0.0450	3.43
IV	0.1	0.0088	0.0088	0.0088	0.0087	0.0086	0.0084	4.54
	0.2	0.0180	0.0179	0.0177	0.0175	0.0175	0.0171	5.00
	0.3	0.0271	0.0269	0.0268	0.0268	0.0262	0.0257	5.17
	0.4	0.0359	0.0363	0.0360	0.0356	0.0353	0.0349	3.86

Table 5.13(continued)

Combinations	u_{in} /(m·s^{-1})	Flow rate of the pulsating flow outlet M/(kg·s^{-1})						Maximum relative error /%
		1	2	3	4	5	6	
V	0.1	0.0134	0.0134	0.0134	0.0134	0.0133	0.0131	2.24
	0.2	0.0268	0.0268	0.0267	0.0268	0.0265	0.0267	1.12
	0.3	0.0408	0.0402	0.0409	0.0407	0.0402	0.0395	3.42
	0.4	0.0555	0.0554	0.0553	0.0541	0.0542	0.0535	3.60
VI	0.1	0.0103	0.0103	0.0102	0.0103	0.0102	0.0100	2.91
	0.2	0.0211	0.0208	0.0209	0.0208	0.0207	0.0207	1.90
	0.3	0.0320	0.0313	0.0311	0.0315	0.0314	0.0309	3.44
	0.4	0.0430	0.0424	0.0428	0.0416	0.0418	0.0419	3.26

In a word, this distributed pulsating flow generator can ensure that the flowrate of the pulsating flow outlet of each branch tube is consistent, which is the premise for the formation of pulsating flow with basically the same frequency and intensity in each pulsating flow domain.

5.4.2 Distribution of Pulsating Flow

Fig. 5.27 shows the velocity isosurface diagrams of each branch domain of the distributed pulsating flow generator under the conditions of $d_1 = 40$ mm, $d_2 = 25$ mm (combination I as shown in Table 5.11) and $u_{in} = 0.1$ m/s, where the value of the isosurface is 0.2 m/s.

Fig. 5.28 shows the velocity cloud diagrams of each branch domain of the distributed pulsating flow generator under the conditions of $d_1 = 40$ mm, $d_2 = 25$ mm (the combination I as shown in Table 5.11) and $u_{in} = 0.1$ m/s.

From Fig. 5.27 and Fig. 5.28, it can be concluded as follows.

(1) Asymmetrical pulsating flow can be formed in each branch tube of the pulsating flow generator. The pulsating flow is mainly distributed at the bottom of the wake of the vortex element due to the diversion effect of the branch elbow and the diversion tube.

(2) This distributed pulsating flow generator is easy to form a consistent pulsating flow in each branch tube.

To analyze the consistency of the pulsating flow formed by each branch tube, it is necessary to further analyze the frequency and intensity of the pulsating flow.

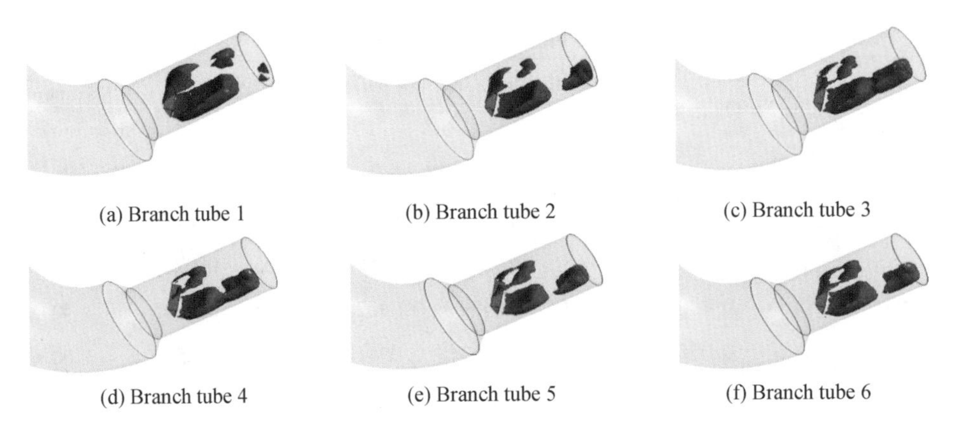

Fig. 5. 27 Velocity isosurface diagrams of each branch domain of the distributed pulsating flow generator

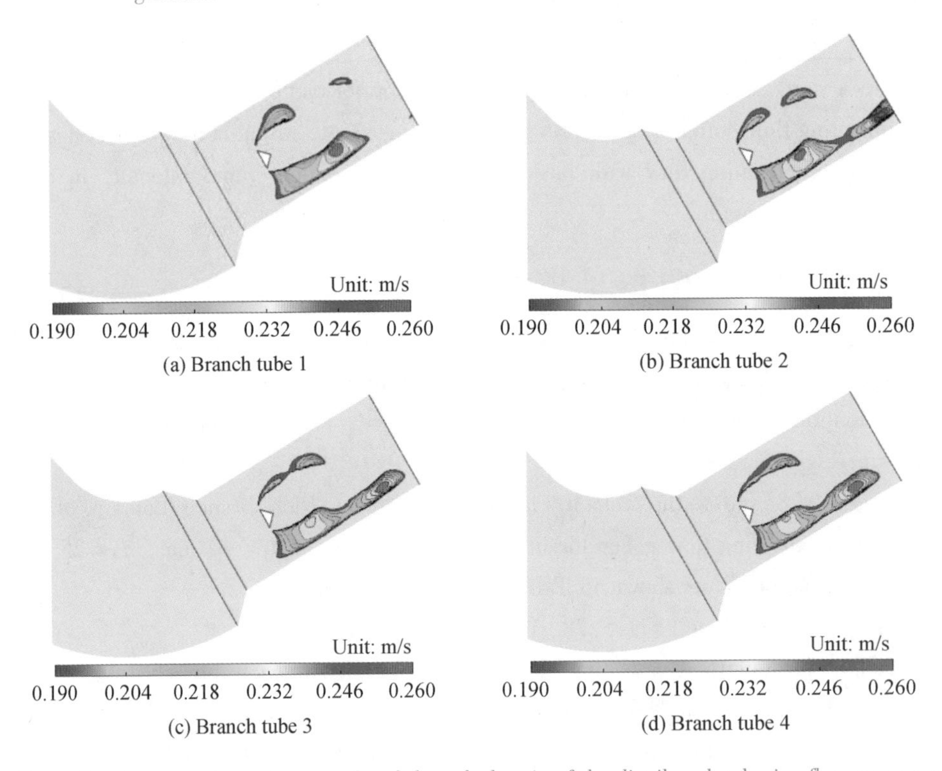

Fig. 5. 28 Velocity cloud diagrams of each branch domain of the distributed pulsating flow generator

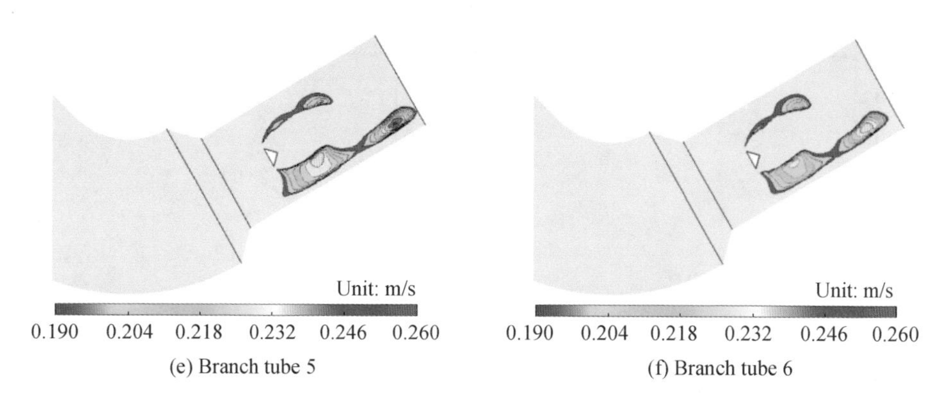

(e) Branch tube 5 (f) Branch tube 6

Fig. 5. 28 (continued)

5.4.3 Frequency Analysis

To study the frequency of the pulsating flow in each branch tube, based on the monitoring points set in each branch tube in Fig. 5. 25, the FFT calculation is performed on the velocity data of each monitoring point to obtain the frequency of velocity, that is, the frequency of the pulsating flow.

Fig. 5. 29 shows the velocity spectrums of the monitoring point in branch tube 6 under different inlet velocities (u_{in} = 0. 1 m/s, 0. 2 m/s, 0. 3 m/s and 0. 4 m/s) when d_1 = 40 mm and d_2 = 25 mm (combination Ⅰ as shown in Table 5. 11).

From Fig. 5. 29, the following conclusions can be drawn.

(1) Under different inlet velocity conditions, the monitoring point velocity varies with time with a main frequency, indicating that pulsating flow can be formed in the branch tube 6 of the distributed pulsating flow generator.

(2) With the increase of the inlet velocity, the frequency (f_v) and amplitude (A_v) of the monitoring point velocity both increase. Within the selected calculation parameter range, both show an approximate linear relationship.

(3) When the inlet velocity is low (u_{in} = 0. 1 m/s), the frequency of the monitoring point velocity changing with time has an obvious high harmonic frequency, which gradually disappears with the increase of the inlet velocity.

To further study the frequency of the pulsating flow, based on different inlet velocities, the frequencies of velocity changes over time of all monitoring points under different combination conditions are calculated, as shown in Table 5. 14.

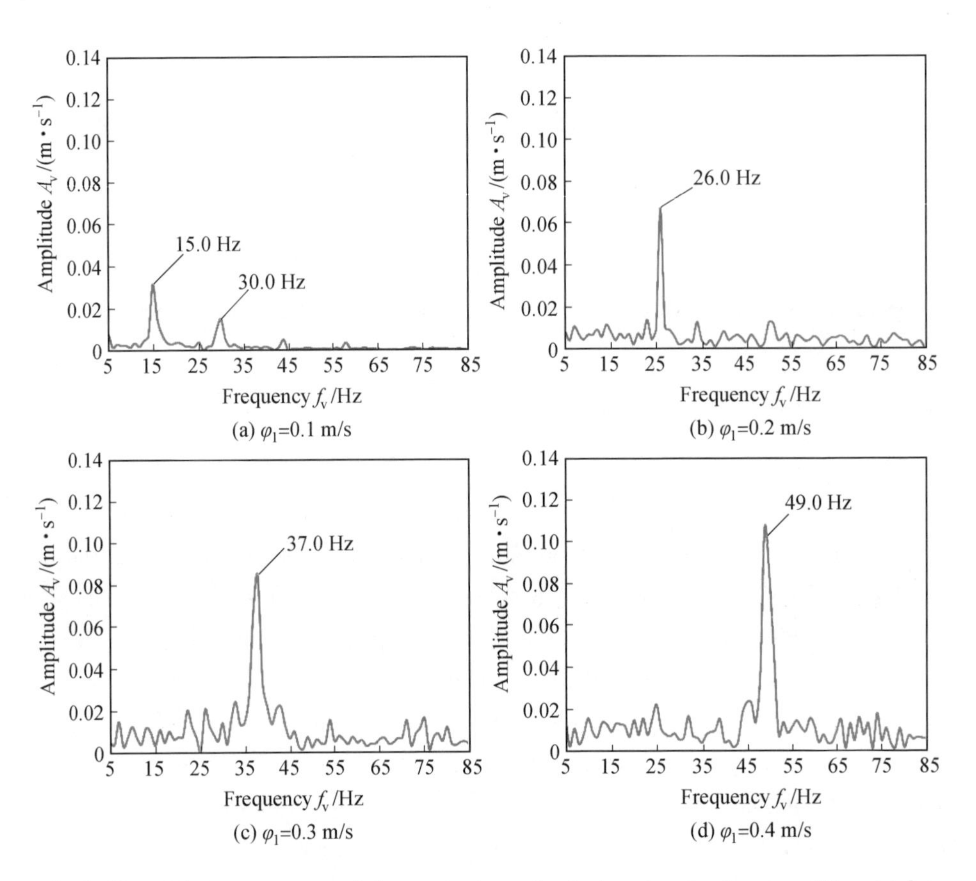

(a) φ_1=0.1 m/s (b) φ_1=0.2 m/s (c) φ_1=0.3 m/s (d) φ_1=0.4 m/s

Fig. 5.29 Velocity spectrums of the monitoring point in branch tube 6 under different inlet velocities

Table 5.14 Pulsating flow frequencies of each combination under different inlet velocities

Combinations	u_{in} /(m·s^{-1})	Frequency f_v/Hz						Maximum relative error /%
		1	2	3	4	5	6	
I	0.1	15.5	15.5	15.5	15.5	15.0	15.0	3.23
	0.2	26.5	26.5	26.5	26.5	26.0	26.0	1.87
	0.3	39.0	39.0	39.0	39.0	38.0	37.0	5.13
	0.4	51.0	51.0	51.0	50.0	50.0	49.0	3.92

Table 5.14(continued)

Combinations	u_{in} /(m · s^{-1})	Frequency f_v/Hz						Maximum relative error /%
		1	2	3	4	5	6	
Ⅱ	0.1	11.5	11.0	11.5	11.0	11.0	11.0	4.35
	0.2	21.5	22.0	22.0	21.5	21.0	20.5	6.82
	0.3	30.5	30.0	30.0	29.5	29.0	29.0	4.92
	0.4	40.0	40.0	39.0	38.5	38.0	38.5	5.00
Ⅲ	0.1	19.5	19	19	19.5	19	19	2.56
	0.2	36	34	33.5	33.5	33	33	8.33
	0.3	57	49.5	49	49.5	49	48	15.79
	0.4	—	65	64	64	63	62.5	—
Ⅳ	0.1	15.5	15	15	15	15	15	3.23
	0.2	27	27	26.5	27.5	27.5	25.5	7.27
	0.3	39.5	39.5	39	38.5	38	37.5	5.06
	0.4	51.5	51.5	50.5	50.5	50	49	4.85
Ⅴ	0.1	21.5	21.5	21	20.5	20.5	22	6.82
	0.2	39	39	39	39	38.5	38.5	1.28
	0.3	58	58	58.5	58	56	56	4.27
	0.4	83	79.5	82	74.5	73.5	73	12.05
Ⅵ	0.1	17.5	17.5	17.5	17.5	17.5	17	2.86
	0.2	31	31	31	30.5	30.5	30.5	1.61
	0.3	45	45.5	45	45	45	44.5	2.20
	0.4	59.5	59.5	59.5	59.5	59.5	58.5	1.68

From Table 5.14, the following conclusions can be drawn.

(1) Under the condition that the inlet velocity of a certain combination is determined, the frequency of velocity changes over time decreases with the increase of the number of branch tubes, but the change range is small, which is consistent with the change of the flow rate of the pulsating flow outlet.

(2) For a certain inner diameter ofthe vertical tube, when the inner diameter of the horizontal tube is large, the fluid flow into each branch tube is small, and the frequency of velocity changes over time is relatively low.

(3) For the inner diameter of a certain horizontal tube, when the inner diameter of the vertical tube is larger, the fluid flow into each branch tube is larger, and the frequency of velocity changes over time is relatively higher.

(4) When the ratio of the inner diameter of the vertical tube to the inner diameter of the horizontal tube d_1/d_2 is small, the frequency of pulsating flow formed in each branch tube is low under the condition of the same inlet velocity. With the increase of d_1/d_2, the frequency of pulsating flow gradually increases.

(5) When d_1/d_2 is large (combination Ⅲ and Ⅴ as shown in Table 5.11, $d_1/d_2 = 1.8$) and the inlet velocity is high ($u_{in} = 0.3$ m/s or 0.4 m/s), no pulsating flow can be formed in some branch tubes or the pulsating flow is abnormal.

5.4.4 Intensity Analysis

To study the intensity of pulsating flow in each branch tube, based on the monitoring points set in each branch tube in Fig. 5.25, the RMS calculation of the velocity data at each monitoring point is carried out to obtain the RMS of the velocity under different conditions, that is, the intensity of the pulsating flow.

Based on different inletvelocities, the RMS of the velocity at each monitoring point is calculated under different combination conditions, as shown in Table 5.15.

Table 5.15 Pulsating flow intensity of each combination under different inlet velocities

Combinations	u_{in} /(m·s^{-1})	RMS of velocity/(m·s^{-1})						Maximum relative error /%
		1	2	3	4	5	6	
Ⅰ	0.1	0.186	0.183	0.179	0.185	0.183	0.179	3.76
	0.2	0.333	0.343	0.345	0.342	0.345	0.336	3.48
	0.3	0.508	0.508	0.521	0.517	0.509	0.501	3.84
	0.4	0.685	0.684	0.697	0.692	0.681	0.659	5.45
Ⅱ	0.1	0.149	0.148	0.146	0.150	0.148	0.147	2.67
	0.2	0.272	0.263	0.269	0.272	0.270	0.258	5.15
	0.3	0.406	0.403	0.405	0.398	0.407	0.391	3.93
	0.4	0.555	0.539	0.555	0.549	0.550	0.527	5.05

Table 5.15(continued)

Combinations	u_{in} /$(m \cdot s^{-1})$	RMS of velocity/$(m \cdot s^{-1})$						Maximum relative error /%
		1	2	3	4	5	6	
Ⅲ	0.1	0.231	0.236	0.240	0.243	0.233	0.239	4.94
	0.2	0.431	0.443	0.447	0.443	0.437	0.436	3.58
	0.3	0.674	0.651	0.665	0.671	0.658	0.646	4.15
	0.4	0.946	0.862	0.886	0.879	0.866	0.852	9.94
Ⅳ	0.1	0.195	0.194	0.198	0.199	0.190	0.193	4.52
	0.2	0.345	0.357	0.358	0.360	0.347	0.344	4.44
	0.3	0.513	0.533	0.531	0.535	0.518	0.511	4.49
	0.4	0.689	0.706	0.716	0.706	0.695	0.680	5.03
Ⅴ	0.1	0.253	0.255	0.250	0.262	0.263	0.255	4.94
	0.2	0.492	0.491	0.495	0.501	0.505	0.497	2.78
	0.3	0.707	0.705	0.710	0.735	0.730	0.728	3.81
	0.4	0.979	0.936	0.969	0.962	0.965	0.962	4.41
Ⅵ	0.1	0.209	0.210	0.209	0.212	0.207	0.208	2.36
	0.2	0.401	0.406	0.396	0.408	0.394	0.385	5.64
	0.3	0.613	0.602	0.592	0.611	0.600	0.597	3.43
	0.4	0.800	0.799	0.790	0.810	0.799	0.792	2.47

From Table 5.15, the following conclusions can be drawn.

(1) Under the condition that the inlet velocity of a certain combination is determined, the intensity of pulsating flow decreases with the increase of the number of branch tubes, but the change range is small, which is consistent with the change of the flow rate and the frequency of the pulsating flow outlet.

(2) For a certain inner diameter of the vertical tube, when the inner diameter of the horizontal tube is larger, the fluid flow into each branch tube is smaller and the pulsating flow intensity is lower.

(3) For a certain inner diameter of the horizontal tube, when the inner diameter of the vertical tube is larger, the fluid flow into each branch tube is larger, and the pulsating flow intensity is higher.

(4) When some branch tubes cannot form pulsating flow, the RMS of the

velocity is abnormal. For example, when $u_{in} = 0.4$ m/s, the RMS of the velocity calculated in branch tube 1 of Combination Ⅲ is significantly different from that of other branch tubes.

(5) Under the conditions of different combinations and different inlet velocities, the pulsating flow intensity of each branch tube has little difference, indicating that the pulsating flow intensity of each branch tube has a good consistency.

5.5 Design Principles of Distributed Pulsating Flow Generator

Based on the analysis of the pulsating flow in each branch tube of the above six tube inner diameter combinations, and combined with the calculation results of the flow around the right triangular cylinder with different structural sizes of the improved branch domain, the following principles should be followed in the design of the distributed pulsating flow generator.

(1) The ratio of the inner diameter of the vertical tube to the inner diameter of the horizontal tube d_1/d_2 should not be too large or too small. When d_1/d_2 is large, pulsating flow cannot be formed in some branch tubes at high inlet velocity or the pulsating flow is abnormal. When d_1/d_2 is small, the fluid flow into each branch tube is small, the intensity of the pulsating flow is low, and then the excitation force to induce the vibration of the ETB is small.

(2) The selection of the vertical tube size should consider the size of the ETB heat exchanger. A larger vertical tube size will not only increase the manufacturing cost of the pulsating flow generator, but also affect its layout in the heat exchanger.

(3) The frequency and intensity of pulsating flow formed in each branch tube have greater flexibility. When the size of each part is determined, the pulsating flow formed can be adjusted by changing the shape and/or size of the disturbing fluid block.

(3) The flexibility of adjusting the frequency and intensity of pulsating flow in each branchtube is greater. Once the size of each part is determined, the pulsating flow can be adjusted by changing the shape and/or the size of the vortex element.

In summary, for the ETB heat exchanger shown in Fig. 4. 1, a six branch distributed pulsating flow generator with a vertical tube diameter $d_1 = 45$ mm and a horizontal tube diameter $d_2 = 30$ mm is selected. Based on the six branch distributed pulsating flow generator, further research will be conducted in this book.

References

[1]GAO H, LIU J F. Experimental analysis of heat transfer enhancement by using Helmholtz oscillator [J]. Energy technology, 2009, 30(3): 141-144.

[2]JI J D, GE P Q, LIU P, et al. Design and application of a new distributed pulsating flow generator in elastic tube bundle heat exchanger [J]. International journal of thermal sciences, 2018, 130: 216-226.

[3]JIANG B. Analysis on mechanism of heat transfer enhancement by vibration and experimental research on a new type of vibrational heat transfer component [D]. Jinan: Shandong University, 2010.

[4]MENG H T. Study on the pulsating flow generating device in elastic tube bundle heat exchanger [D]. Jinan: Shandong University, 2012.

[5]HEMMAT E M, BAHIRAEI M, TORABI A, et al. A critical review on pulsating flow in conventional fluids and nanofluids: Thermo-hydraulic characteristics [J]. International communications in heat and mass transfer, 2021, 120: 104859.

[6]XU C, XU S L, WANG Z Y, et al. Experimental investigation of flow and heat transfer characteristics of pulsating flows driven by wave signals in a microchannel heat sink [J]. International communications in heat and mass transfer, 2021, 125: 105343.

[7]DUAN D R, CHENG Y J, GE M R, et al. Experimental and numerical study on heat transfer enhancement by flow-induced vibration in pulsating flow [J]. Applied thermal engineering, 2022, 207: 118171.

[8]LIU J Q, TIAN M C, WANG H X, et al. Experimental study on flow-induced vibration of compound bent beam under pulsating flow [J]. Journal of hydrodynamics, 1998, 13(4): 467-472.

[9]TIAN M C, CHENG L, LIN Y Q, et al. Experimental investigation of heat transfer enhancement by crossflow induced vibration [J]. Journal of engineering thermophysics, 2002, 23(4): 485-487.

[10]NORBERG C. Flow around a circular cylinder: aspects of fluctuating lift [J]. Journal of fluids and structures, 2001, 15: 459-469.

[11]ZHANG P F, WANG J J , HUANG L X. Numerical simulation of flow around cylinder with an upstream rod in tandem at low Reynolds numbers [J]. Applied ocean research, 2006, 28: 183-192.

[12]GOPALKRISHNAN R. Vortex induced forces on oscillating bluff cylinders [D]. Cambridge City: Massachusetts Institute of Technology, 1993.

Chapter 6 Effect of Pulsating Flow Generator on Vibration and Heat Transfer of ETBs

The design of distributed pulsating flow generator has changed the traditional design idea that the shell-side fluid of the ETB heat exchanger flows into the heat exchanger through the shell-side inlet tube on the bottom shell cover[1-2]. For the distributed pulsating flow generator, one end is welded to the upper shell cover, and the other end is suspended at the bottom of the heat exchanger. Part of the fluid flows into the branch tube and then flows around the vortex element, forming the pulsating flow and impacting the stainless steel connector of the ETB. The opening is located on the horizontal tube at the bottom, which impacts the inner wall of the bottom shell cover, causing the fluid to flow approximately spirally upward inside the heat exchanger.

The vibration of the ETB of the heat exchanger installed with the distributed pulsating flow generator is the response induced by the coupling of the shell-side fluid and the pulsating flow[3]. Compared with the vibration of the ETB inside the TETB heat exchanger, the internal fluid domain is more complex, the fluid flow characteristics are more disordered, and the vibration characteristics of the ETB are also significantly different.

In this chapter, the six-row ETB heat exchanger installed with the distributed pulsating flow generator was taken as the research object, and the vibration and heat transfer characteristics of the ETB under the coupling induction of the shell-side fluid and the pulsating flow were analyzed numerically and experimentally, to verify the actual efficacy of the distributed pulsating flow generator in the ETB heat exchanger.

In this chapter, the overall shell-side fluid domain of the heat exchanger installed with the distributed pulsating flow generator was established first, and the mesh of the overall shell-side fluid domain is formed by fluid domain partitioning, segmented fluid domain meshing and shell-side fluid domain mesh assembly. Then, numerical analysis was conducted on the vibration and heat transfer characteristics of each row of ETBs under different inlet velocity conditions induced by coupling fluids. Secondly, based on the ETB heat exchanger used in numerical analysis, a test bench was built for the coupling of shell-side fluid and pulsating fluid to induce the vibration of ETBs. Then, experimental research was conducted on the vibration response of each row of ETBs under the coupling of shell-side fluid and pulsating fluid.

6.1 Shell-side Fluid Domain and Its Mesh

6.1.1 Heat Exchanger with Distributed Pulsating Flow Generator

Fig. 6.1 shows the schematic diagram of the ETB heat exchanger with the distributed pulsating flow generator and its shell-side fluid domain.

To facilitate manufacturing and comparison with the subsequent experimental data, the two stainless steel connectors of the ETB are replaced by two connecting copper tubes with a size of $\phi 18$ mm×3.0 mm, as shown in Fig. 6.2.

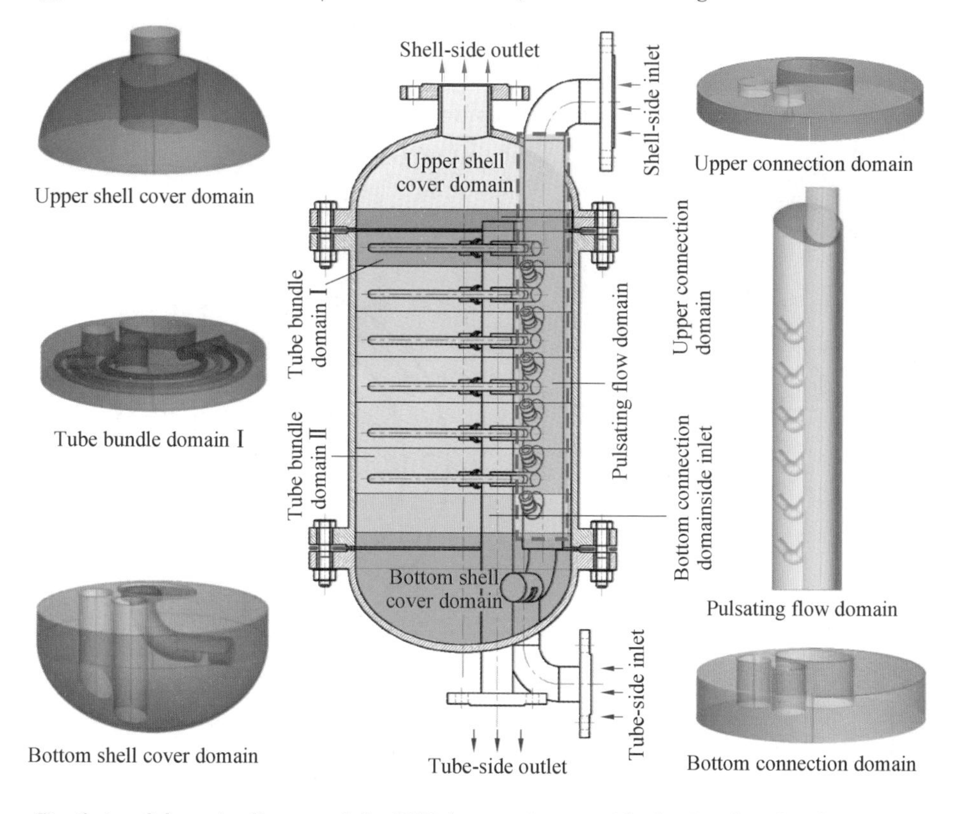

Fig. 6.1 Schematic diagram of the ETB heat exchanger with the distributed pulsating flow generator and its shell-side fluid domain

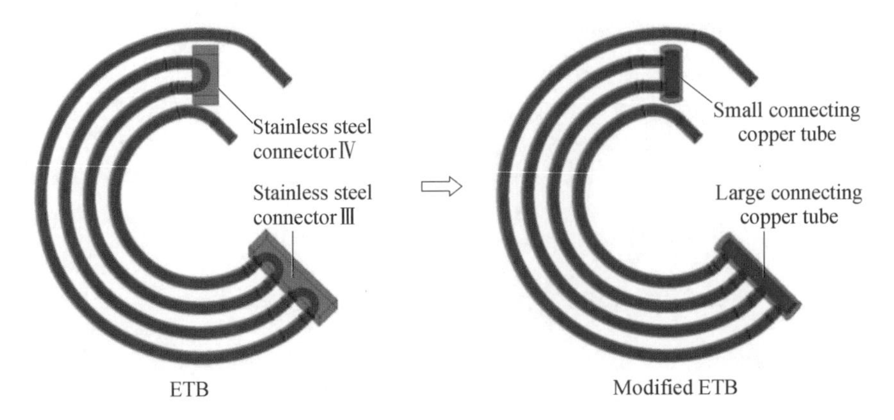

Fig. 6.2 Modify of the ETB

It is calculated that the vibration response of the two kinds of ETBs under the shell-side fluid induction is the same except that the vibration equilibrium position of the two stainless steel connectors is lower in the x-direction[4-5].

Fig. 6.3 shows the installation of the distributed pulsating flow generator on the upper shell cover. The pulsating flow generator is fixed on the upper shell cover of the heat exchanger by welding. In the calculation process, if not specified, $m=112$ mm, $n=15$ mm, $s=417.5$ mm, $\theta=70°$.

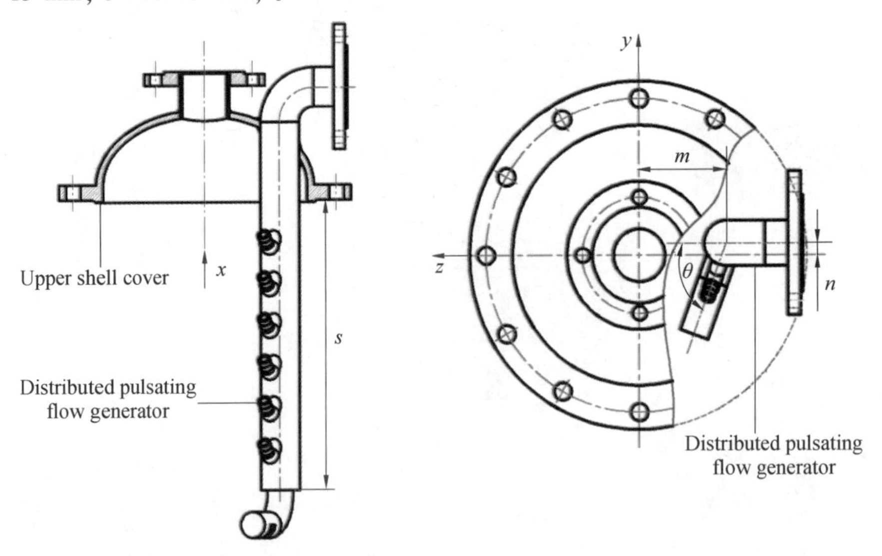

Fig. 6.3 Installation of the distributed pulsating flow generator on the upper shell cover

Due to the complex structure of the shell-side fluid domain in the heat exchanger, it is difficult and inefficient to divide the overall mesh, and it is not easy to adjust and modify the mesh. Therefore, the overall fluid domain is segmented based on its structural characteristics. As shown in Fig. 6.1, the segmented fluid domain includes seven parts: upper shell cover domain, bottom shell cover domain, tube bundle domain I, tube bundle domain Ⅱ, upper connection domain, bottom connection domain and pulsating flow domain.

The tube bundle domain Ⅱ is generated by stretching based on the tube bundle domain Ⅰ (Fig. 6.4), and the stretching height depends on the spacing between ETBs, both of which have similar structural characteristics. When the row spacing increases or decreases, the top surface of tube bundle Ⅱ is stretched or compressed, while keeping the height of tube bundle domain Ⅰ unchanged, so as not to affect the distance between the top row ETB and the shell-side outlet.

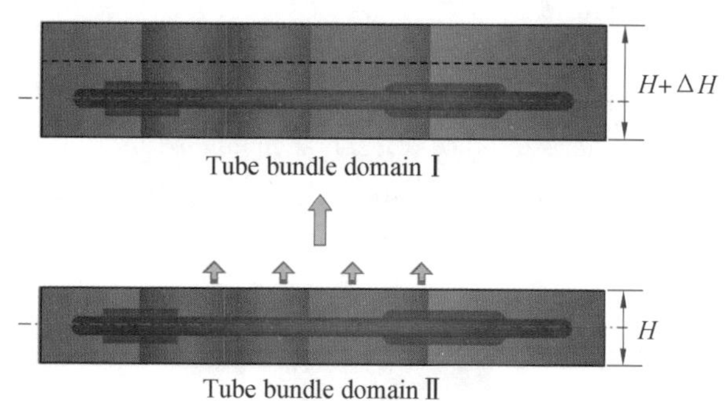

Fig. 6.4 Schematic diagram of stretching in the tube bundle domain

Relatively speaking, the structure of pulsating flow domain is relatively complex, and the mesh division is difficult. Therefore, further segmentation is carried out, as shown in Fig. 6.5. The divided pulsating flow domain includes three parts: external domain, branch tube domain and vertical tube domain.

The specific dimensions of the ETB heat exchanger used in the numerical analysis can be found in "4.1.1 ETB Heat Exchanger and Its Shell Side Fluid Domain". Except for the two connecting copper tubes with a size of $\phi18$ mm×3.0 mm, the remaining dimensions of the ETB heat exchanger are consistent with the dimensions of the ETB heat exchanger used in Chapter 4 of this book.

The distributed pulsating flow generator is shown in Fig. 5.20, and its installation inside the heat exchanger is shown in Fig. 5.21. The diameter of the vertical tube of the pulsating flow generator $d_1 = 45$ mm, the diameter of the horizontal tube $d_2 = 30$ mm, the vortex element is a right triangular cylinder, and the hypotenuse length is 2.0 mm.

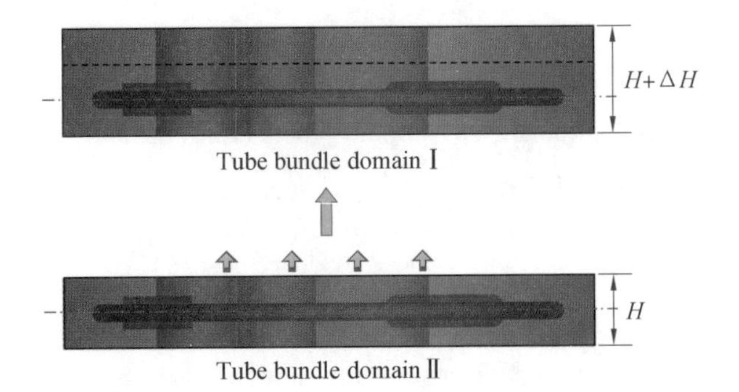

Fig. 6. 5 Schematic diagram of the pulsating flow domain segmentation

6.1.2 Mesh Division and Assembly

Based on the segmentation of the shell-side fluid domain mentioned above, and based on the structural characteristics of each segmented fluid domain, different mesh strategies are adopted for mesh partitioning. The schematic diagrams of the meshes in each segmented fluid domain are shown in Fig. 6. 6.

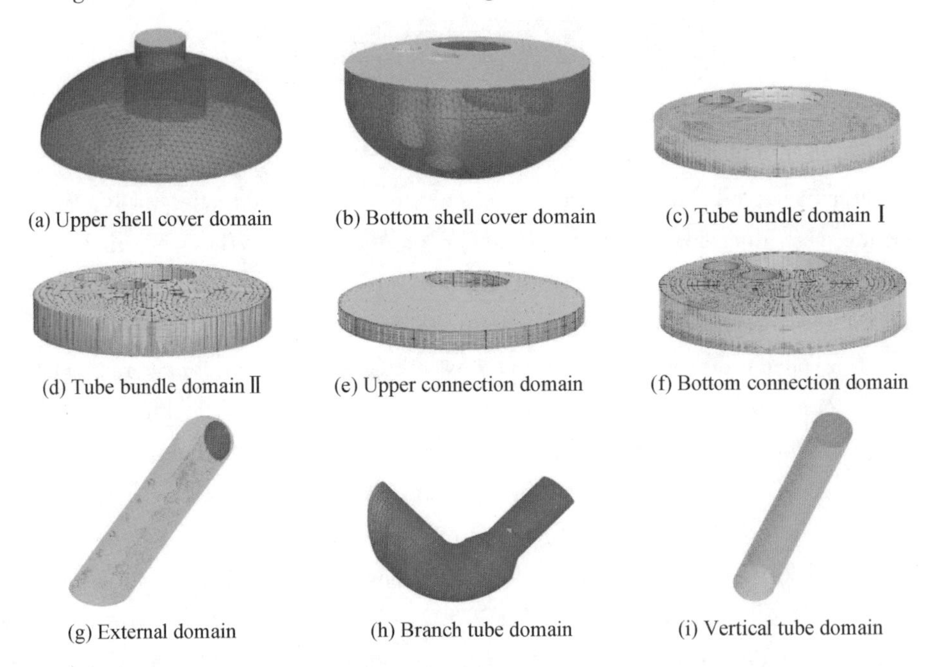

Fig. 6. 6 Schematic diagrams of the meshes in each segmented fluid domain

In Fig. 6. 6, the meshes of each segmented fluid domain are completed using the meshing software ICEM. As shown in Fig. 6. 6, tetrahedral meshes are used for the upper shell cover domain, bottom shell cover domain and external domain, and hexahedral meshes are used for other segmented fluid domains. To improve the accuracy of the calculation, the mesh size of the wall connected with the ETB inside the tube bundle domain Ⅰ and tube bundle domain Ⅱ is smaller and the density is larger, and a 6-layer Boundary layer mesh is set. The mesh size set on the walls connecting the ETB inside the tube bundle domain Ⅰ and tube bundle domain Ⅱ is smaller and the density is larger, and 6 layers of boundary layer mesh are used.

Table 6.1 shows the statistics of the mesh quantity and quality in each segmented fluid domain. From Table 6.1, it can be seen that the lowest minimum mesh quality is 0.35, which occurs in the external domain (non-vibration region). Tube bundle domain Ⅰ, tube bundle domain Ⅱ and branch tube domain are key areas for calculation, with high grid density and a large number.

Table 6.1 Statistics of the mesh quantity and quality in each segmented fluid domain

Segmented fluid domain	Number ofmeshes		Minimummesh quality
	Elements	Nodes	
Upper shell cover domain	140,983	24,973	0.38
Bottom shell cover domain	210,245	37,077	0.39
Tube bundle domain Ⅰ	215,410	196,692	0.50
Tube bundle domain Ⅱ	261,125	241,462	0.50
Upper connection domain	22,456	19,183	0.54
Bottom connection domain	51,233	45,550	0.58
External domain	294,727	53,042	0.35
Branch tube domain	90,160	83,961	0.48
Vertical tube domain	80,332	75,750	0.60

The mesh assembly of the shell-side fluid domain is done in ANSYS CFX software, and the implementation steps are shown in Fig. 6.7. It should be noted that, for the sake of observation, the mesh lines are not shown in Fig. 6.7. The specific steps for mesh assembly are as follows.

(1) Import the branch tube domain and vertical tube domain meshes, copy 5 copies of the branch tube domain mesh along the x-direction, and establish connections between the meshes of each branch tube domain and the vertical tube domain.

(2) Import the external domain mesh, and establish a connection between the

meshes of the branch tube domain and the vertical tube domain.

(3) Import the bottom shell cover domain, and establish connections between the meshes of the external domain and the bottom shell cover domain, as well as the meshes of the bottom shell cover domain and the vertical tube domain, respectively.

(4) Import the bottom connection domain mesh, and establish a connection between the meshes of the bottom connection domain and the bottom shell cover domain.

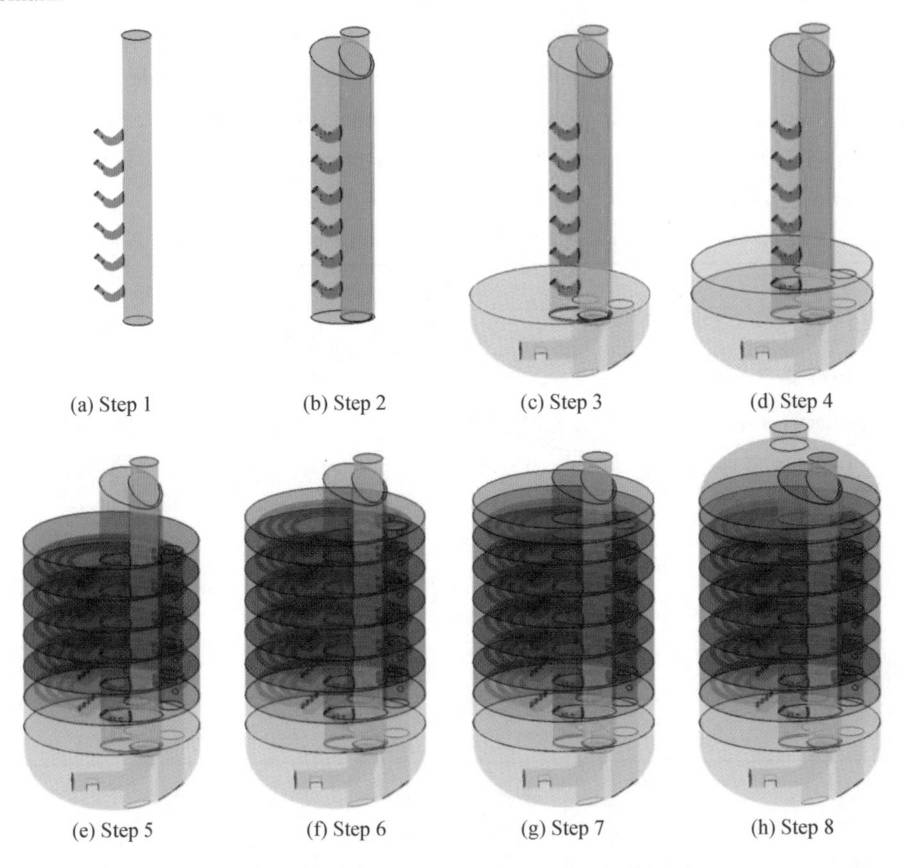

(a) Step 1 (b) Step 2 (c) Step 3 (d) Step 4

(e) Step 5 (f) Step 6 (g) Step 7 (h) Step 8

Fig. 6.7 Assembly steps of the overall shell-side fluid domain

(5) Import the mesh to the tube bundle domain Ⅱ, copy the tube bundle domain Ⅱ mesh 4 times in the x-direction, and establish connections between the meshes of the tube bundle domain Ⅱ and the bottom connection domain, as well as the meshes of each replicated tube bundle domain Ⅱ.

(6) Import the mesh of tube bundle domain Ⅰ, and establish a connection between the meshes of tube bundle domain Ⅰ and the adjacent replicated tube bundle

domain Ⅱ.

(7) Import the upper connection domain mesh, and establish a connection between the meshes of the upper connection domain and the tube bundle domain Ⅰ.

(8) Import the upper shell cover domain, and establish a connection between the meshes of the upper shell cover domain and the upper connection domain.

The mesh connections between the above segmented fluid domains are made by the General Grid Interface (GGI) method. At this point, the overall shell-side fluid domain mesh assembly is completed.

Thus, take the shell-side fluid domain mesh of a six-row ETB heat exchanger with tube row spacing $H=60$ mm as an example.

The number of elements in the shell-side fluid domain (n_c) is

$$\begin{aligned} n_c &= 140,983 + 210,245 + 215,410 + 261,125 \times 5 + 22,456 + 51,233 + \\ &\quad 294,727 + 90,160 + 80,332 \\ &= 2,411,171 \end{aligned} \tag{6.1}$$

The number of nodes in the shell-side fluid domain (n_n) is

$$\begin{aligned} n_n &= 24,973 + 37,077 + 196,692 + 241,462 \times 5 + 19,183 + 45,550 + \\ &\quad 53,042 + 83,961 + 75,750 \\ &= 1,743,538 \end{aligned} \tag{6.2}$$

The above mesh division strategy significantly reduces the number of meshes, improves the quality of meshes, and facilitates the adjustment of the structure and meshes of various segmented fluid domains, thereby improving the efficiency of mesh division.

Based on the mesh duplication, translation and connection functions of ANSYS CFX software, the overall fluid domain mesh with any number of ETB rows and any row spacing can be generated, and the structure and/or size of the distributed pulsating flow generator in the pulsating flow domain can be easily adjusted.

Similar to the analysis of the vibration response of ETBs induced by shell-side fluid in Chapter 4 of this book, the tube-side inlet tube, the tube-side outlet tube and the distributed pulsating flow generator are steel structures, and there is basically no vibration under fluid induction. Therefore, to improve calculation efficiency, the structural domain only retains the ETBs without affecting the calculation results.

The mesh division of the ETB is carried out using the Workbench platform's mesh division module, as shown in Fig. 6.8.

As can be seen from Fig. 6.8, the mesh of the ETB consists of tetrahedral mesh (large and small connecting copper tubes) and hexahedral mesh (copper bend tube).

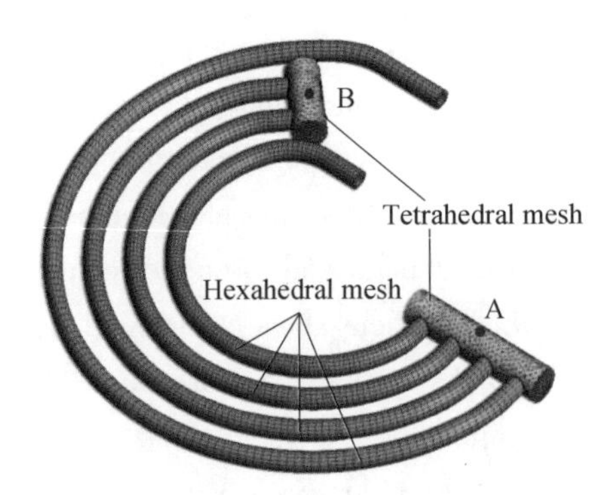

Fig. 6. 8 Mesh of the ETB

The minimum mesh quality is 0. 34, with a total of 20,142 elements and 80,355 nodes. In Fig. 6. 8, two points A and B are monitoring points set on two connecting copper tubes to monitor the vibration response of the ETB under fluid induction.

For the convenience of analysis, the ETB in the heat exchanger is numbered from bottom to top as 1, 2, ..., 6. The monitoring points A_n and B_n (n = 1, 2, ..., 6) are set on the two connecting copper tubes of each row of ETBs to monitor the vibration of the ETBs under fluid induction.

6.1.3 Boundary Conditions and Mesh Independence Analysis

The boundary condition settings are similar to "4. 1. 3 Boundary Conditions" of this book, with the basic settings as follows.

The boundary conditions of the structural domain are set as follows.

(1) The cross sections at the two fixed ends of each row of ETBs are set as "Fixed Support".

(2) The outer surfaces of each row of ETBs are set as "Fluid Solid Interface", and are sequentially named from the bottom row of ETBs as FSI-1, FSI-2, ..., FSI-6.

(3) The direction of gravitational acceleration (Standard Earth Gravity) is set as vertical downward (x-direction), and its value is 9. 8066 m/s^2.

The boundary conditions of the fluid domain are set as follows.

(1) Set the shell-side inlet boundary type to "Inlet", and give the inlet velocity and temperature.

(2) Set the shell-side fluid outlet boundary type to "Outlet", and give the outlet relative static pressure to 0 Pa.

(3) Set the inner surfaces of all tube bundle domains are "Fluid Solid Interface",

and give the wall temperature. The FSI interfaces in the fluid domain correspond one-to-one with the FSI interfaces in the structural domain.

(4) The boundary types of the other surfaces are set to "Wall", and the boundary details included: the "Mesh Motion" option is set as "Stationary", and the "Heat Transfer" option is set as "Adiabatic".

As with the numerical simulation in Chapter 2 of this book, due to the large number of elements in the fluid and structural domains, the FSI calculation requires high requirements for computer hardware. To obtain the vibration response of the ETBs under fully developed flow conditions, the step-by-step calculation strategy of rough calculation plus actuarial calculation is adopted. To ensure the full development of the shell-side fluid, the total time of the rough calculation is defined as 500 s, with a time step of 0.1 s. In addition, the total time of the actuarial calculation is defined as 1.5 s, with a time step of 0.001 s.

To analyze the mesh independence of the overall shell-side fluid domain and the structural domain, based on three different mesh division schemes, the vibration response of the ETB is calculated when the inlet velocity of the shell-side inlet u_{in} = 0.4 m/s. Among them, the fluid medium is water.

Table 6.2 shows the comparison of the first vibration frequency f_x and its amplitude A_x at the monitoring point A_1 on the lowest ETB in the x-direction (vertical direction) under different mesh division schemes.

Table 6.2 Comparison of calculation results in different grid division schemes

Case	Nodes	Calculation results		Relative error/%		Calculation time/h
		f_x/Hz	A_x/mm	f_x	A_x	
1	1,663,292	22.67	0.0825	6.21	7.82	86
2	2,225,668	24.17	0.0895	—	—	124
3	3,950,507	25.00	0.0917	3.43	2.46	215

Note: The relative error is calculated based on the calculation result of case 2.

In Table 6.2, the number of nodes refers to the total number of nodes in the overall shell-side fluid domain and the structural dimain. Case 2 refers to the mesh division strategy mentioned above. Cases 1 and 3 are obtained by decreasing or increasing the mesh density of the structural domain and each segmented fluid domain based on case 2. The number of nodes in case 1 is about 72.6% of that in case 2, and the number of nodes in case 3 is about 1.84 times that in case 2.

As can be seen from Table 6.2, when the number of nodes is increased (case

3), it has little impact on the calculation results. The maximum relative error is only 3.43%, but the calculation time is about 1.73 times that of case 2. However, when the number of nodes is reduced (case 1), it has a bigger impact on the calculation results. The maximum relative error is 7.82%, and the calculation time is about 69.4% of that of case 2. Considering the calculation accuracy and efficiency, the mesh division scheme in case 2 is selected.

6.2 Numerical Analysis of Vibration and Heat Transfer

6.2.1 Vibration Analysis of ETBs

Fig. 6.9 shows the change of the displacement (A_x) of monitoring point A_1 in the x-direction on the bottom row of ETB with the actuarial calculation time (t_a) when the inlet velocity u_{in} = 0.4 m/s. Among them, the fluid medium is water.

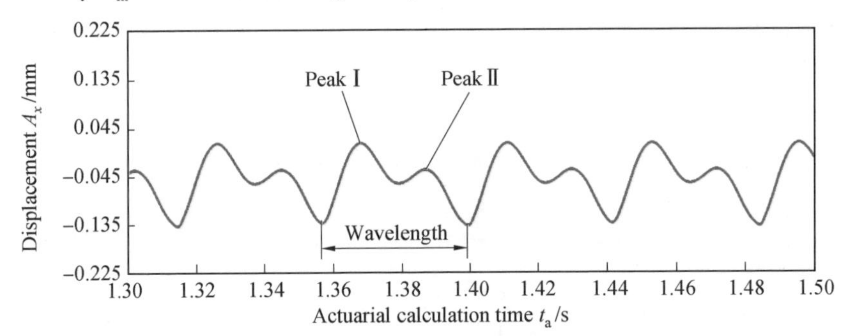

Fig. 6.9 The change of the displacement of monitoring point A_1 in the x-direction on the bottom row of ETB with the actuarial calculation time

From Fig. 6.9, it can be seen that the displacement of the monitoring points changes with the actuarial calculation time, and there is a "double peak" phenomenon within each wavelength. Within the calculation time shown in Fig. 6.9, the vibration of the monitoring points reaches stability, which also verifies the rationality of the above calculation time and time step settings.

To facilitate further analysis of the amplitude and frequency of the monitoring points, FFT calculations are performed on the displacement time history curve after vibration stabilization. Table 6.3 shows the frequency and amplitude of the two monitoring points on the bottom ETB in the x-direction when u_{in} = 0.1 m/s and 0.4 m/s.

Table 6.3 Frequency and amplitude of the two monitoring points on the bottom ETB in the x-direction

Monitoring point	$u_{in}/(\mathrm{m \cdot s^{-1}})$	f_c		f_v	
		f/Hz	A/mm	f/Hz	A/mm
A_1	0.1	24.17	0.0281	12.33	0.0138
	0.4	24.17	0.0895	43.67	0.0579
B_1	0.1	24.17	0.0212	12.33	0.0144
	0.4	24.17	0.0514	43.67	0.0328

It can be seen from Table 6.3 as follows.

(1) After FFT calculation of the displacement signals of the monitoring points, the vibration of the two monitoring points under different inlet velocity conditions has two vibration frequencies. The size of one frequency does not change with the increase of the inlet velocity, which is called constant frequency (f_c). The size of the other fequency increases with the inlet velocity and is called variable frequency (f_v).

(2) The displacement amplitude of monitoring points increases with the increase ofthe inlet velocity. The amplitude of constant frequency is higher than that of the variable frequency at the same monitoring point, and the amplitude of monitoring point A_1 is higher than that of monitoring point B_1 at the same inlet velocity.

Table 6.4 shows the frequency and amplitude of the two monitoring points on the bottom ETB in each direction when $u_{in} = 0.4$ m/s. As each row of ETBs mainly vibrates with constant frequency, only the constant frequency and its amplitude in each direction are listed here.

Table 6.4 Frequency and amplitude of the two monitoring points on the bottom ETB in each direction

Monitoring point	Direction	Calculation result		Monitoring point	Direction	Calculation result	
		f/Hz	A/mm			f/Hz	A/mm
A_1	x	24.17	0.0895	B_1	x	24.17	0.0514
	y	24.17	0.0081		y	24.17	0.0018
	z	24.17	0.0020		z	24.17	0.0013

As can be seen from Table 6.4, compared with the amplitude in the x-direction, the amplitude in the y- and z-directions is small, with a difference of about one order of magnitude. This indicates that the vibration of the monitoring point induced by the coupling fluids is mainly in the x-direction, and mainly manifested as out-plane vibration.

Fig. 6. 10 and Fig. 6. 11 show the variation of amplitude A_x of constant frequency and variable frequency of two monitoring points in the x-direction along with the number of ETBs at different inlet velocities (u_{in}=0. 1 m/s and 0. 4 m/s).

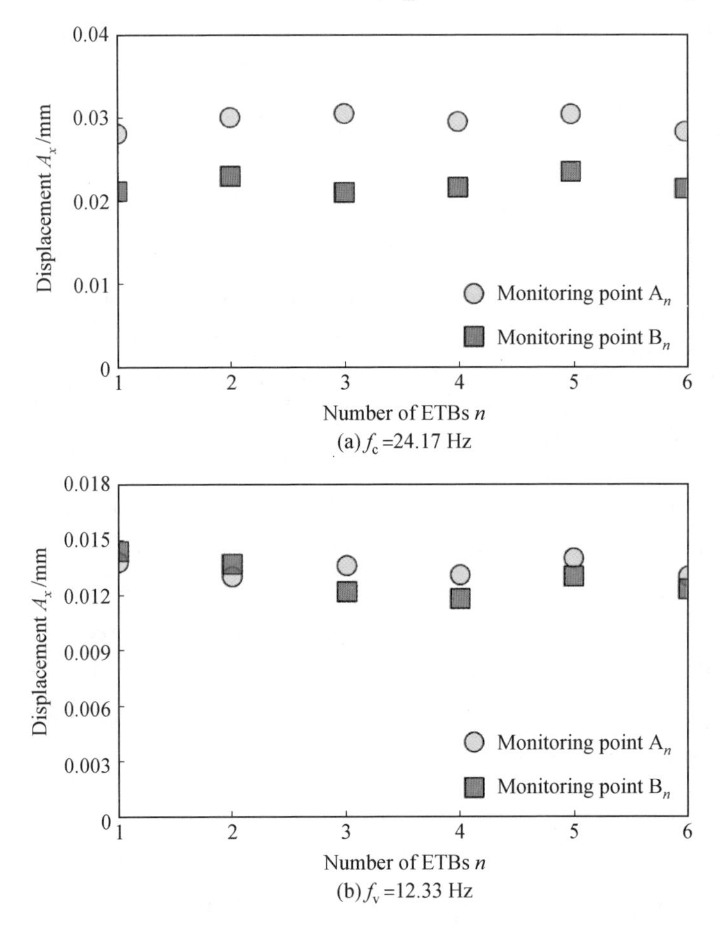

Fig. 6. 10 Amplitude variation of monitoring points with the number of ETBs (u_{in}=0. 1 m/s)

From Fig. 6. 10 and Fig. 6. 11, it can be seen as follows.

(1) Under different inlet velocities, the amplitude corresponding to constant frequency increased first and then decreased with the change of ETB number, which is consistent with the changing trend of ETB induced by shell-side fluid.

(2) When the inlet velocity is low (u_{in} = 0. 1 m/s), the maximum amplitude relative errors of the constant frequency of the two monitoring points for each row of

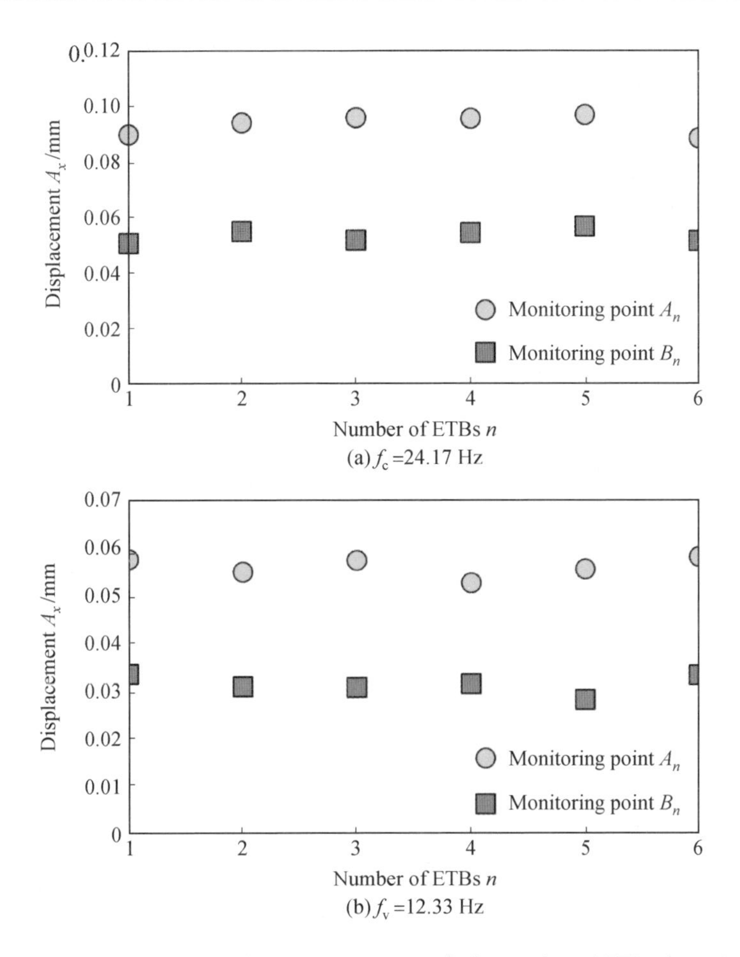

Fig. 6. 11 Amplitude variation of monitoring points with the number of ETBs (u_{in} = 0. 4 m/s)

ETBs are A_n—7. 87% and B_n—9. 01%, respectively. And also, the maximum amplitude relative errors of the variable frequency are A_n—7. 14% and B_n—18.06%, respectively. When the inlet velocity is high (u_{in} = 0. 4 m/s), the maximum amplitude relative errors of the constant frequency of the two monitoring points for each row of ETBs are A_n—8. 66% and B_n—9. 82%, respectively. And also, the maximum amplitude relative errors of the variable frequency are A_n—9. 28% and B_n—15. 41%, respectively. It indicates that the amplitude uniformity of the constant frequency is relatively good, while that of variable frequency is relatively poor.

(3) Under differentinlet velocity conditions, the amplitude of the constant frequency is significantly higher than that of the variable frequency, indicating that the

vibration of each row of ETBs is dominated by constant frequency. Compared with the vibration of each row of ETBs at the same inlet velocity in Chapter 4 (the maximum amplitude relative errors of the monitoring points are higher than 10%), the vibration uniformity is significantly enhanced.

(4) Compared with the amplitude of the monitoring point under the same inlet velocity condition in Chapter 4, the amplitude of the monitoring point of the ETB under the induction of the coupling fluids is significantly enhanced. It is calculated that when the inlet velocity $u_{in} = 0.1$ m/s, the amplitudes of A_n and B_n at the monitoring points increase by 4.2 times and 3.1 times, respectively. In addition, when the inlet velocity $u_{in} = 0.4$ m/s, the amplitudes of A_n and B_n at the monitoring points increase by about 2.5 times and 2.1 times, respectively. This shows that the vibration intensity of each row of ETBs increases under the coupling induction of the shell-side fluid and the distributed pulsating fluid.

6.2.2 Heat Transfer Analysis of ETBs

According to the previous research, the flow-induced vibration of ETBs in the heat exchanger can greatly improve the heat transfer coefficient of each row of ETBs at a low velocity (or low *Re*), but the effect of heat transfer enhancement is not obvious at a high velocity. Here, the heat transfer coefficient of each row of ETBs in the heat exchanger is calculated only when the inlet velocity $u_{in} = 0.1$ m/s, as shown in Fig. 6.12. The inlet fluid temperature of the pulsating flow generator is set as 293.15 K, and the wall temperature of the ETB is set as 353.15 K.

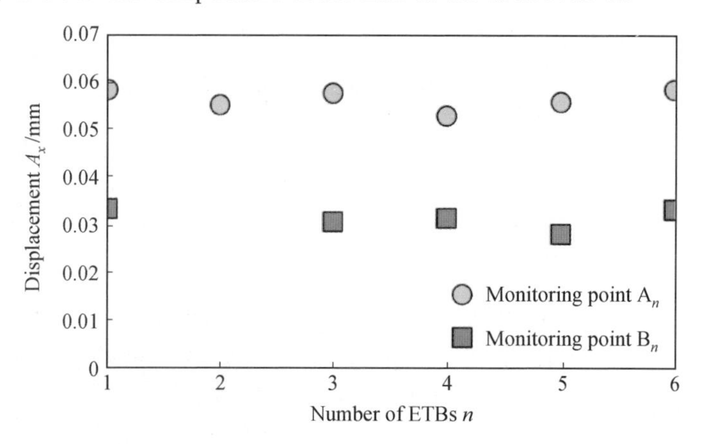

Fig. 6.12 Heat transfer coefficient of each row of ETBs when $u_{in} = 0.1$ m/s

From Fig. 6.12, it can be seen as follows.

(1) Compared with the heat transfer coefficient in the condition of no vibration, the heat transfer coefficient of each row of ETBs in the heat exchanger under the condition of low velocity induced by coupling fluids is significantly increased. Under the condition of inlet velocity $u_{in} = 0.1$ m/s, the heat transfer coefficient increases about 5.3 times on average.

(2) Compared with the heat transfer coefficient under the same inlet velocity in Chapter 4 of this book, the heat transfer coefficient of the ETBs without vibration is basically the same, and the average heat transfer coefficient of each row of ETBs only increases by 0.463 $W \cdot m^{-2} \cdot K^{-1}$ under the coupling fluids.

To sum up, under the coupling induction of the shell-side fluid and the distributed pulsating fluid, the vibration frequency of each row of ETBs in the heat exchanger includes the constant frequency and variable frequency. The vibration intensity and vibration uniformity are enhanced, and the heat transfer performance of each row of ETBs is greatly improved.

6.3 Vibration Test Bench and Shell Vibration Test

6.3.1 Construction of Vibration Test Bench

Based on the numerical study of the vibration response of the ETB under the shell-side fluid and the distributed pulsating fluid, the vibration of the ETB is mainly manifested as out-plane vibration. Therefore, only the out-plane vibration response of the ETB is tested during the experimental study.

Fig. 6.13 shows the experimental principle diagram of the vibration test of ETB induced by the coupling of the internal shell-side fluid and the distributed pulsating fluid. Fig. 6.14 shows the vibration test bench of the ETB induced by the coupling of the internal shell-side fluid and the distributed pulsating fluid of heat exchanger.

The vibration test platform mainly consists of the ETB heat exchanger, water circulation system, and signal acquisition and analysis system.

The water circulation system mainly includes water tank, high-pressure hose, water pump and pump speed regulation system. Where, the water pump adopts a three-phase oil-filled submersible pump, the model is QY15-26-2.2C, the rated flow rate is 15 m^3/h and the synchronous speed is 3,000 r/min. By adjusting the working

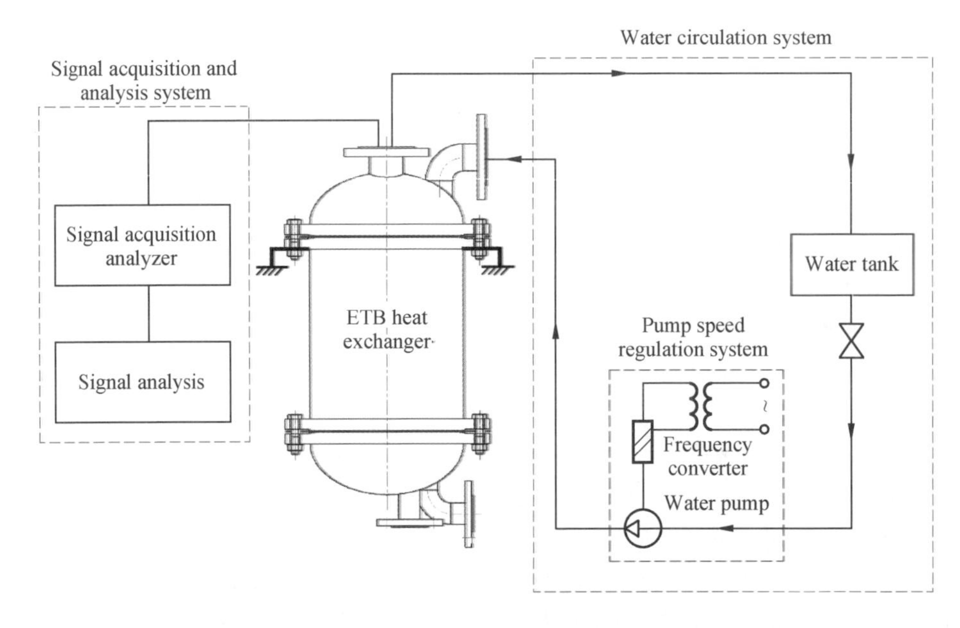

Fig. 6. 13 Experimental principle diagram of the vibration test

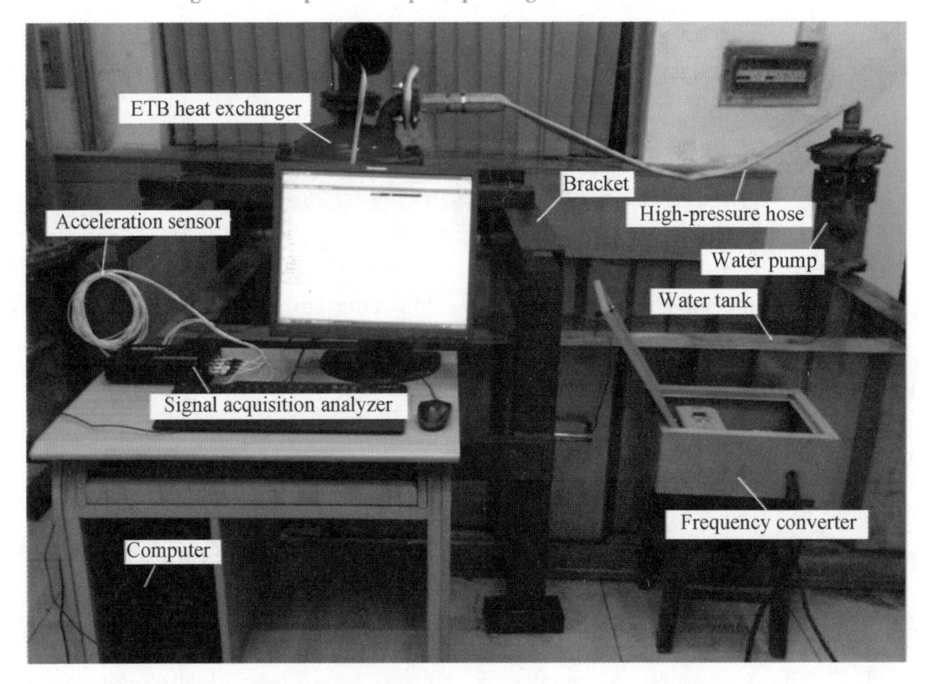

Fig. 6. 14 Vibration test bench of the ETB

speed of the water pump through a frequency converter, the purpose of controlling the flow rate is achieved.

The signal acquisition and analysis system mainly includes sensor, signal acquisition analyzer and computer, where the sensor adopts acceleration sensor PCB-W352C65/002P20. The signal acquisition analyzer INV3018A is used for signal acquisition and analysis, and Coinv-DASP-V10 software is used to process the measured acceleration signal.

Fig. 6. 15 shows the physical diagram of the ETB heat exchanger with the distributed pulsating flow generator. Fig. 6. 16 shows the physical diagram of the installation of the distributed pulsating flow generator and the ETBs in the heat exchanger.

Fig. 6. 15 Physical diagram of the ETB heat exchanger with the distributed pulsating flow generator

The schematic diagram of the ETB heat exchanger used in the experiment is shown in Fig. 6. 1. The structural dimensions of the ETB heat exchanger used in the experiment are consistent with those of the ETB heat exchanger used in the numerical analysis mentioned above.

As the three-phase oil-filled submersible pump will cause additional vibration during operation, to reduce its impact on the vibration response of the ETBs in the heat exchanger, two vibration reduction measures are taken.

(1) A silicone gasket with a thickness of 10 mm is laid at the bottom of the water pump to reduce the vibration transmission between the pump and the water tank.

(2) Fix the heat exchanger on the bracket (Fig. 6. 14) and separate the bracket from the water tank at a certain distance. In this way, the vibration caused by the water pump is only transmitted to the heat exchanger through the ground, and the vibration intensity is weak.

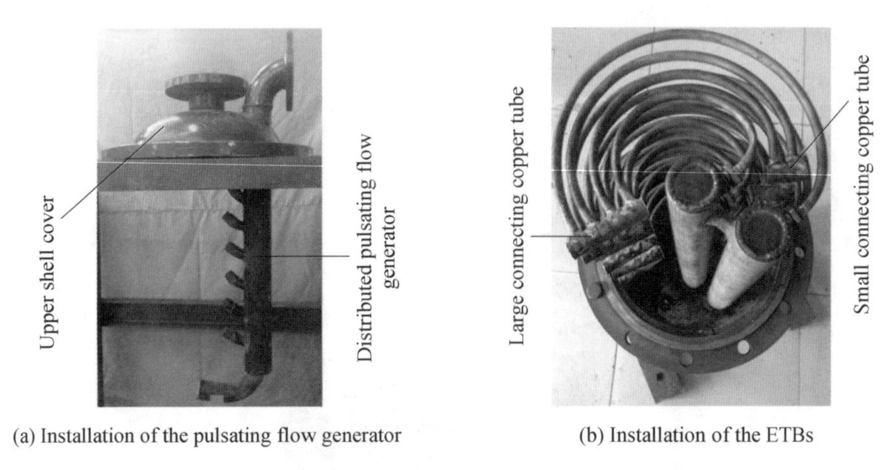

Fig. 6. 16 Physical diagram of the installation of the distributed pulsating flow generator and the ETBs in the heat exchanger

6. 3. 2 Vibration Test of Heat Exchanger Shell

Before the vibration test of the ETBs, to eliminate the interference vibration caused by environmental factors and the three-phase oil-filled submersible pump and increase the rationality of the analysis, two acceleration sensors are arranged at the proper position of the heat exchanger shell to analyze the vertical vibration response of the shell under different inlet velocity conditions[6]. Fig. 6. 17 shows the arrangement position of the acceleration sensor on the heat exchanger shell.

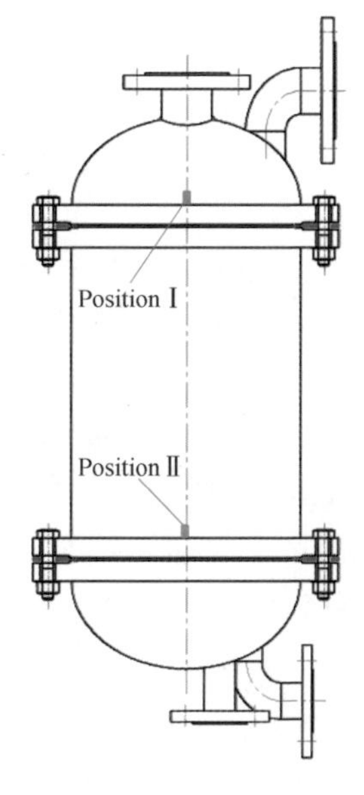

Fig. 6. 17 Arrangement position of the acceleration sensor on the heat exchanger shell

Fig. 6. 18 and Fig. 6. 19 show the vertical acceleration spectrum at different positions of the heat exchanger shell under different inlet velocities of the distributed pulsating flow generator (u_{in} = 0. 1 m/s and 0. 4 m/s). In Fig. 6. 18 and Fig. 6. 19, A_a represents the acceleration amplitude and f_a represents the frequency.

From Fig. 6. 18 and Fig. 6. 19, the following conclusions can be drawn.

(1) When u_{in} = 0. 1 m/s, the vibration at shell position Ⅰ with a frequency of 50 Hz is more pronounced, while when u_{in} = 0. 4 m/s, the vibration at shell position Ⅱ with a frequency of 29 Hz is more pronounced, and

the vibration intensity (acceleration amplitude) of the shell increases with the increase of inlet velocities.

(2) Within the testing inlet velocity range (u_{in}) in this chapter, the main vibration frequencies of the shell of the ETB heat exchanger include 16 Hz, 29 Hz and 50 Hz, and the 50 Hz frequency only appears at the position Ⅰ.

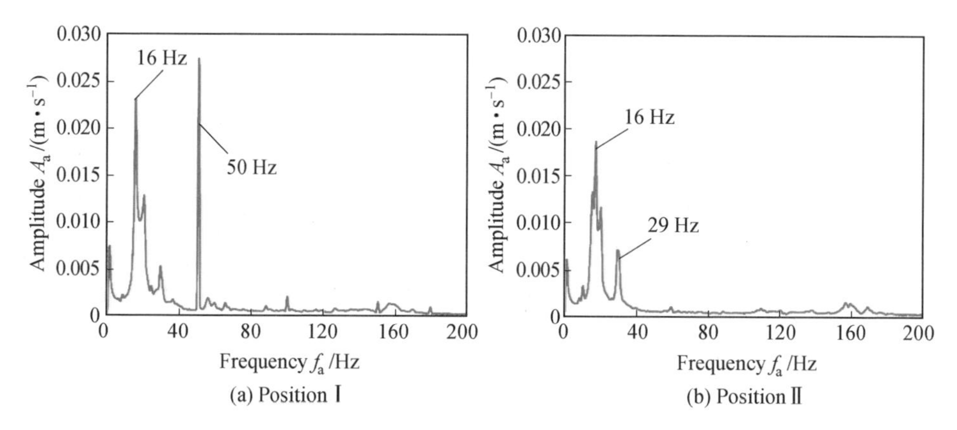

Fig. 6. 18 Acceleration spectrum at different positions of the heat exchanger shell (u_{in} = 0.1 m/s)

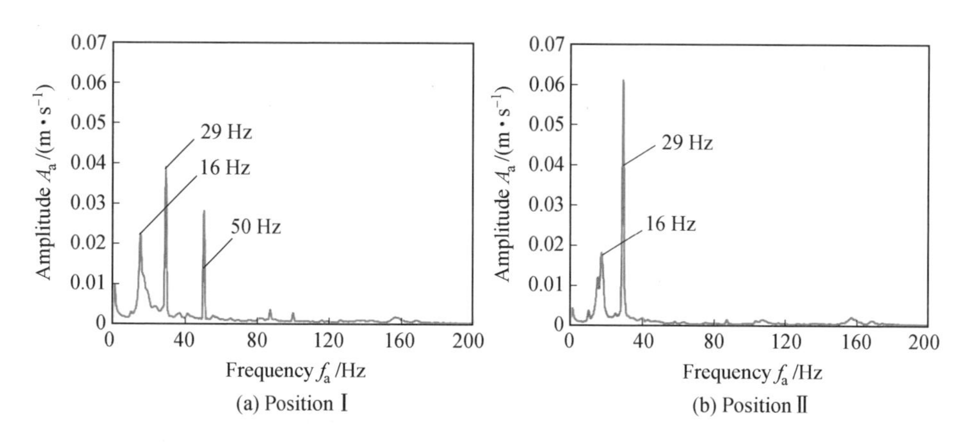

Fig. 6. 19 Acceleration spectrum at different positions of the heat exchanger shell (u_{in} = 0.4 m/s)

6.4 Experimental Study on Vibration of ETBs

6.4.1 Vibration Test of ETBs Induced by Shell-side Fluid

To verify the role of the distributed pulsating flow generator in the ETB heat exchanger, the outlet of each branch tube of the pulsating flow generator is sealed without changing the structural size. In this way, the fluid only enters the heat exchanger from the opening located on the horizontal tube. Based on the vibration test bench of the ETBs built in the previous section, the out-plane vibration responses of each row of ETBs under shell-side fluid induction are tested.

During the testing process, the placement position of the acceleration sensor is shown in Fig. 6.8. Each row of ETBs is numbered from bottom to top as 1, 2, 3,..., 6, and the acceleration sensors are numbered as A_n and B_n ($n=1, 2, 3, \ldots, 6$) from bottom to top.

Fig. 6.20 and Fig. 6.21 show the vertical acceleration spectrum of sensors A_1, A_6, B_1 and B_6 on the first and sixth rows of ETBs under the conditions of inlet velocity $u_{in}=0.1$ m/s and 0.4 m/s, respectively.

It can be seen from Fig. 6.20 and Fig. 6.21 as follows.

(1) After excluding the interference frequency of the shell at 16 Hz, under different inletvelocity conditions, the acceleration signals detected by sensors A_1 and A_6 are more prominent in the spectrum of 24 Hz or 25 Hz and 42 Hz or 44 Hz, while the acceleration signals detected by sensors B_1 and B_6 are more prominent in the vibration spectrum of 26 Hz, 43 Hz or 44 Hz and 89 Hz.

(2) Under different inletvelocity conditions, the vibration of the bottom row ETB (ETB 1) is more severe than that of the top row ETB (ETB 6), and the vibration frequency of the top row ETB is slightly higher than the corresponding vibration frequency of the low row ETB. For example, when $u_{in}=0.1$ m/s, sensor A_1 detects a vibration frequency of 24 Hz, while sensor A_6 detects a corresponding vibration frequency of 25 Hz.

(3) For the same ETB, when the inlet velocity is high, the vibration of the ETB is more severe (with higher acceleration amplitude), but the frequencies remain unchanged. This indicates that within the parameter range tested in this book, the influence of the inlet velocity on the frequency values of each monitoring position of

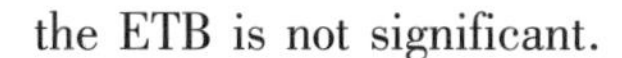

the ETB is not significant.

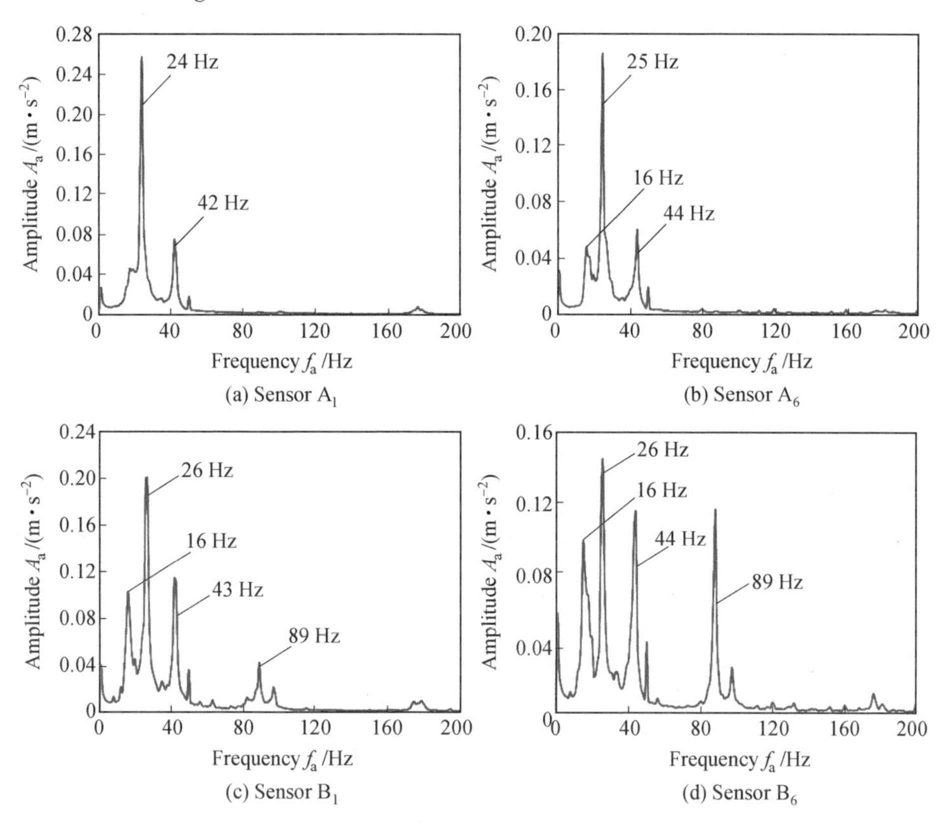

(a) Sensor A_1 (b) Sensor A_6

(c) Sensor B_1 (d) Sensor B_6

Fig. 6. 20 Acceleration spectrum of sensors on the first and sixth rows of ETBs (u_{in} = 0. 1 m/s)

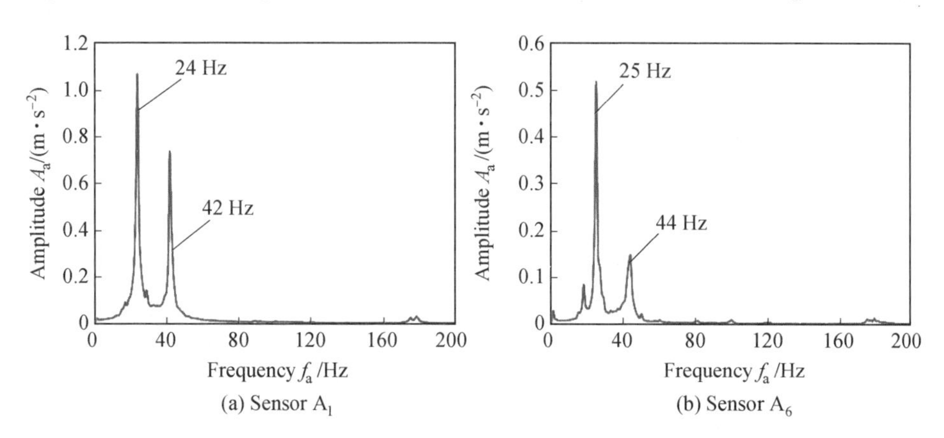

(a) Sensor A_1 (b) Sensor A_6

Fig. 6. 21 Acceleration spectrum of sensors on the first and sixth rows of ETBs (u_{in} = 0. 4 m/s)

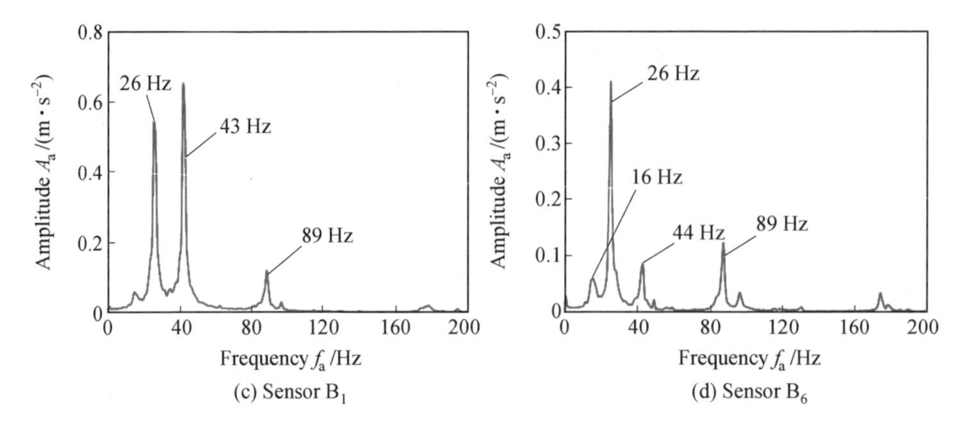

Fig. 6. 21 (continued)

(4) On the same ETB, the acceleration signal monitored by the sensor A_n is strong, indicating that the vibration of the small connecting copper tube (corresponding to stainless steel connector Ⅲ) is relatively small, and the vibration of the large connecting copper tube (corresponding to stainless steel connector Ⅳ) is severe. In addition, the vibration frequency of the small connecting copper tube is slightly higher than that of the large connecting copper tube. For example, when u_{in} = 0. 1 m/s, the vibration frequency with the highest amplitude detected by sensor A_1 is 24 Hz, and the corresponding amplitude is 0. 26 m/s^2. And also, the vibration frequency with the highest amplitude detected by sensor B_1 is 26 Hz, and the corresponding amplitude is 0. 20 m/s^2.

Fig. 6. 22 shows the variation of acceleration amplitude on the large connecting copper tube of ETB with the number of ETBs under the conditions of inlet velocity u_{in} = 0. 1 m/s and 0. 4 m/s, where the acceleration amplitude is the average value of multiple measurements.

It can be seen from Fig. 6. 22 as follows.

(1) After the FFT calcualtion of the acceleration signal monitored by each acceleration sensor, there are multiple vibration frequencies, and the vibration frequency increases with the increase of the number of ETBs, but the increase is very small.

(2) The acceleration amplitudes of each row of ETBs vary greatly under different conditions, and the maximum relative errors of 24 Hz and 25 Hz under different inlet velocity conditions are about 65. 6% and 68. 2%, respectively, indicating that the vibration uniformity of each row of ETBs is poor.

In summary, the vibration response of each row of ETBs in the heat exchanger

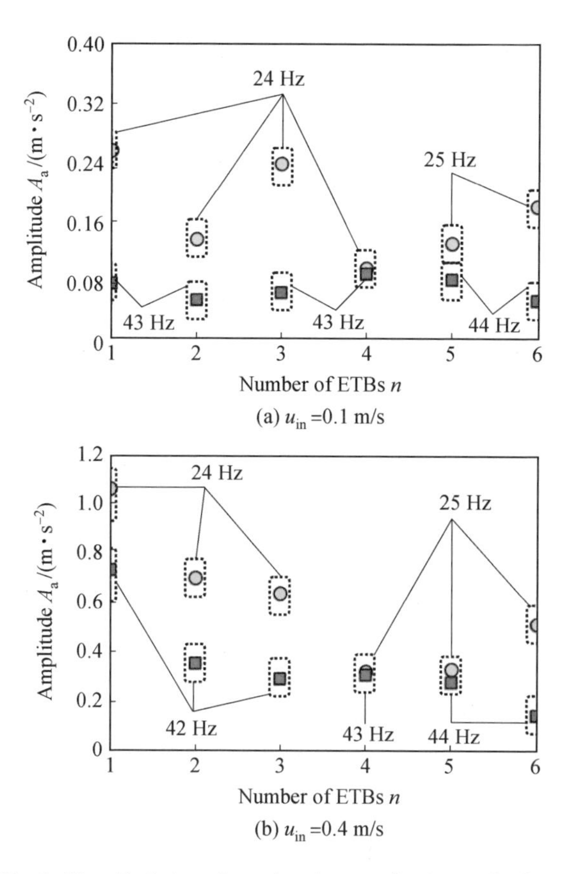

Fig. 6. 22 Variation of acceleration amplitude on the large connecting copper tube of ETB with the number of ETBs under different inlet velocities

induced by the shell side fluid is experimentally tested by sealing the outlet of each branch tube of the distributed pulsating flow generator. Through testing, it is found that the vibration of each row of ETBs is uneven under the induction of the shell-side fluid. In this way, during the actual operation of the heat exchanger, ETBs with severe vibration are prone to fatigue damage, while ETBs with less obvious vibration have poor heat transfer enhancement effects. This further demonstrates the necessity of designing a distributed pulsating flow generator in this book.

6.4.2 Vibration Test of ETBs Induced by Coupling Fluids

In order to test the vibration response of each row of ETBs in the heat exchanger under the coupling of shell-side fluid and distributed pulsating fluid, the outlet of each branch tube of the pulsating flow generator is opened, allowing some of the fluid flowing into the pulsating flow generator to flow out from each branch tube. Through flow around the vortex elements located on the pulsating flow tube, the pulsating flows with basically the same frequency and intensity are formed at each pulsating flow outlet, impacting the large connecting copper tubes of the ETBs and inducing the vibration of the ETBs.

During the testing process, the installation position of the acceleration sensor is shown in Fig. 6.8. Similarly, the number the acceleration sensors from bottom to top as A_1, A_2, ..., A_6 and B_1, B_2, ..., B_6.

Due to the structural differences in each row of ETBs during the manufacturing process, to eliminate the impact of the ETB structure on vibration, the arrangement position of the ETB in the heat exchanger is changed multiple times during the testing process, and the measured signals are compared and analyzed to eliminate unreasonable data.

Fig. 6.23 and Fig. 6.24 show the vertical acceleration spectrum of sensors A_1, A_6, B_1 and B_6 on the first and sixth rows of ETBs at inlet velocities of $u_{in}=0.1$ m/s and 0.4 m/s, respectively.

It can be seen from Fig. 6.23 and Fig. 6.24 as follows.

(1) Under the coupling induction of shell-side fluid and distributed pulsating fluid, there are two main vibration frequencies of ETBs. One of which is a constant frequency ($f_c=24$ Hz or 25 Hz), and its frequency value is not affected by the inlet velocity. The other is a variable frequency ($f_v=13$ Hz or 14 Hz, 46 Hz or 47 Hz), its frequency value increases with the increase of the inlet velocity. This is consistent with the previous analysis based on Fig. 6.10 and Fig. 6.11.

(2) The constant frequency is independent of whether the ETB is impacted by pulsating flow. When the branch tube outlets of the pulsating flow generator are blocked (Fig. 6.20 and Fig. 6.21), this frequency still exists. Therefore, this frequency is caused by the impact of the shell-side fluid and the structure of the ETB itself. The variable frequency under different inlet velocity conditions is close to the frequency of pulsating flow in each branch tube of the pulsating flow generator, so the variable frequency is generated by the induction of the pulsating flow.

(3) When $u_{in}=0.1$ m/s, the variable frequency $f_v=13$ Hz or 14 Hz. Under this inlet velocity condition, the frequency of pulsating flow formed by each branch tube of the pulsating flow generator is 15 Hz to 15.5 Hz. Similarly, when $u_{in}=0.4$ m/s, the variable frequency $f_v=46$ Hz or 47 Hz, and under this inlet velocity condition, the frequency of pulsating flow formed by each branch tube of the pulsating flow generator

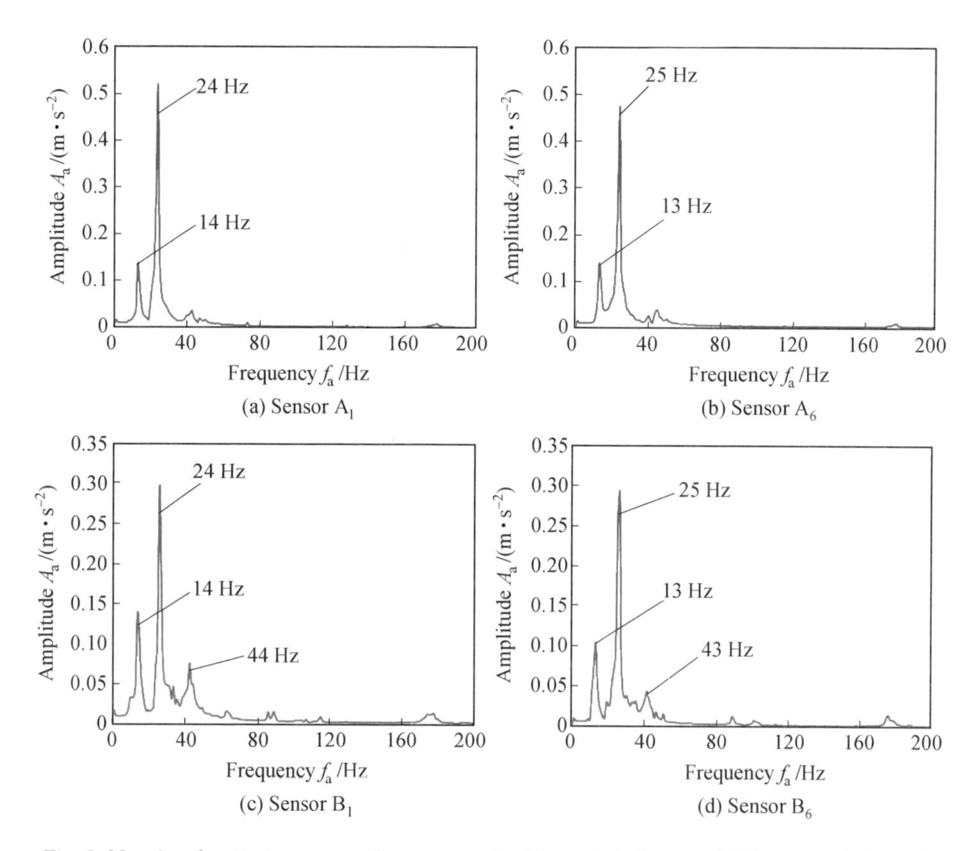

Fig. 6.23 Acceleration spectrum of sensors on the first and sixth rows of ETBs (u_{in} = 0.1 m/s)

is 49 Hz to 51.5 Hz. It can be seen that the variable frequency is smaller than the pulsating flow frequency. Through previous analysis, it is known that this is related to the shell-side fluid flowing outside the branch tube (Table 5.9).

(4) Compared with Fig. 6.20 and Fig. 6.21, the vibration intensity of ETBs induced by the coupling fluids is significantly increased, and the vibration intensity of ETBs induced by the coupling fluids is basically the same. This also reflects the effectiveness of pulsating flow generatosr in vibration induction and control.

(5) For the analysis of the signals of the two acceleration sensors on the same ETB, it can be seen that the intensity of the acceleration signal detected by the acceleration sensor A_n is higher than that detected by the acceleration signal B_n, indicating that the vibration intensity of the large connecting copper tube is higher.

Fig. 6.25 and Fig. 6.26 show the variation of the amplitudes corresponding to different frequencies of the acceleration signals of the two sensors on the ETB with the number of ETBs under the conditions of inlet velocity u_{in} = 0.1 m/s and 0.4 m/s,

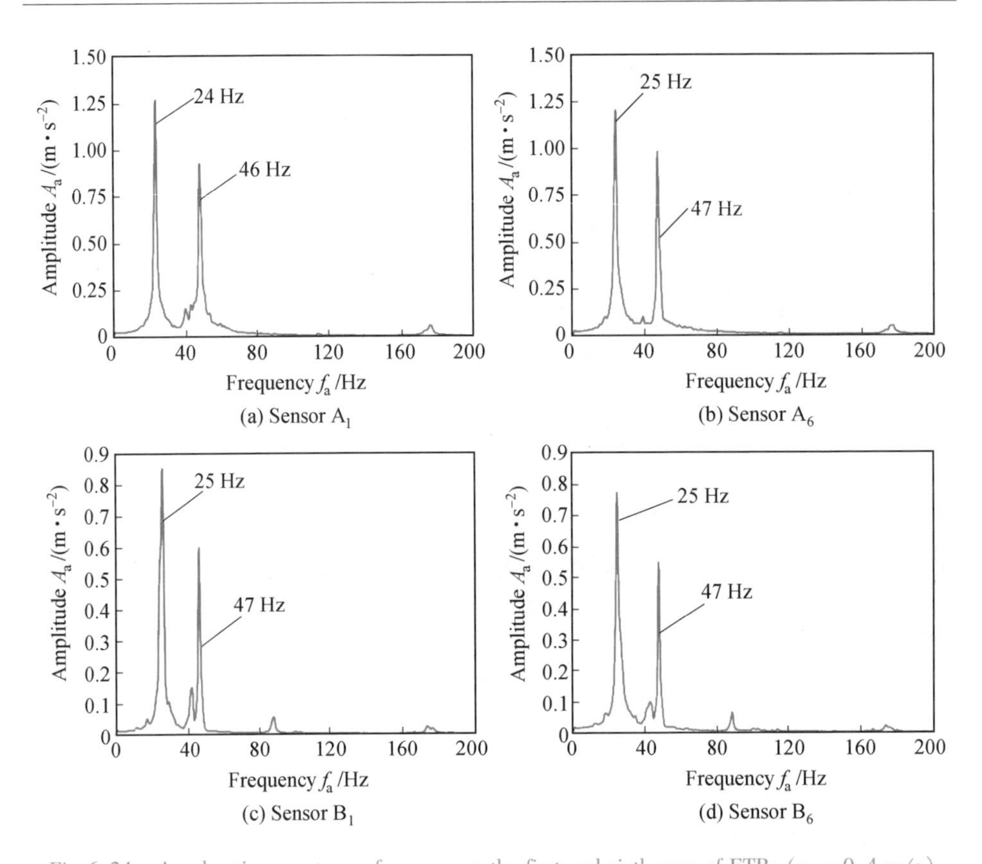

Fig. 6.24 Acceleration spectrum of sensors on the first and sixth rows of ETBs (u_{in} = 0.4 m/s)

where the acceleration amplitude is the average value of multiple measurements.

It can be seen from Fig. 6.25 and Fig. 6.26 as follows.

(1) The vibration frequency and intensity of each row of ETBs under different conditions are basically the same, indicating that the distributed pulsating flow generator can induce uniform vibration of each row of ETBs. For example, when u_{in} = 0.1 m/s, the maximum errors of the acceleration amplitudes of the two sensors at frequencies of 24 Hz and 25 Hz are approximately 10.7% and 12.8%, respectively, which is significantly reduced compared to the maximum relative error shown in Fig. 6.22.

(2) Compared with the results in Fig. 6.22, it can be seen that the vibration intensity of the ETB, especially the vibration intensity of the upper ETBs, has been significantly improved under the coupling of the shell-side fluid and distributed pulsating fluid. Taking the acceleration amplitude measured by the sensors A_n at an inlet velocity u_{in} = 0.1 m/s as an example, the average amplitude of the constant

frequency (24 Hz or 25 Hz) of each row of ETBs increases by about 2.9 times.

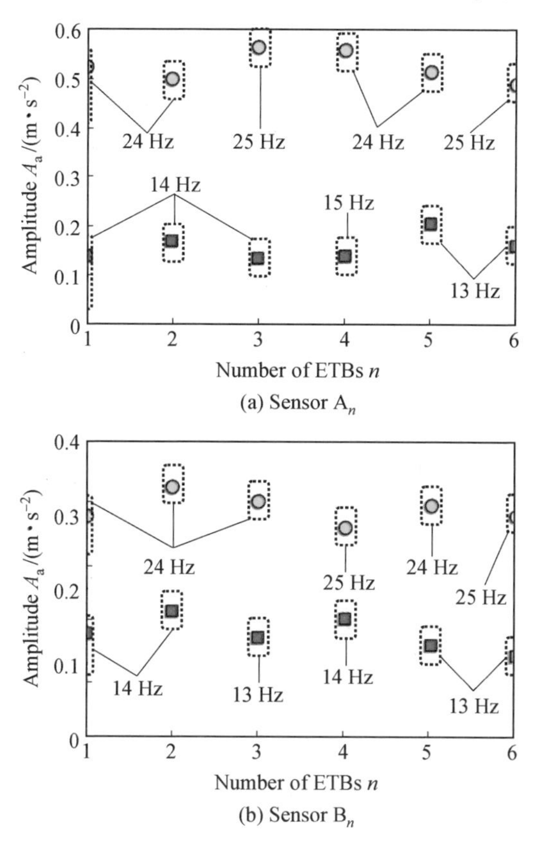

Fig. 6.25 Variation of acceleration amplitude with the number of ETBs (u_{in} = 0.1 m/s)

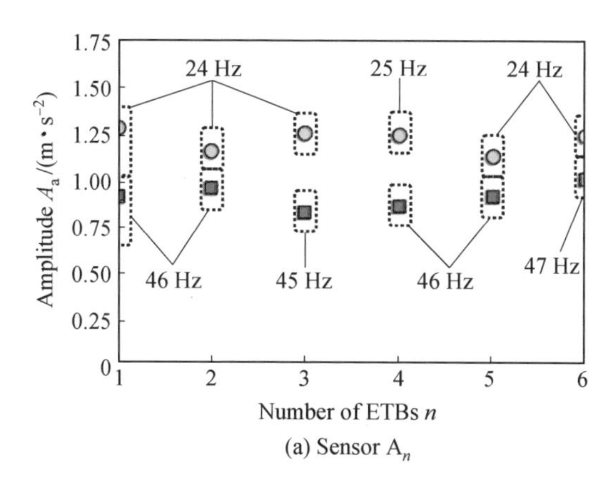

Fig. 6.26 Variation of acceleration amplitude with the number of ETBs (u_{in} = 0.4 m/s)

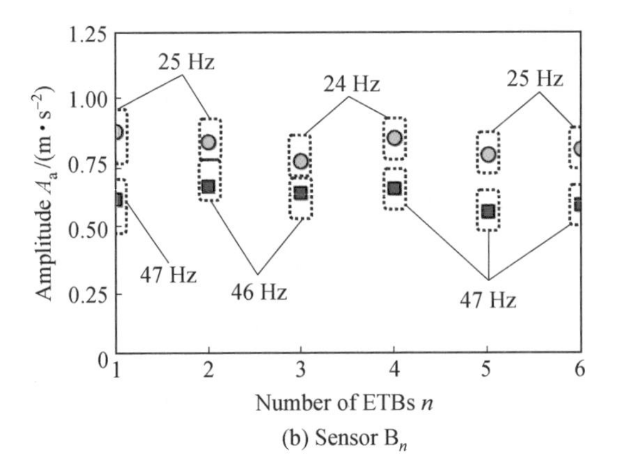

(b) Sensor B_n

Fig. 6. 26(continued)

Table 6.5 shows the variation of variable frequency and constant frequency with the inlet velocity. For comparison, the corresponding pulsating flow frequencies in Table 5.14 are also listed in Table 6.5, where the research object is the acceleration signal measured by sensor A_1.

Table 6.5 Variation of virable frequency and constant frequency with the inlet velocity

u_{in}/(m · s^{-1})	f_s/Hz	f_d/Hz	f_v/Hz
0.1	24	13	15.5
0.2	25	23	27.0
0.3	24	36	39.5
0.4	24	46	51.5

From Table 6. 5, it can be seen that the constant frequency is basically not affected by the inlet velocity, while the variable frequency increases with the increase of the inlet velocity, and its trend is consistent with the pulsating flow frequency. In addition, the variable frequency is slightly lower than the pulsating flow frequency, based on the research of the flow around the vortex element in the branch tube in Chapter 5 of this book, this phenomenon is caused by the shell-side fluid with a certain inlet velocity (Table 5.9).

To sum up, through testing the vibration response of each row of ETBs in the heat exchanger under the coupling induction of shell-side fluid and pulsating fluid, it is found as follows.

(1) The vibration frequency and intensity of each row of ETBs induced by the coupling of shell-side fluid and pulsating fluid are basically the same, and the

vibration intensity of the ETBs is significantly higher than that induced by shell-side fluid alone.

(2) Under the coupling induction of shell-side fluid and pulsating fluid, the ETB mainly vibrates at two frequencies, one of which remains basically unchanged, and the other increases with the increase of inlet velocity. And also, the frequency of the variable frequency is basically close to the frequency of the pulsating flow formed by the branch tube of the pulsating flow generator.

In summary, the distributed pulsating flow generation device designed in this chapter. On the one hand, generates variable frequencies and significantly increases the vibration intensity of the ETB. On the other hand, it has to some extent solved the phenomenon of uneven vibration of each row of ETBs in TETB heat exchangers, achieving effective excitation and control of the vibration.

6.4.3 Comparison of Numerical Results and Experimental Data

To verify the correctness of the numerical simulation method and the accuracy of the calculation results in this chapter, the vibration frequency (f_v) and acceleration amplitude (A_v) of the monitoring points on the two connecting copper tubes of the bottom row ETB in the x-direction (vertical direction) are calculated under the condition of inlet velocity $u_{in}=0.4$ m/s. The numerical simulation results are compared with the experimental data, as shown in Table 6.6.

From Table 6.6, it can be seen that for the vibration frequency, the numerical calculation results are basically consistent with the experimental data, with a maximum relative error of only 3.43%. The acceleration error is slightly larger, and the maximum relative error is about 12%. The main reasons for the relative error include the followings.

(1) The difference between the experimental equipment and geometric modeling.

(2) The interference of factors such as water pump and environmental noise during the experimental testing process.

Table 6.6 Comparison between numerical simulation results and experimental data

Monitoring point	Vibration frequency f_v/Hz			Acceleration amplitude A_v/(m · s^{-2})		
	Numerical result	Experimental data	Relative error	Numerical result	Experimental data	Relative error
A_1	24.17	24	0.7%	1.491	1.312	12.01%
B_1	24.17	25	3.43%	0.976	0.869	10.96%

Note: Relative error = |Numerical result−Experimental data|/Numerical result×100%.

References

[1]MENG H T. Study on the pulsating flow generating device in elastic tube bundle heat exchanger [D]. Jinan: Shandong University, 2012.

[2]JI J D, GAO R M, SHI B J, et al. Improved tube structure and segmental baffle to enhance heat transfer performance of elastic tube bundle heat exchanger [J]. Applied thermal engineering, 2022, 200: 117703.

[3]JI J D, LU Y, SHI B J, et al. Numerical research on vibration and heat transfer performance of a conical spiral elastic bundle heat exchanger with baffles [J]. Applied thermal engineering, 2023, 232: 121036.

[4]JI J D, CHEN W Q, GAO R M, et al. Research on vibration and heat transfer in heat exchanger with vortex generator [J]. Journal of thermophysics and heat transfer, 2021, 35(1): 164-170.

[5]JI J D. Study on flow-induced vibration of elastic tube bundle with shell-side distributed pulsating flow in heat exchanger [D]. Jinan: Shandong University, 2016.

[6]YAN K. A study on the vibration and heat transfer characteristics of conical spiral tube bundle in heat exchanger [D]. Jinan: Shandong University, 2012.

Chapter 7 Effect of Baffles on Vibration and Heat Transfer of TETBs

The shell-side distributed pulsating flow generator designed in Chapter 6 of this book significantly improves the vibration intensity and the heat transfer performance of the ETBs in the heat exchanger. And also, the shell-side distributed pulsating flow generator solves the problem of uneven vibration of each row of ETBs in the heat exchanger to a certain extent[1]. However, achieving precise control of vibration and heat transfer without changing the size and arrangement of internal ETBs and affecting the overall heat transfer performance of the heat exchanger is a technically challenging. The main reasons include the followings.

(1) Due to the arrangement of distributed pulsating flow generator in the ETB heat exchanger, the size parameters of each row of ETBs are limited. This further reduces the heat transfer coefficient per unit volume, which further affects the overall heat transfer performance of the heat exchanger[2-3].

(2) For distributed pulsating flow generators of specific sizes, pulsating flow with basically the same intensity and frequency only occurs under certain specific flow velocity conditions. This device cannot effectively adjust the vibration intensity at the specific vibration frequency of the ETB, nor can it effectively adjust the vibration frequency at the specific vibration intensity of the ETB[4].

Therefore, installing baffles in the ETB heat exchanger to optimize the flow path of the shell-side fluid is also an effective means of regulating vibration and heat transfer[5-9].

In this chapter, for improving the performance of heat exchangers by effectively utilizing the flow-induced vibration, two types of heat exchangers: TETB heat exchanger with baffles and TETB heat exchanger without baffles were proposed. Based on mesh division and boundary condition settings, by comparing and analyzing the vibration and heat transfer characteristics of the inner traditional ETBs, the effects of baffles on the ETB vibration responses and vibration-enhanced heat transfer performances under different conditions were investigated.

7.1 Heat Exchangers with or Without Baffles

7.1.1 Three-dimensional Model and Fluid Domain Model

Fig. 7. 1 shows a three-dimensional cross-section of two types of heat exchangers: TETB heat exchanger with baffles (markd as TETB-HB heat excahnger) and TETB heat exchanger without baffles (markd as TETB-NB heat exchanger).

Fig. 7. 1 clearly shows demonstrate the different structures and working principles of two heat exchangers. It should be noted that the structure of the heat exchanger has been modified, and its shape is consistent with that of the traditional shell-and-tube heat exchangers, and it is placed horizontally. In Fig. 7. 1, the y-direction is the vertical direction. In addition, to improve the comprehensive performance of the heat exchanger, an appropriate adjustment is made to the ETB based on the ETB shown in Fig. 3. 1, and the length of the copper tube connected to the stainless steel connector Ⅳ is appropriately stretched and bent according to the characteristics of the heat exchanger.

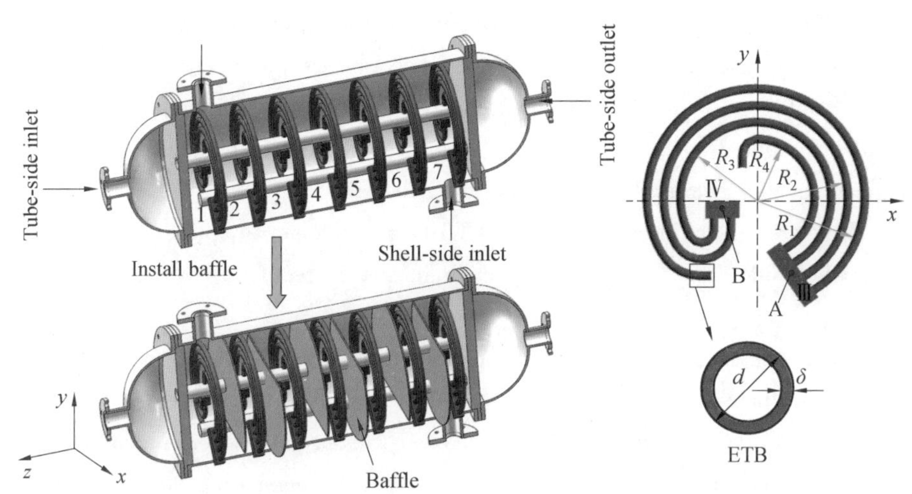

Fig. 7. 1 The three-dimensional cross-section view of TETB-NB and TETB-HB heat exchangers

In Fig. 7. 1, seven rows of ETBs are present in each heat exchanger, and each row of ETBs has two stainless steel connectors, one large and one small. For analysis, the ETBs in the heat exchanger are numbered 1, 2, ... , 6 from bottom to top.

During operation, low-temperature fluid works on shell-side, which enters through shell-side inlet and exits through shell-side outlet. And high-temperature fluid works on tube-side, which enters tube-side inlet and exits through tube-side outlet. The two different fluids achieve opposite flows in heat exchanger. Heat is exchanged between two fluids of different temperatures through wall of ETB. The TETB – HB heat exchanger is formed by adding baffles to TETB–NB heat exchanger, which changes the flow path of fluid. To study vibration characteristics of ETB, monitoring points A_n, B_n(n=1, 2, ... , 6) are set on the two stainless steel connectors of each row of ETBs to monitor the vibration responses of the ETBs induced by the actual shell-side fluid.

Since the tube-side fluid has little effect on heat transfer and vibration characteristics of ETB, this book only studies shell-side fluid. Fig. 7. 2 shows the shell-side fluid domains of the TETB – NB heat exchanger and the TETB – HB heat exchanger.

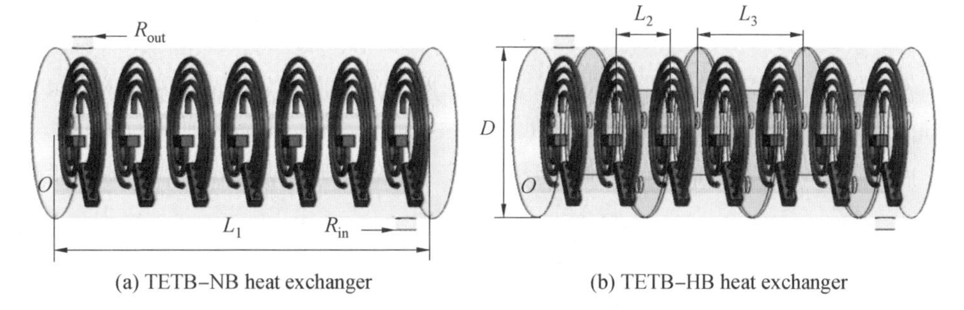

(a) TETB–NB heat exchanger (b) TETB–HB heat exchanger

Fig. 7. 2 Shell-side fluid domains of the TETB – NB heat exchanger and the TETB – HB heat exchanger

Table 7. 1 shows the shell-side fluid domain and ETB structure parameters used in the following calculation.

Table 7. 1 Shell-side fluid domain and ETB structure parameters

Project	Parameter	Value
Shell-side fluid domain	Length L_1/mm	630
	Length L_3/mm	170
	Diameter D/mm	300
	Shell-side inlet radius R_{in}/mm	17
	Shell-side outlet radius R_{out}/mm	17

Table 7.1(continued)

Project	Parameter	Value
ETB structure	Length L_2/mm	90
	Tube outer diameter d/mm	10
	Tube wall thickness δ/mm	1
	Bend radius: R_1, R_2, R_3, R_4/mm	70, 90, 110, 130
	Stainless steel connector Ⅲ/mm	80×20×20
	Stainless steel connector Ⅳ/mm	40×20×20

7.1.2 Meshes and Boundary Conditions

Numerical calculations on a variety of meshing methods are conducted, and finally adopted a more reliable meshing method, as shown in Fig. 7.3.

In this chapter, the tetrahedral mesh method is used to divide the shell-side fluid domain and the two stainless steel connectors, and the hexahedral mesh method is used to divide ETBs. At the same time, a 7-layer boundary is set near the FSI wall to obtain more reliable and accurate calculation results.

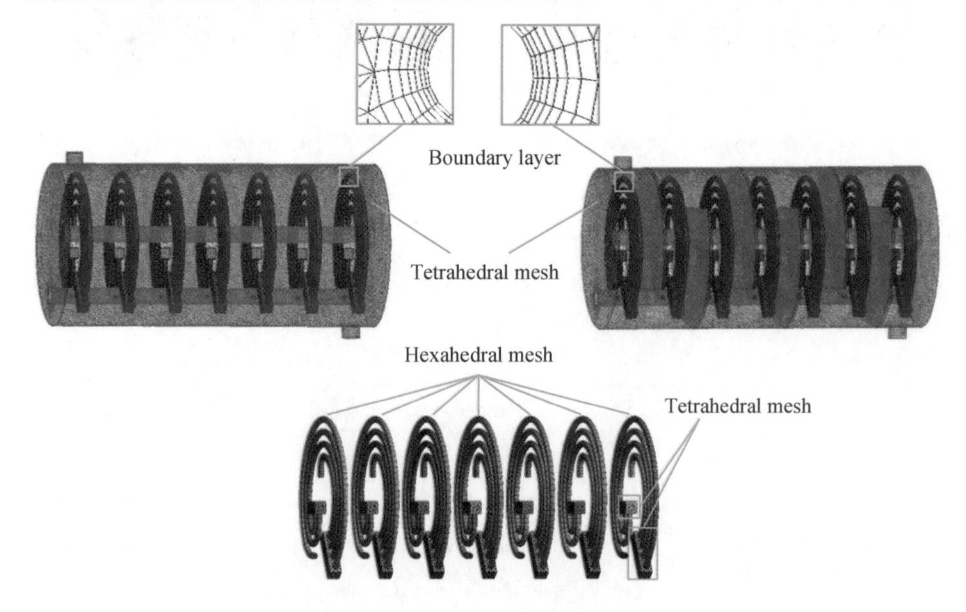

Fig. 7.3 Mesh delineation of the structural and fluid domains

The number of structural and fluid domain elements and nodes is shown in Table 7.2.

The following are boundary conditions simulated in this chapter.

(1) Inlet temperature T_{in} = 293.15 K. Outlet pressure P_{out} = 0 Pa (this pressure is relative to atmospheric pressure (101,325 Pa)). Inlet velocity u_{in} = 0.7 m/s and 1.0 m/s.

(2) Temperature T_{wall} = 333.15 K at the FSI surfaces.

(3) The wall surface of the fluid domain is set to be a non-slip and adiabatic surface.

(4) The two ports of the ETBs are fixed supports. The gravity direction is set to the "-y" direction with a magnitude of 9.807 m/s^2.

During the calculation, the calculation time and time step of rough calculation are set to 300 s and 0.1 s respectively, and the calculation time and time step of precise calculation are set to 1.2 s and 0.001 s respectively.

Table 7.2 Number of structual and fluid domain elements and nodes

Project	Area	Elements	Nodes
TETB-NB heat exchanger	Structural domain	29,393	124,453
	Fluid domain	7,766,253	1,412,928
TETB-HB heat exchanger	Structural domain	29,393	124,453
	Fluid domain	7,720,560	1,410,768

7.1.3 Mesh Independence Analysis

Table 7.3 shows the mesh independence analysis. TETB-NB heat exchanger is selected for analysis, and the effects of different element numbers on the calculation results are analyzed from two perspectives of heat transfer characteristics (pressure drop ΔP) and vibration characteristics (vibration amplitude A and frequency f) at the u_{in} = 1.0 m/s.

Table 7.3 Mesh independence analysis

Case	Element number	Numerical results			Relative error/%		
		ΔP/Pa	A/mm	f/Hz	ΔP	A	f
1	6,864,965	759.41	0.01203	22.98	2.79%	4.45%	2.50%
2	7,795,646	738.79	0.01259	23.57	—	—	—
3	15,953,124	730.94	0.01292	23.81	1.06%	2.62%	1.02%

Note: The relative error is calculated based on the calculation results of case 2.

From Table 7.3, it can be seen that the variation of the element number did not have a large effect on the numerical results. Therefore, it is verified that the mesh delineation method used in this chapter achieves mesh independence.

7.2 Vibration Analysis

7.2.1 Deformation and Displacement Analysis

Fig. 7.4 shows the streamlines of the shell-side fluid domains of the TETB-NB heat exchanger and the TETB-HB heat exchanger with $u_{in} = 1.0$ m/s.

From Fig. 7.4, it is not difficult to see that, compared with the TETB-HB heat exchanger, the flow path of the fluid in the TETB-NB heat exchanger heat exchanger is very confusing and relatively short. This illustrates that adding the baffle helps to increase the heat absorption time of fluid within the ETB heat exchanger and improves the consistency of the fluid flow path.

To specifically analyze the effect of the with/without of baffles on vibration characteristics, vibration displacements of two monitoring points on ETB 1 are selected for qualitative analysis, while the vibration displacements of the monitoring points on each ETB at vibration stabilization are then quantitatively analyzed.

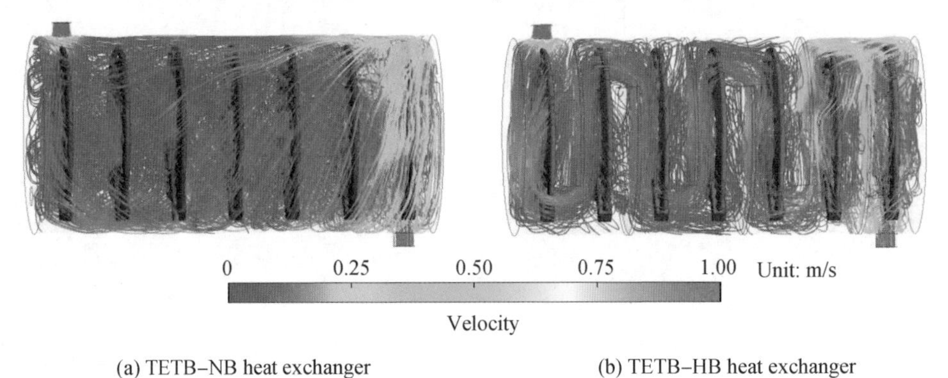

Fig. 7.4 Streamlines of the shell-side fluid domains of the TETB-NB heat exchanger and the TETB-HB heat exchanger

Fig. 7.5 shows the vibration displacement distribution of the ETB in the TETB-NB heat exchanger and the TETB - HB heat exchanger, the total vibration displacement S of A_1 and B_1, and the average total vibration displacement S_a of A_n and B_n on each tube bundle during vibration stabilization, where $u_{in} = 1.0$ m/s.

From Fig. 7.5, it can be seen as follows.

(1) Vibration displacement distribution of each ETB in the ETB heat exchanger is relatively approximate. The vibration displacement of monitoring point B is larger than monitoring point A. Meanwhile, the vibration displacement of the inner copper tube of is greater than that of the outer copper tube.

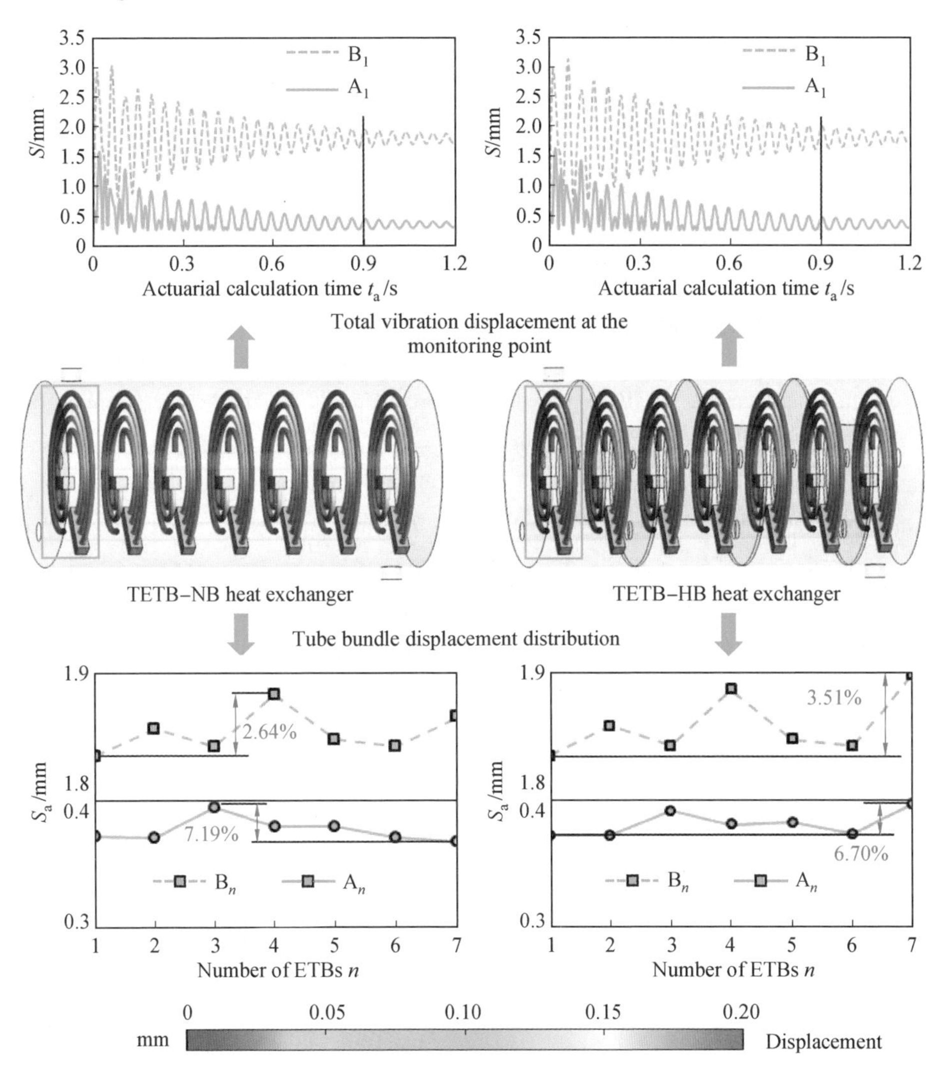

Fig. 7.5 Vibration displacement distribution of the ETB in the TETB-NB heat exchanger and the TETB-HB heat exchanger

(2) As actuarial calculation time t_a increases, S tends to stabilize, and when t_a = 0.9 s, the vibration reaches stability. Meanwhile, the total vibration displacement of

monitoring point B_1 is much larger than that of monitoring point A_1. And the average total vibration displacement illustrates the average total vibration displacements S_a of the monitoring points A_n and B_n on each ETB for vibration stabilization (t_a>0.9 s) in the TETB–NB heat exchanger and TETB–HB heat exchanger.

(3) The difference in the vibration displacement of each ETB in the TETB–NB heat exchanger is small. The maximum relative errors of the average total vibration displacements at monitoring points A_n and B_n are 7.19% and 2.67%, respectively. The difference in vibration displacements of each ETB within the TETB–HB heat exchanger is similarly small. The maximum relative errors of the average total vibration displacements at monitoring points A_n and B_n are 6.70% and 3.51%, respectively.

(4) Moreover, the average total vibration displacement at the two monitoring points on the ETB 1 inside the TETB–HB heat exchanger is elevated compared to the TETB–NB heat exchanger due to the baffle.

To facilitate the specific analysis of the amplitude and frequency of each monitoring point on the TETB–NB heat exchanger and TETB–HB heat exchanger, the displacement time curves in three directions (x, y, z) of the four monitoring points after the vibration stabilization (t_a=0.9–1.2 s) at the inlet velocity u_{in}=1.0 m/s are subjected to FFT calculation, as shown in Fig. 7.6 and Fig. 7.7.

From the comparison of Fig. 7.6 and Fig. 7.7, it can be seen that the vibration of the monitoring point B_n on each ETB is mainly in the y-direction, and the vibration of A_n is mainly in the x-direction, which indicates that the in-plane vibration intensity of ETB is much greater than the out-plane vibration intensity. Meanwhile, several conclusions can be drawn.

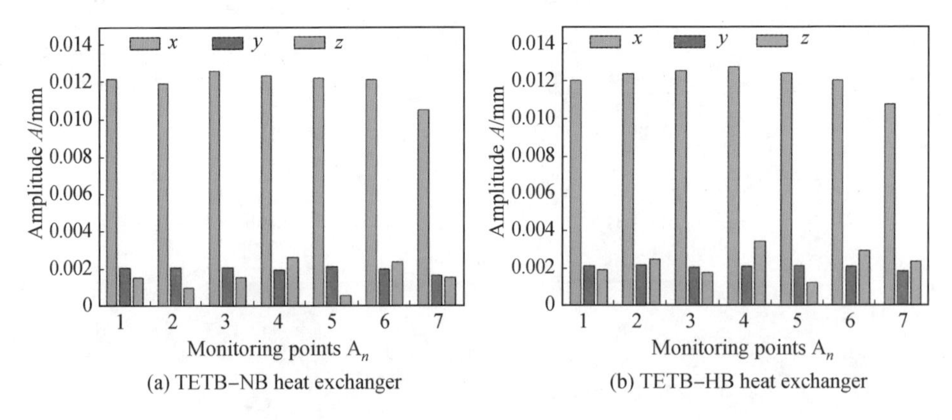

Fig. 7.6 Three directions (x, y, z) amplitude of monitoring point A_n

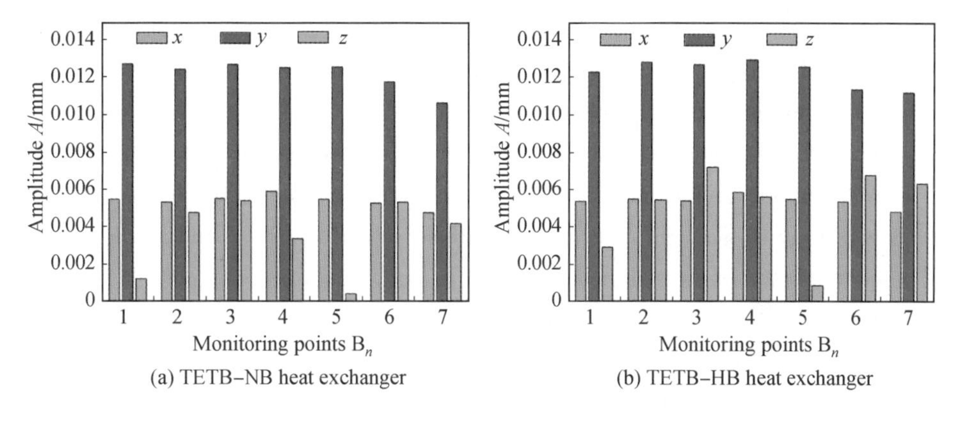

Fig. 7.7 Three directions (x, y, z) amplitude of monitoring point B_n

(1) The installation of baffle insidethe ETB heat exchanger has a small influence on the variation of amplitudes at each row of ETB monitoring points A_n and B_n in x- and y-directions, and there is both growth and reduction. For monitoring point A_n, the maximum growth rate of x-direction amplitude is 3.53%, the minimum is −1.22%, the maximum growth rate of y-direction amplitude is 8.85%, and the minimum is −1.06%. In addition, for monitoring point B_n, the maximum growth rate of x-direction amplitude is 4.57%, the minimum is −0.88%, the maximum growth rate of y-direction amplitude is 5.96%, and the minimum is −2.88%. However, the z-directional amplitude of the two monitoring points on each row of ETB has a significant increase. The maximum growth rate of the z-directional amplitude of monitoring point A_n is 146.05%, and the minimum is 13.43%; the maximum growth rate of the z-directional amplitude of monitoring point B_n is 144.13%, and the minimum is 15.39%.

(2) For each row of monitoring points A_n on the ETB, the maximum relative errors of vibration amplitude changes in the three directions are 27.72% (x-direction), 27.53% (y-direction) and 351.10% (z-direction) when there is no baffle in the ETB heat exchanger. The maximum relative errors of the vibration amplitude variations in the three directions are 18.52% (x-direction), 17.66% (y-direction) and 184.39% (z-direction) when the baffle is provided inside the ETB heat exchanger. For monitoring point B_n on each row of ETBs, the maximum relative error in the change of vibration amplitude in the three directions when there is no baffle in the ETB heat exchanger is 23.54% (x-direction), 19.68% (y-direction) and 1,160.19% (z-direction), respectively. The maximum relative errors in the change of vibration amplitude in the three directions are 21.15% (x-direction),

15.93% (y-direction) and 710. 58% (z-direction) for the TETB - HB heat exchanger. The above shows that the baffle is beneficial to improving the vibration uniformity of the monitoring points on each row of ETBs.

Because each row of ETBs in the fluid domain is induced by the performance of in-plane vibration, and vibration amplitude in the y-direction of each row of ETBs monitoring point B_n is much larger than the other two directions (x- and z-directions), vibration amplitude in the x-direction of each row of ETBs monitoring point A_n is much larger than other two directions (y- and z-directions). Therefore, in the later analysis, for monitoring points B_n, the focus is on the vibration in the y-direction, and for monitoring points A_n, the focus is on the x-direction.

7.2.2 Analysis of Vibration

From the above, it can be seen that the vibration characteristics of the monitoring points at the same location on each row of ETBs are relatively close to each other. A qualitative analysis of A_1 and B_1 on ETB 1 is carried out to analyze the effect of shell-side inlet velocity on the vibration characteristics of ETBs in two different heat exchangers.

Fig. 7. 8 and Fig. 7. 9 show the vibration displacement spectrum diagram of A_1 and B_1 for u_{in} =0. 7 m/s and 1. 0 m/s respectively, where NB stands for TETB-NB heat exchanger and HB stands for TETB-HB heat exchanger.

The following conclusions can be drawn from Fig. 7. 8 and Fig. 7. 9.

(1) Compared to the TETB-NB heat exchanger, the amplitudes of the A_1 and B_1 on the ETB in the TETB-HB heat exchanger change in all three directions. Among them, the amplitude of B_1 is almost unchanged in x- and y-directions (amplitude is reduced by 0. 50% in the x-direction, and amplitude is reduced by 2. 88% in the y-

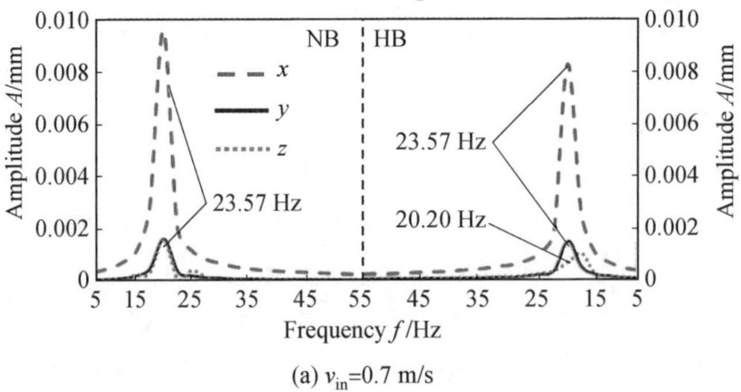

(a) v_{in}=0.7 m/s

Fig 7. 8 Vibration displacement spectrum diagram of A_1 at different inlet velocities

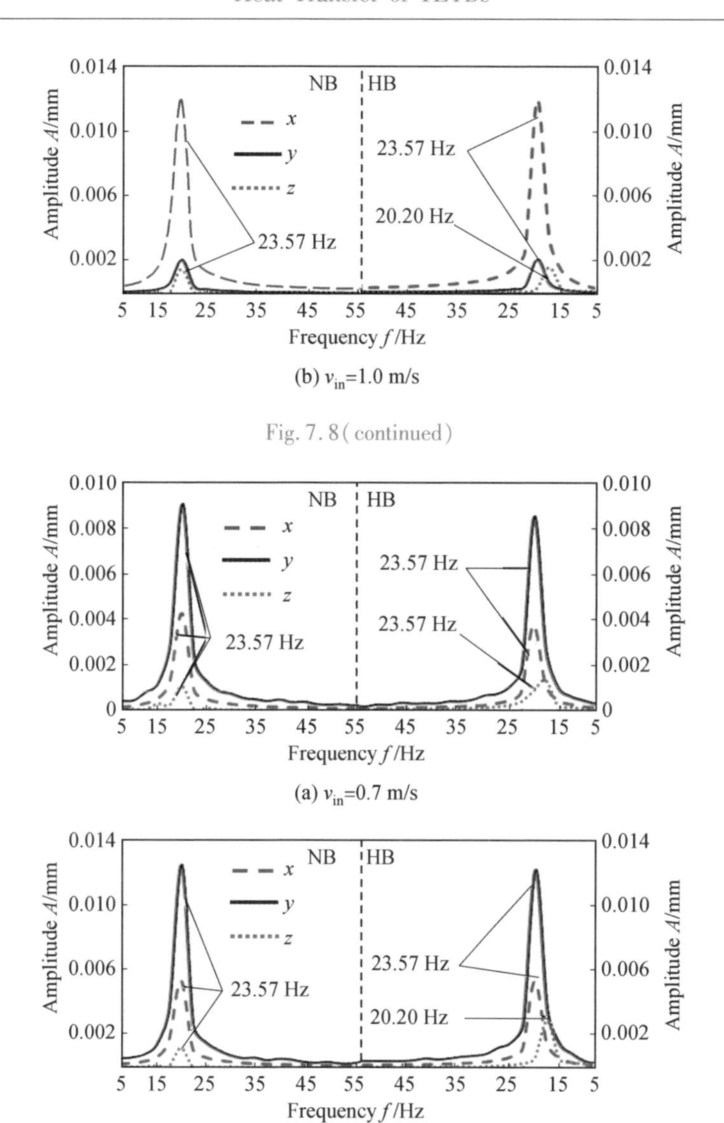

Fig. 7.9 Vibration displacement spectrum diagram of B_1 at different inlet velocities

direction), but the amplitude in the z-direction has a large change (increased by 144.13%) and the vibration is more easily excited (frequency is changed from 23.57 Hz to 20.20 Hz); amplitude changes of A_1 are also very weak in x- and y-directions (amplitude is decreased 1.21% in the x-direction, and amplitude is decreased 0.80% in the y-direction), but there is a large change in the z-direction (27.01% increase) and the vibration is easier to excite (the frequency changes from

23.57 Hz to 20.20 Hz).

(2) The inlet velocity is different while thefrequency of A_1 and B_1 on the same ETB is the same size (f=23.57 Hz), the amplitude is similar in size but different in the main vibration direction, the amplitude of B_1 in the y-direction is the largest, and the amplitude of A_1 in the x-direction is the largest (at u_{in} = 0.7 m/s, A_y = 0.0100 mm of B_1 and A_x = 0.0097 mm of A_1; at u_{in} = 1.0 m/s, A_y = 0.0125 mm of B_1 and A_x = 0.0121 mm of A_1).

(3) To 1.0–0.7 m/s, the amplitude of A_1 and B_1 on ETB in all three directions are enhanced to different degrees after the inlet velocity is increased. Among them, the amplitude of B_1 in three directions increased by 26.15% (x-direction), 25.40% (y-direction) and 12.03% (z-direction), and the amplitude of A_1 in three directions increased by 25.92% (x-direction), 26.53% (y-direction) and 6.17% (z-direction), respectively.

The above shows that the increase of inlet velocity will further strengthen the vibration of ETB, to specifically study the increase of vibration intensity of each ETB by flow velocity. Based on different inlet velocities, the vibration response of each row of ETBs under the induced flow field of the shell-side is investigated. Fig. 7.10 and Fig. 7.11 show the vibration displacements of monitoring points A_n in the x-direction and B_n in the y-direction for u_{in} = 0.7 m/s and 1.0 m/s.

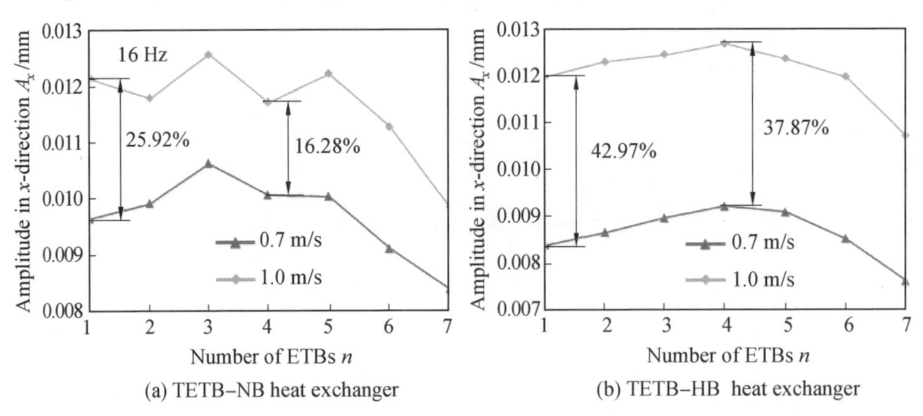

Fig. 7.10 The vibration amplitude of monitoring point A_n in the x-direction

From Fig. 7.10 and Fig. 7.11, the following conclusions can be drawn.

(1) The inlet velocity has a large effect on the vibration response of the ETB, and when u_{in} increases from 0.7 m/s to 1.0 m/s, the amplitude of each monitoring point on the ETB within both heat exchangers increases to different degrees. In the

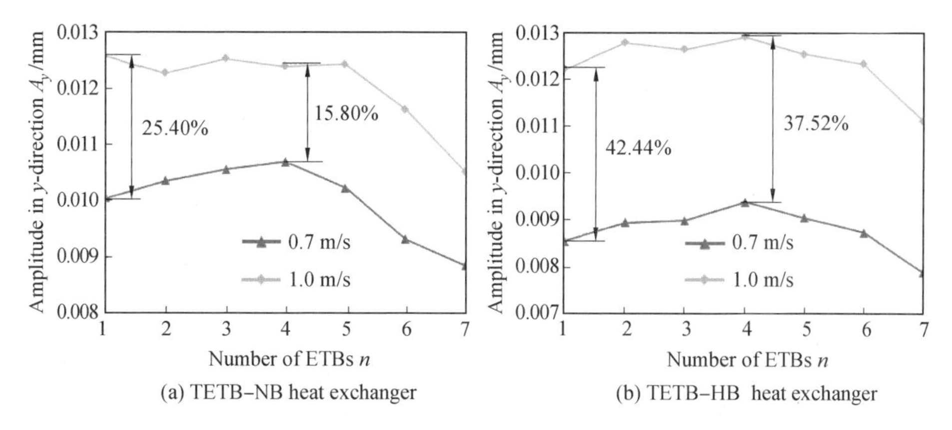

Fig. 7. 11 The vibration amplitude of monitoring point B_n in the y-direction

TETB-NB heat exchanger, the increase in amplitude of A_1 and B_1 on the first row of ETBs is the largest (25.92% and 25.40%, respectively), while the increase in the amplitude of monitoring points A_4 and B_4 on the fourth row of ETBs is the smallest (16.29% and 15.79%, respectively). Within the TETB-HB heat exchanger, the amplitude growth of monitoring points A_1 and B_1 on the first row of ETBs is the largest (42.97% and 42.44%, respectively), while the amplitude growth of monitoring points A_4 and B_4 on the fourth row of ETBs is the smallest (37.87% and 37.52%, respectively). This indicates that there is a more pronounced effect of increased inlet velocity in the TETB-HB heat exchanger on the vibration enhancement of the ETB.

(2) When u_{in} = 0.7 m/s, the maximum relative errors of amplitude changes of monitoring points on each ETB in TETB-NB heat exchanger are 21.09% (A_n) and 20.92% (B_n), and the maximum relative errors of amplitude changes of monitoring points on each ETB in TETB-HB heat exchanger are 19.04% (A_n) and 18.75% (B_n), respectively. When u_{in} = 1.0 m/s, the maximum relative errors of amplitude changes of monitoring points on each ETB in TETB-NB heat exchanger are 27.72% (A_n) and 19.68% (B_n), and the maximum relative errors of amplitude changes of monitoring points on each ETB in TETB-HB heat exchanger are 18.52% (A_n) and 15.93% (B_n), respectively. This shows that adding baffles can effectively improve the uniformity of ETB vibration.

7.3 Heat Transfer Analysis

7.3.1 Temperature Field Analysis

Fig. 7. 12 shows the average outlet temperature changes of shell-side fluid with two different structures of ETB heat exchangers at $u_{in}=1.0$ m/s. And Fig. 7. 13 shows the temperature distribution of the shell-side flow field for two different configurations of heat exchangers at $u_{in}=1.0$ m/s, where the temperature cloud in the $y-z$ plane is shown on the right side. The effect of the baffle on the overall performance of the ETB heat exchanger is shown by shell-side outlet temperature variation shown in Fig. 7. 12, and Fig. 7. 13 shows the shell-side flow field temperature distribution.

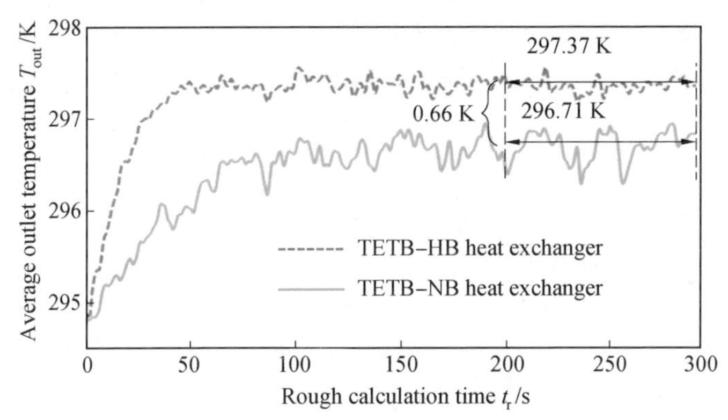

Fig. 7. 12 Average outlet temperature changes of the shell-side fluid with two different stnutures of ETB heat exchangers

From Fig. 7. 12, the average outlet temperature enters dynamic stability after the time of iterative calculations reaches 200 s, at which time the average outlet temperature of the TETB–HB heat exchanger is 297. 37 K, which is higher than the average outlet temperature of TETB–NB heat exchanger, 296. 71 K, and the former outlet temperature reaches dynamic stability in a shorter time than the latter.

From Fig. 7. 13, the temperature of each radial section of the flow field in the TETB–HB heat exchanger gradually increases, and its final outflow temperature is higher than the final outflow temperature of the TETB–NB heat exchanger. From this, it is clear that the entire heat exchange process within the ETB heat exchanger is essentially a process of heat accumulation. The condensate enters the ETB heat exchanger and continuously absorbs the heat emitted by the ETB. The longer the path of a given volume of condensate flowing through the ETB heat exchanger, the more heat it will absorb. Therefore, the longer the path of condensate flows, the better the ETB heat exchanger can exchange heat.

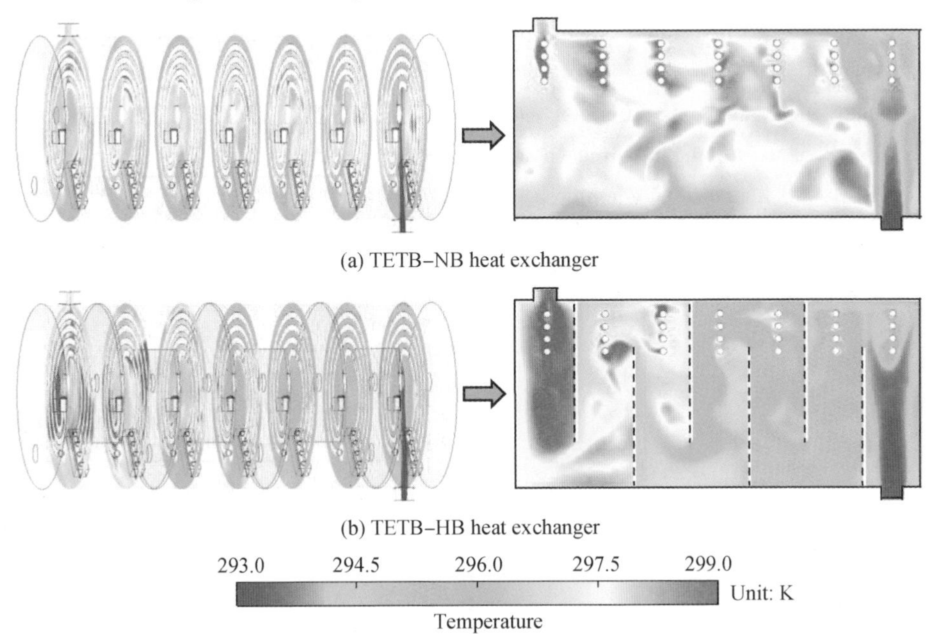

Fig. 7. 13 Temperature distribution of the shell-side flow field for two different configurations of heat exchangers

The above description shows that the baffle can make heat transfer between the tube-side and shell-side more adequate, and fluid temperature after absorbing heat is significantly increased. Therefore, adding a baffle can significantly improve the overall heat transfer performance of the heat exchanger.

7.3.2 Analysis of Heat Transfer

Fig. 7. 14 shows the heat transfer enhancement with the vibration of ETB in the TETB-NB heat exchanger and TETB-HB heat exchanger at two inlet velocities. h stands for heat transfer coefficient of ETB without vibration and h_v stands for heat transfer coefficient of ETB with vibration.

The following conclusions can be drawn from Fig. 7. 14.

(1) The increase in inlet velocity significantly improves the h and h_v of the ETB in both heat exchangers. When inlet velocity increases from 0.7 m/s to 1.0 m/s, for

the TETB-NB heat exchanger, h and h_v of ETB increased by 32.46% and 30.48% on average; for the TETB-HB heat exchanger, h and h_v of ETB increased by 31.16% and 31.36% on average.

(2) Heat transfer coefficients of all ETBs in both heat exchangers are improved to different degrees by vibration. When $u_{in}=0.7$ m/s, h of ETB in the TETB-NB heat exchanger increased by 3.88% on average, and those of ETBs in the TETB-HB heat exchanger increased by 3.36% on average. When $u_{in}=1.0$ m/s, h of ETB in the TETB-NB heat exchanger increased by 3.65% on average, and the h of the ETB in the TETB-HB heat exchanger increased by 4.13% on average.

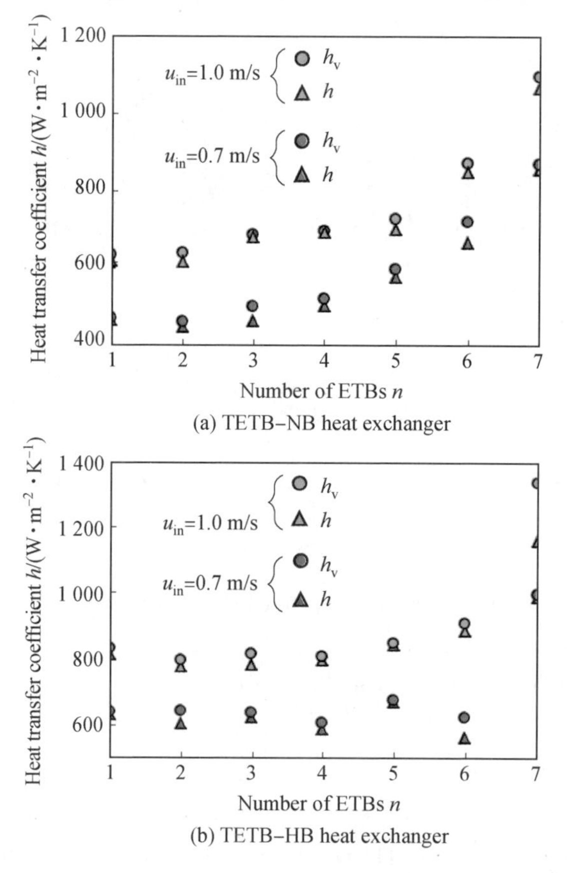

Fig. 7.14 Heat transfer enhancement with the vibration of ETB in the TETB-NB heat exchanger and TETB-HB heat exchanger at inlet velocities

(3) Average h and average h_v of the ETB in the TETB-HB heat exchanger are higher and the heat transfer performance is stronger than the TETB - NB heat

exchanger. h and h_v of each ETB in the TETB–HB heat exchanger are more uniform in size. This indicates that the addition of baffles has an important role in improving heat transfer and stabilizing the flow field.

According to the above analysis, the heat transfer coefficient of the heat exchanger is not only related to the inlet velocity, but also related to the strength of the ETB vibration and the heat exchanger structure. The vibration analysis shows that the strength of the ETB vibration is affected by inlet velocity and baffle. Therefore, the effect of vibration-enhanced heat transfer can be used to reflect the influence of the above factors on the heat transfer performance of the heat exchanger.

7.3.3 Comprehensive Heat Transfer Analysis

To effectively measure the effectiveness of vibration-enhanced, two sets of heat transfer data (u_{in} = 0.1 m/s and 0.4 m/s) are extended and the performance evaluation criteria (PEC) are used in this chapter. The expression can be found in Eq. (2.25).

Fig. 7.15 to Fig. 7.17 shows the trend of average h, $Nu_v/\Delta P$ (Nusselt number with vibration of unit pressure drop) and PEC at different inlet velocities, where NB stands for TETB–NB heat exchanger and HB stands for TETB–HB heat exchanger.

It can be seen in Fig. 7.15 to Fig. 7.17 as follows.

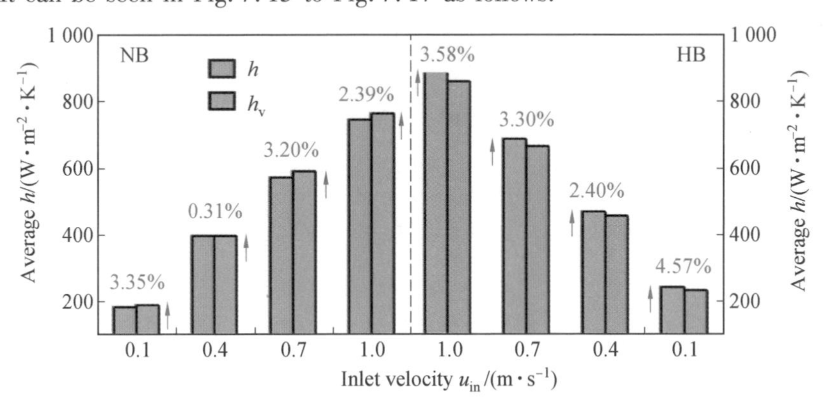

Fig. 7.15 The trend of average h at different inlet velocities

(1) Consistent with the heat transfer analysis above, with the gradual increase of inlet velocity, the average h of ETB within the two different heat exchangers gradually increases.

(2) The average h of the ETBs in the two different structures of heat exchangers are enhanced to different degrees due to vibration at four different inlet velocities. For

the TETB–NB heat exchanger, the vibration of the ETB resulted in an increase of the average h by 3.35%, 0.31%, 3.20% and 2.39%, respectively; for the TETB–HB heat exchanger, the vibration of the ETB resulted in an increase of the average h by 4.57%, 2.40%, 3.30% and 3.58%, respectively.

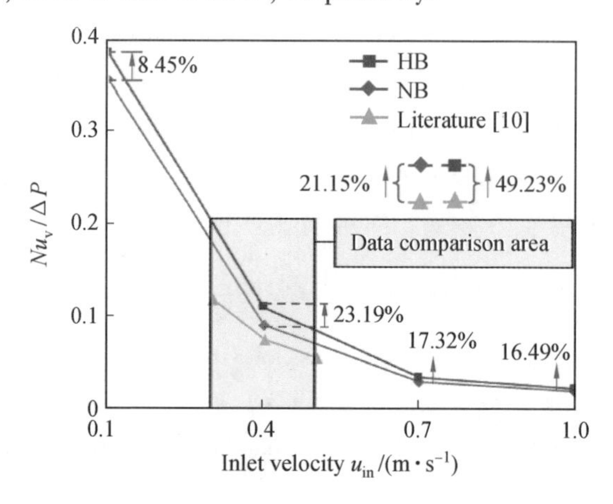

Fig. 7.16 The trend of $Nu_v/\Delta P$ at different inlet velocities

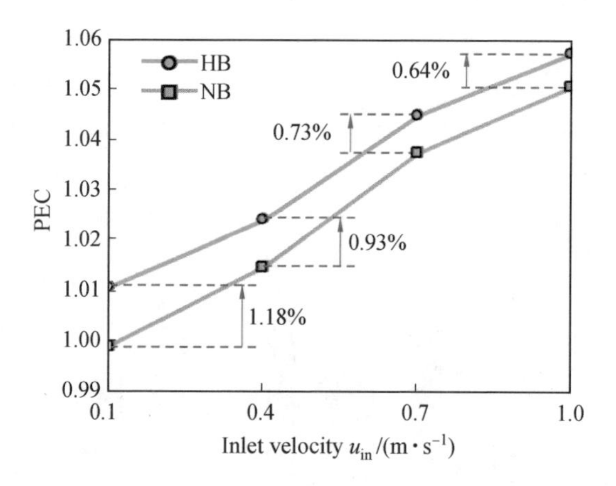

Fig. 7.17 The trend of PEC at different inlet velocities

(3) Fluid transport requires energy consumption, so the $Nu_v/\Delta P$ decreases with increasing inlet velocity. However, the TETB–HB heat exchanger has higher $Nu_v/\Delta P$ at the same inlet velocity. Compared with the $Nu_v/\Delta P$ of the TETB–NB heat exchanger, that of the TETB–HB heat exchanger increases by 8.45%, 23.19%, 17.32% and 16.49% for the four different inlet velocities, respectively. It is proved that the TETB–HB heat exchanger has better heat exchange performance than the TETB–NB heat exchanger.

(4) At the same inlet velocity, the $Nu_v/\Delta P$ of both the TETB-NB and the TETB-HB heat exchangers are greater than those studied in literature [10]. Compared with the $Nu_v/\Delta P$ in the literature, when $u_{in}=0.4$ m/s, the TETB-NB heat exchanger is 21.15% higher, the TETB-HB heat exchanger is 49.23% higher. This shows that the heat exchanger studied in this chapter has better heat transfer performance, and also confirms the positive effect of baffle on the heat exchanger performance.

(5) The PECs are 0.999, 1.015, 1.038 and 1.051 for the TETB-NB heat exchanger, and 1.011, 1.024, 1.045 and 1.058 for the TETB-HB heat exchanger at different inlet velocities. And for the TETB-NB heat exchanger, all values are greater than 1.0, except for $u_{in}=0.1$ m/s, where PEC is less than 1.0 (The impact of fluid does not excite the ETB vibration better when inlet velocity is small, so there is a certain amount of calculation error). This indicates that the vibration of ETB can substantially enhance the overall heat transfer performance of heat exchangers and achieve enhanced heat transfer.

(6) Compared with the TETB-NB heat exchanger, the PEC of the TETB-HB heat exchanger improves by 1.18%, 0.93%, 0.73% and 0.63% at four different inlet velocities. As flow velocity increases, the percentage increase of PEC decreases. Combined with the results of vibration analysis, it shows that, at low inlet velocities, the improvement of ETB vibration intensity of baffle is more obvious. It is further proved that flow-induced vibration is easier to achieve enhanced heat transfer at low velocity.

The above shows that adding baffles and increasing the inlet velocity can improve the overallheat transfer coefficient of the heat exchanger. Increasing flow velocity and installing baffles also enhance the vibration of ETB, which improves the enhanced heat transfer effect and further improves overall heat transfer performance.

References

[1] JI J D, GE P Q, LIU P, et al. Design and application of a new distributed pulsating flow generator in elastic tube bundle heat exchanger [J]. International journal of thermal sciences, 2018, 130: 216-226.

[2] CHENG L, QIU Y. Complex heat transfer enhancement by fluid-induced vibration [J]. Journal of hydrodynamics, Ser, B. 2003, 15(1): 84-89.

[3] SU Y C, LI M L, LIU M L, et al. A study of the enhanced heat transfer of flow-

induced vibration of a new type of heat transfer tube bundle—the planar bending elastic tube bundle [J]. Nuclear engineering and design, 2016, 309: 294-302.

[4] JI J D, GAO R M, SHI B J, et al. Improved tube structure and segmental baffle to enhance heat transfer performance of elastic tube bundle heat exchanger [J]. Applied thermal engineering, 2022, 200: 117703.

[5] CHEN J, LI N Q, DING Y, et al. Experimental thermal-hydraulic performances of heat exchangers with different baffle patterns [J]. Energy, 2020, 205: 118066.

[6] WANG K, LIU J Q, LIU Z C, et al. Fluid flow and heat transfer characteristics investigation in the shell side of the branch baffle heat exchanger [J]. Journal of applied fluid mechanics, 2021, 14: 1775-1786.

[7] SUN Y R, JI J D, HUA Z S. Hua, et al. Flow-induced vibration and heat transfer analysis for a novel hollow heat exchanger [J]. Journal of thermophysics and heat transfer, 2023, 37(1): 94-103.

[8] JI J D, LI F Y, SHI B J, et al. Analysis of the effect of baffles on the vibration and heat transfer characteristics of elastic tube bundles [J]. International communications in heat and mass transfer, 2022, 136: 106206.

[9] JI J D, LU Y, SHI B J, et al. Numerical research on vibration and heat transfer performance of a conical spiral elastic bundle heat exchanger with baffles [J]. Applied thermal engineering, 2023, 232: 121036.

[10] DUAN D R, GE P Q, BI W B. Numerical investigation on heat transfer performance of planar elastic tube bundle by flow-induced vibration in heat exchanger [J]. International journal of heat and mess transfer, 2016, 103: 868-878.

Chapter 8 Research on Vibration-enhanced Heat Transfer of IETB Heat Exchanger

Based on the research in Chapter 4 of this book, the IETB improves the flexibility of ETB vibration by dividing the stainless steel connector Ⅲ into two identical stainless steel connectors (Fig. 4. 13), making it easier to achieve flow-induced vibration at low flow rates[1]. If the TETB in Chapter 7 of this book is replaced with the IETB in the two types of improved heat exchangers, it is possible to further improve the comprehensive performance of the heat exchanger.

In this chapter, considering the advantage of IETB that is easy to achieve vibration under low fluid velocity, based on the TETB heat exchangers with or without baffles, two types of heat exchangers: IETB heat exchanger with baffles (marked as IETB–HB heat excahnger) and IETB heat exchanger without baffles (marked as IETB–NB heat exchanger) were proposed. The influence of inlet velocity and baffles installation on the IETB vibration and heat transfer features was studied. In addition, the heat transfer performances of the four heat exchangers (containing two types of heat exchangers in Chapter 7) were compared under different conditions.

8.1 IETB Heat Exchanger and Computional Domain

8.1.1 IETB–HB Heat Exchanger

The structure of the IETB–HB heat exchanger is shown in Fig. 8. 1. It should be noted that the IETB shown in Fig. 8. 1 is proposed based on the TETB in Chapter 7 of this book (Fig. 7. 1), taking into account the effective utilization of heat exchanger space. In addition, the IETB used in this chapter is different from the ETB shown in Fig. 4. 13, mainly reflected in the further bending and stretching of copper tubes connected to stainless steel connectors Ⅲ and Ⅳ.

Same as the TETB–HB heat exchanger, seven IETBs are evenly arranged on the tube-side inlet and outlet tubes in the IETB–HB heat exchanger. Baffles (dashed lines shown in Fig. 8. 1) are alternately placed between each IETB. The shell-side low-temperature fluid enters from the shell-side inlet, then flows fully inside the heat exchanger, and flows out from the shell-side outlet. The tube-side high-temperature fluid enters the tube-side inlet tube from the tube-side inlet, flows into the tube-side outlet tube through the IETBs, and flows out from the tube-side outlet finally. The IETBs are impacted by the fluids during the working process, which is beneficial to excite the vibration of the IETBs, thereby improving the turbulence characteristics of the near-wall fluid inside and outside the IETBs, and then realizing the vibration-enhanced heat transfer.

For the convenience of explanation, the IETBs are marked as n ($n=1, 2, \ldots, 7$). In addition, to study the vibration characteristics of the IETBs, monitoring points (marked as A_n, B_n, C_n) are set to monitor the vibration of the stainless steel connectors under different research conditions as shown in Fig. 8. 1. In the calculation of this chapter, Table 8. 1 shows the partial specific structural parameters, while the remaining parameters are shown in Table 8. 1.

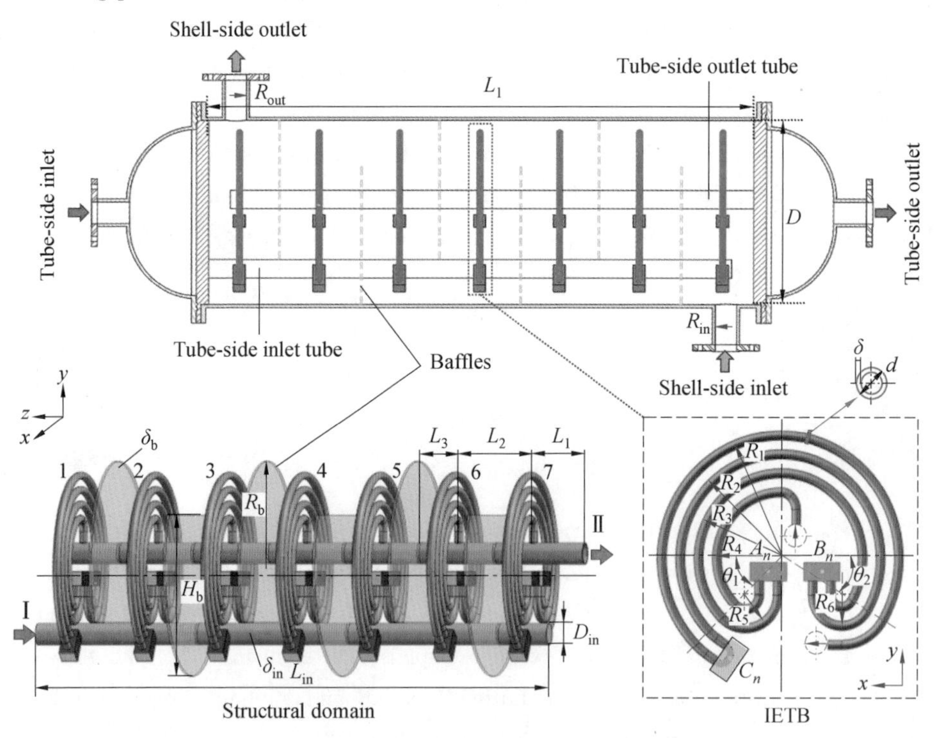

Fig. 8. 1 Structure of the IETB-HB heat exchanger

Table 8.1 Partial specific structural parameters

Parameters	Value	Parameters	Value
S_1/mm	45	H_b/mm	225
S_2/mm	90	δ_b/mm	5
S_3/mm	42.5	R_5/mm	15
L_{in}/mm	600	R_6/mm	35
D_{in}/mm	30	θ_1/(°)	45
δ_{in}/mm	1.5	θ_2/(°)	30
R_b/mm	150		

8.1.2 Meshes and Boundary Conditions

Since flow-induced vibration mainly occurs on the IETBs, only seven IETBs are retained in the structural domain. In addition, because the shell-side fluid is the main factor to induce the vibration of the IETBs, only the shell-side fluid is retained in the fluid domain[2-3]. Fig. 8.2 shows the mesh distribution of the computational domain.

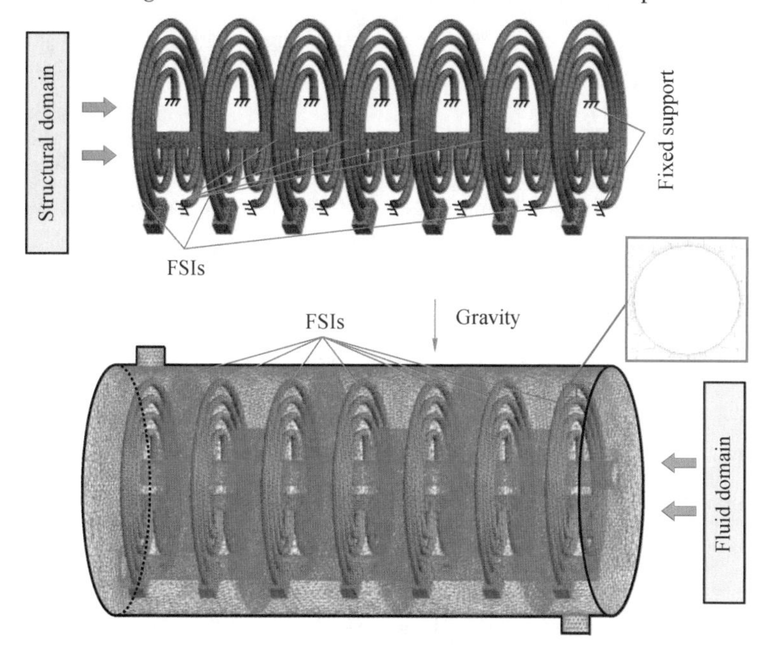

Fig. 8.2 Mesh distribution of the computational domain

As shown in Fig. 8.2, the meshes of all copper bend tubes are hexahedral, and the meshes of all stainless steel connectors are tetrahedral. For the whole structural domain, the number of elements and nodes are 29,729 and 125,482 respectively.

In the process of numerical calculation, the end surfaces of the IETBs are set as the fixed support. The outer surfaces of the IETBs are set as the FSI interface. The gravitational acceleration is set in the opposite y-direction of the structural domain, and the value is 9.8066 N/m^2.

The meshes of the fluid domain are tetrahedral as shown in Fig. 8.2. To improve the exactness of calculation results, seven boundary layers are used in the near-wall domains contacting with the IETBs. The number of elements and nodes are 8,120,090 and 1,483,248 respectively.

In the process of numerical calculation, the "Inlet" and "Outlet" are used as the boundary conditions of the shell-side inlet and outlet. The fluid domain outer wall and contact surfaces of the tube-side inlet and out tubes and the baffles with the shell-side fluid are set as "No slip wall". The inner surfaces of the fluid domains contacting with the IETBs are set as FSI interfaces, which correspond to the FSI interfaces of the structural domain one by one. In addition, the initial boundary conditions are set as the inlet velocity u_{in} (0.1 m/s, 0.4 m/s, 0.7 m/s and 1.0 m/s), inlet fluid temperature T_{in} (293.15 K), the temperature of the FSI interfaces T_F (353.15 K) and outlet relative pressure P_{out} (0 Pa).

The time step of the rough calculation andactuarial calculation is different. In the rough calculation, set the total time (300 s) and the time step (0.1 s). In the precise calculation, set the total time (1.2 s) and the time step (0.001 s). This is consistent with the settings in Chapter 7.

8.1.3 Mesh Independence Analysis

Table 8.2 shows the analysis of mesh independence. Cases 1 and 3 are generated by increasing or decreasing the mesh size and boundary layer number based on case 2. The initial condition is baffles installation (u_{in} = 1.0 m/s). The vibration frequency f_y and amplitude A_y of point A_7 in the y-direction are monitored.

As shown in Table 8.2, the mesh generation method in this chapter is case 2. With the number of elements increased, there is less impact on numerical results. So, it indicates that the mesh generation method used can achieve the mesh independence requirements.

Table 8.2 Analysis of mesh independence

Case	Number of elements	Boundary layer number	Numerical results		Relative error/%		Computing time/h
			f_y/Hz	A_y/mm	f_y	A_y	
1	7,965,862	5	26.10	0.0449	3.08	8.73	41.7
2	8,149,819	7	26.93	0.0492	—	—	50.1
3	8,852,036	12	27.12	0.0501	0.71	1.83	80.6

Note: The relative error is calculated based on the calculation results of case 2.

8.2 Vibration Analysis

8.2.1 Velocity Analysis of Fluid Domain

Fig. 8.3 shows the velocity distribution isosurface on the y–z section fluid domain in the IETB heat exchangers under different inlet velocity conditions. Among them, Fig. 8.3 (a) and (b) respectively show the fluid velocity distribution when there are no baffles (IETB–NB heat exchanger) at different inlet velocities (0.4 m/s and 1.0 m/s). Fig. 8.3 (c) and (d) respectively show the fluid velocity distribution when installing baffles (IETB – HB heat exchanger) at different inlet velocities (0.4 m/s and 1.0 m/s).

It is not difficult to see from Fig. 8.3 that the shell-side fluid flows fully. The contours are dense and the velocity varies greatly around the outer wall of the IETBs. After the impact of the fluid on the IETBs, it is beneficial to generate eddy currents that improve the turbulence characteristics of the fluid. Through comparing Fig. 8.3(a) and (b), and comparing Fig. 8.3(c) and (d), it is not difficult to see that inlet velocity increase has a great influence on the distribution of the fluid velocity of the fluid domain, regardless of whether install baffles or not. Through comparing Fig. 8.3(a) and (b), and comparing Fig. 8.3(c) and (d), it is not difficult to see that under the same inlet velocity, installing baffles in the IETB heat exchanger makes the fluid path change significantly, concentrates the fluid, and helps to increase the fluid velocity. And at the same time better impacts IETBs, and promote the generation of more eddy currents. It is worth noting that the change of fluid velocity directly affects the strength of the impact IETB. Therefore, to further study the vibration of IETBs, inlet velocity and baffles installation are the important factors that need to be considered.

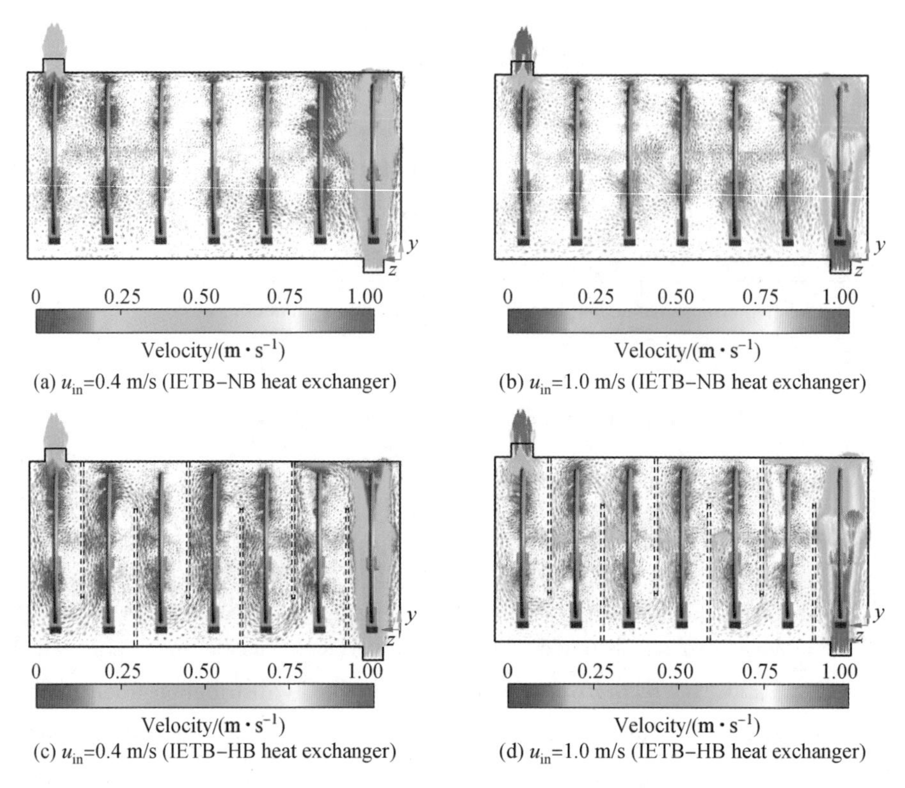

Fig. 8.3 Velocity distribution isosurfaces on the $y-z$ section fluid domain in the IETB heat exchangers under different inlet velocity conditions

8.2.2 Vibration Analysis of the IETBs

To study the vibration of IETBs under different conditions, the overall vibration displacement distribution of IETBs is extracted and analyzed with baffles installation or not, and the overall vibration data of monitoring points A_n, B_n and C_n on each IETB are analyzed, as shown in Fig. 8.4, where $u_{in}=0.4$ m/s.

According to the displacement distribution, the effect of the baffle installation on the overall vibration displacement of IETBs is small. Regardless of whether the baffles are installed or not, the overall vibration displacement distribution of each IETB is the same, so the vibration of the IETBs is uniform. Moreover, for a single IETB, the vibration gradually increases from the center to the periphery. In this regard, the vibration displacement diagrams of monitoring point A_n, B_n and C_n on each IETB are further analyzed. Combind with the displacement distribution, it can be seen that the vibration variation of all monitoring points is periodicity and same, that is, the

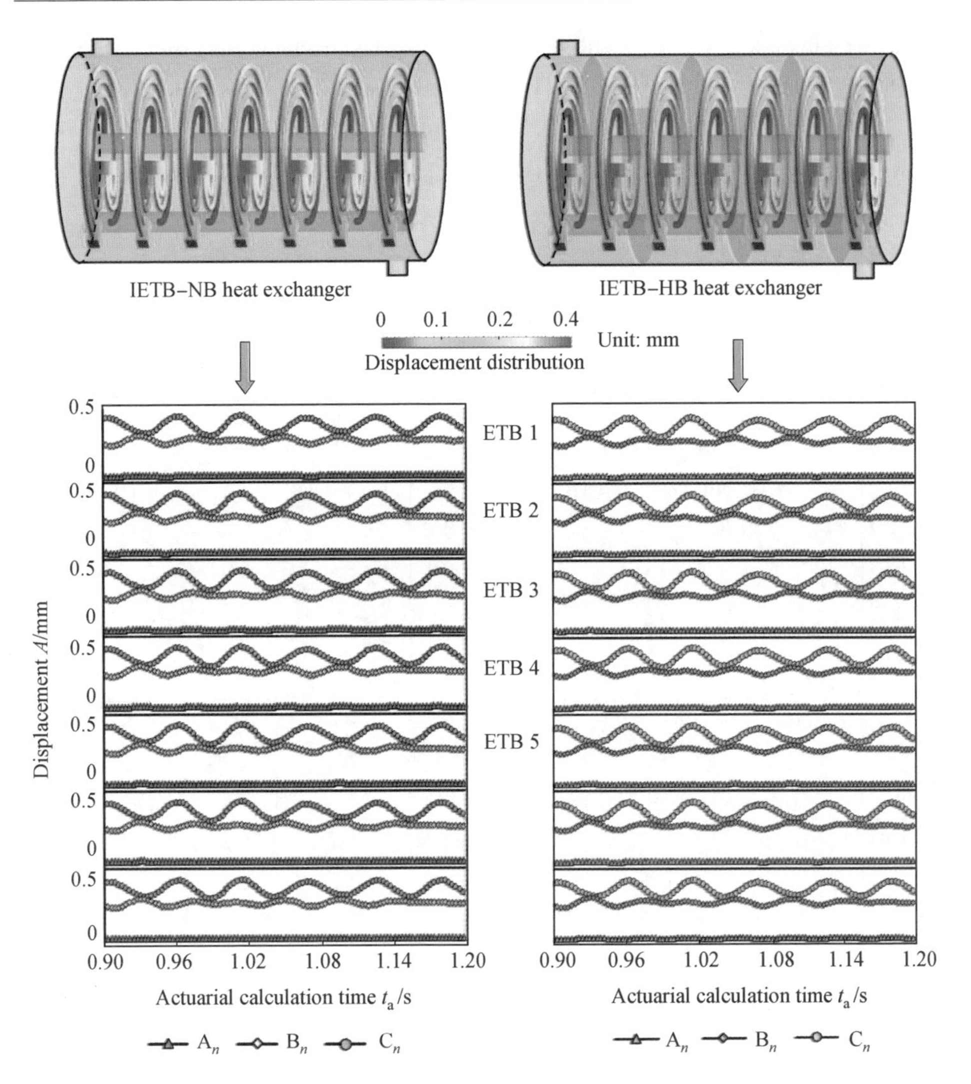

Fig. 8.4 Vibration displacement distribution and vibration time history diagram

vibration of the IETBs has reached dynamic stability. In addition, the vibration of monitoring point C_n is the most intense, followed by B_n and A_n. In general, the vibration of the IETBs is uniform, and the vibration characteristics of monitoring point C_n are representative.

To further study the main manifestations of IETBs vibration, according to the results shown in Fig. 8.4, monitoring point C_n is taken as the research object because of its' most obvious vibration. Based on the FFT calculation, the vibration data obtained by the monitoring point C_n on seven IETBs are processed, and the vibration amplitude A and main frequency f in different directions of the different monitoring points are obtained. Fig. 8. 5 shows the vibration displacement spectrum diagram of monitoring points $C_1 - C_7$ on each IETB at the inlet velocity of 0. 4 m/s.

It is not difficult to see from Fig. 8. 5 as follows.

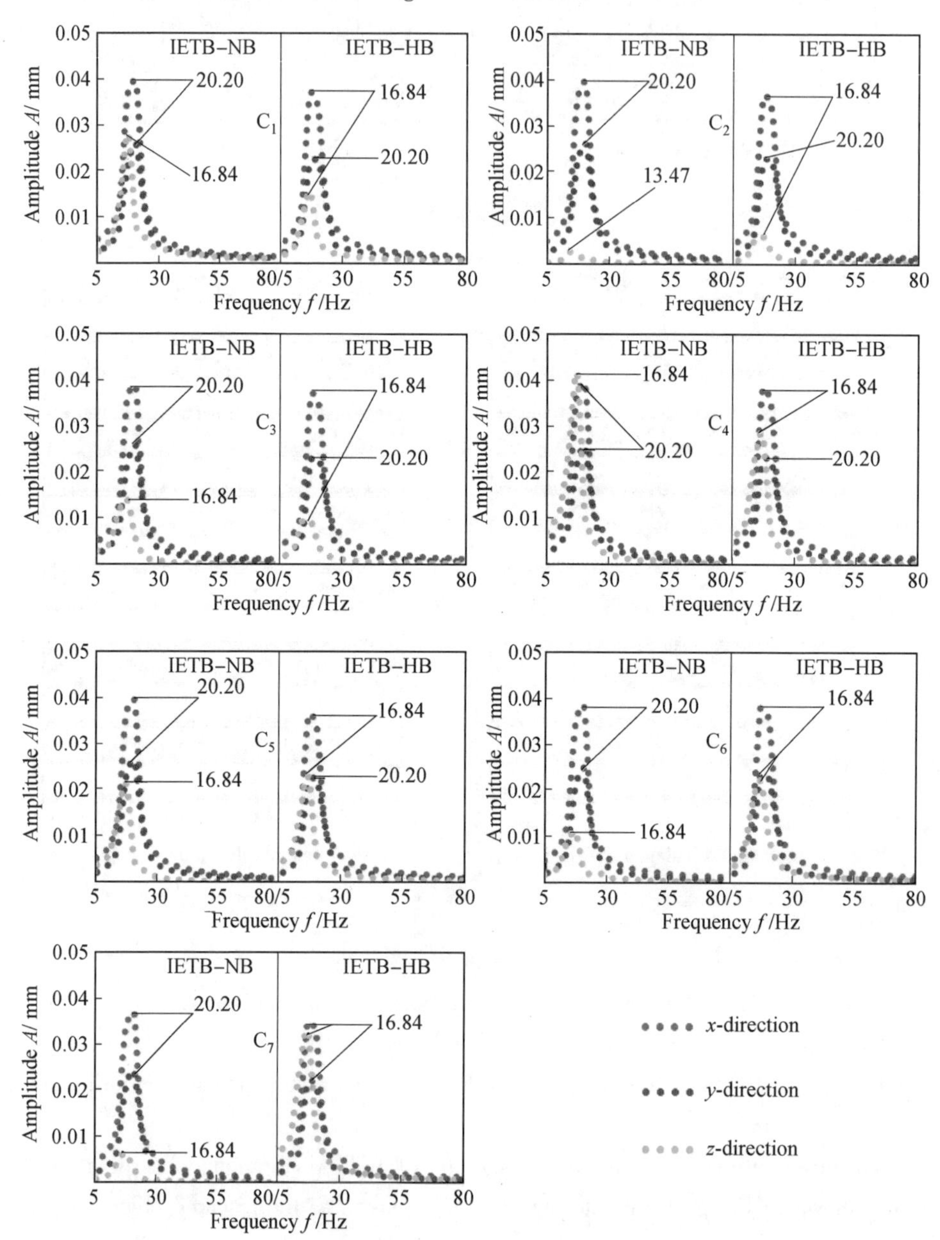

Fig. 8. 5 Vibration displacement spectrum diagram of monitoring points $C_1 - C_7$

(1) No matter whether the baffle is installed or not, there is no regular change of main frequency in all directions of monitoring points C_1-C_7. When there are no baffles, except for monitoring point C_6, all other monitoring points conform to the *x*- and *y*-directions (20.20 Hz) and *z*-direction (16.84 Hz, 13.47 for C_6). After adding baffles, the main frequencies in the three directions changed, among them, the main frequencies in the *x*- and *z*-directions of C_1-C_7 decreased to 16.84 Hz, the main frequencies in the *y*-direction of C_1 and C_2 increased to 16.84 Hz, and the main frequencies in C_3-C_7 increased to 20.02 Hz. It can be seen that the baffles installation reduces the vibration frequency of IETBs in the *x*- and *z*-directions, but increases in the *y*-direction.

(2) Regardless of whether the baffle is installed or not, the maximum amplitude of each monitoring point in the *x*- and *y*-directions has little difference, changes most obvious in the *z*-direction. After the addition of the baffles, the amplitude of each monitoring point in the *x*- and *y*-directions decreased but changed irregularly in the *z*-direction. Generally, the vibration of IETB is mainly reflected in the vibration in the *x*–*y* plane. In addition, because the flow path of the fluid in the heat exchanger affects the direction of the impact of IETB, the amplitude difference in the *z*-direction is large.

By analyzing the spectrogram of different monitoring points, it can be seen that the vibration characteristics of different monitoring points are not very different. So, combined with the conclusions in Fig. 8.4 and Fig. 8.5, it is reliable to use the vibration of monitoring point C_n to reflect the vibration of IETBs. To further study the influence of installing baffles and inlet velocity on the vibration characteristics of the IETBs, the amplitude of monitoring point C_n of each IETB with different inlet velocities (0.4–1.0 m/s) and installing baffles or not is analyzed based on FFT calculation. The average amplitude (A_a, the average vibration amplitude of the monitoring points C_1-C_7 on the seven IETBs) is proposed to study the influence of different conditions on the vibration of IETBs. As shown in Fig. 8.6, the average amplitude comparison of the monitoring point C_n under different research conditions is shown. Among them, "*o*" represents the total vibration displacement direction. In addition, to facilitate a specific analysis of the results and effects, the relative average amplitude is calculated to reflect the specific changes.

In Fig. 8.6, "NB" represents to the IETB–NB heat exchanger, and "HB" represents to the IETB–HB heat exchanger. It is not difficult to see from Fig. 8.6 as follows.

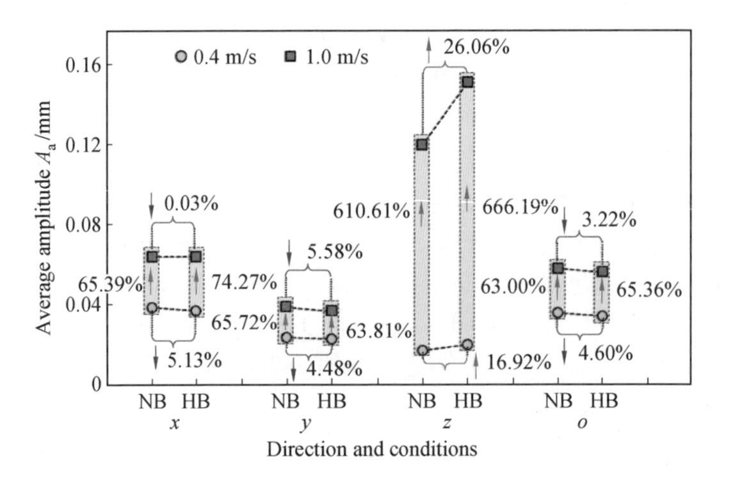

Fig. 8.6 The average amplitude of the monitoring points C_n

(1) Regardless of whether install baffles or not, with the increase of the inlet velocities, the average amplitudes in different directions all increase. The inlet velocity has different effects on the average amplitude in different directions. Among them, the impact on the z-direction is the largest, followed by the x- and y-directions (the impact is roughly the same). When there are no baffles, it is worth noting that while the value of u_{in} rises from 0.4 m/s to 1.0 m/s, the relative average amplitude under each direction is larger, and the maximum is 610.61%, the minimum is 63.00%. After installing baffles, the maximum relative average amplitude is 666.19%, and the minimum is 63.81%. So, it can be seen that at low inlet velocities, increasing the inlet velocity is beneficial to increase the vibration amplitude of the IETBs, and when the inlet velocity is increased to a large value, the vibration of the IETBs transitions from in-plane vibration to out-plane vibration, vibration will not cause excessive vibration of the IETBs.

(2) At different inlet velocities, the installation of baffles has different effects on the average amplitude of the IETBs in different directions. Among them, it has a greater impact on the z-direction, followed by the x- and the y-directions. In the z-direction, the installation of baffles effectively increased the average amplitude, with the maximum relative average amplitude being 26.06% (u_{in} = 0.4 m/s) and the minimum being 16.92% (u_{in} = 1.0 m/s). In other directions, baffles installation reduced the average amplitude by a maximum of 5.58%. The total amplitude decreased by 4.60% (u_{in} = 0.4 m/s) and 3.22% (u_{in} = 1.0 m/s), respectively. This is because adding baffles can effectively improve the flow path of the fluid, so

that its influence on IETBs is more concentrated, which is conducive to improving the out-plane vibration intensity and vibration characteristics of the IETB, and has a suppression effect on the overall vibration.

In conclusion, the following important summaries are included.

(1) When the inlet velocity is low (u_{in} = 0.4 m/s), the amplitude in the x-direction is large, indicating that the vibration is mainly in-plane vibration; when the inlet velocity is high (u_{in} = 1.0 m/s), the amplitude in the z-direction is large, indicating that the vibration is mainly out-plane vibration.

(2) The velocity change has the most obvious effect on the vibration in the z-direction.

(3) After the installation of baffles, the amplitude of the z-direction increased, but the amplitude of x- and y-directions, and the total decreased to a certain extent, indicating that adding baffles has a certain inhibitory effect on vibration.

8.3 Heat Transfer Analysis

8.3.1 Analysis of Temperature Field

To study the effect of different inlet velocities and baffles installation on the heat transfer characteristics of the IETBs, Fig. 8.7 shows the temperature distribution nephogram of the fluid domain of the y-z plane section of the IETB heat exchanger under different research conditions. Among them, Fig. 8.7(a) and (c) typify the temperature field while the u_{in} = 0.4 m/s (IETB-NB heat exchanger or IETB-HB heat exchanger). Fig. 8.7(c) and (d) typify the temperature field while u_{in} = 1.0 m/s (IETB-NB heat exchanger or IETB-HB heat exchanger).

It is not difficult to see from Fig. 8.7 that the fluid temperature around the IETBs is both higher at different inlet velocities. By the way, the IETB 7 is close to the inlet, so the fluid flows faster, the high-temperature area here is less compared to other IETBs. As the inlet velocity increases, the area of low temperature becomes more. This is because the higher the fluid inlet velocities, the less time it stays inside the IETB heat exchangers, the shorter the heat exchange time, and the lower the fluid temperature.

The high-temperature area of the temperature field is concentrated above the heat exchanger when there are no baffles. However, in the case of installing baffles, the

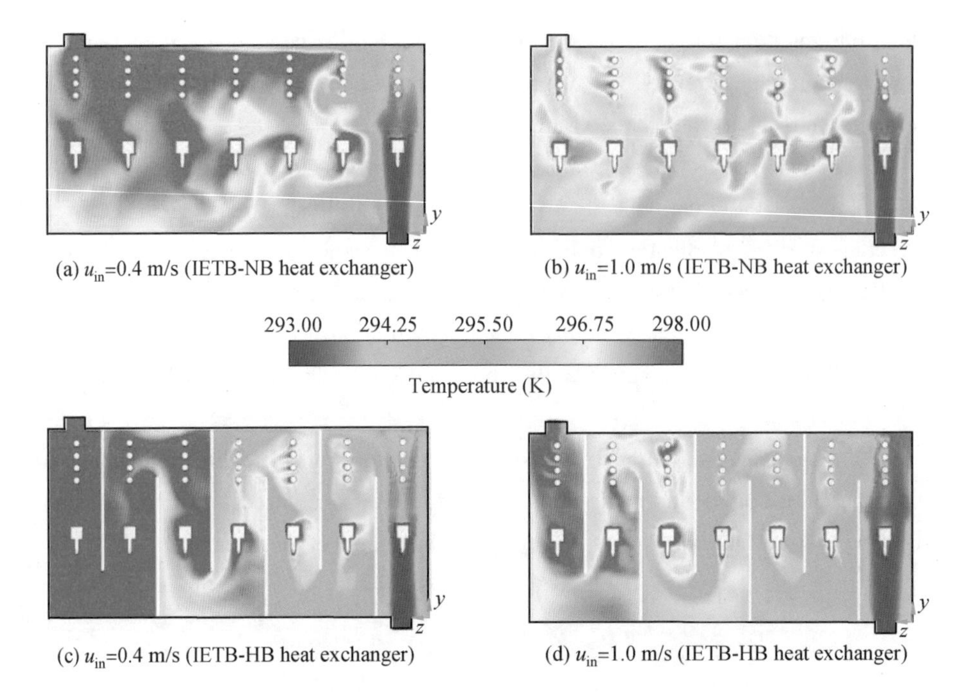

Fig. 8.7 Temperature distribution nephogram of the fluid domain

high-temperature area becomes more hierarchical and progressively fills the inside of the heat exchanger, which shows that installing the baffles has a great impact on the temperature change of the shell-side fluid and are is conducive to increasing the outlet temperature. This is due to the installation of the baffles changing the internal space structure in the IETB heat exchanger and affecting the flow path of the fluid, which makes it stay longer and lengthens the heat exchange process time. The baffles make the fluid oriented, concentrate and impact IETBs in an orderly manner, which directly affects the temperature area around IETBs and the outlet.

So, it is very important to study the two factors that affect the IETB heat exchanger heat transfer capacity.

8.3.2 Analysis of Heat Transfer

To further study IETB heat exchanger heat transfer performance affected by different inlet velocities and baffles installation, the effect on the heat transfer features of each IETB was researched, and the heat transfer coefficient of each IETB under different conditions are compared and analyzed.

Fig. 8.8 (a) shows the heat transfer coefficient of each IETB at different conditions under non-vibration. Fig. 8.8 (b) shows the heat transfer coefficient of each IETB under vibration. Among them, h represents the heat transfer coefficient of the IETB without vibration, h_v represents the heat transfer coefficient of the IETB under vibration, and the relative heat transfer coefficient (relative h and h_v) is calculated to reflect the lifting strength.

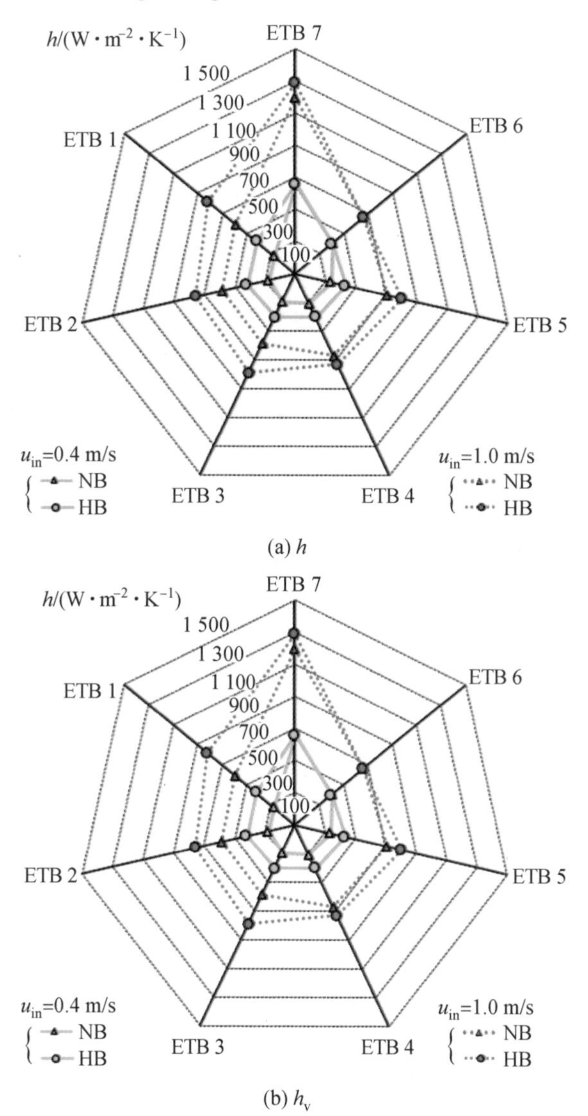

Fig. 8.8 Heat transfer coefficient of seven IETBs at different conditions

It is not difficult to see from Fig. 8.8 (a) and (b) that the heat transfer coefficient of IETB 7 is large, while the heat transfer coefficient of other IETBs has little difference (with a slightly decreasing trend). This is due to the direct impact of the fluid on IETB 7 being stronger than others (Fig. 8.3). By comparing Fig. 8.8(a) and (b), the heat transfer coefficient variation trend of IETBs under vibration and non-vibration conditions is the same. In addition, comparing the heat transfer coefficient under vibration and non-vibration conditions, the maximum value is 17.33% and the minimum value is 0.52%. So, vibration is beneficial to reinforce the heat transfer coefficient of the IETB, that is, vibration-enhanced heat transfer.

As u_{in} rises from 0.4 m/s to 1.0 m/s, while there are no baffles, the maximum relative h is 118.23% and the minimum is 64.84%; the maximum relative h_v is 135.16% and the minimum is 52.88%. When baffles are installed, the maximum relative h is 66.89% and the maximum relative h_v is 106.72%. In addition, comparing the heat transfer coefficient between baffles installation or not, it can be found that the maximum relative h is 52.24% and the maximum relative h_v is 46.65% (u_{in} = 0.4 m/s). The maximum relative h is 38.39% and the maximum relative h_v is 37.81% (u_{in} = 1.0 m/s). So, increasing the inlet velocity and installing baffles can greatly improve the heat transfer coefficient of the IETBs.

8.3.3 Comprehensive Heat Transfer Analysis

Based on the above analysis of the heat transfer coefficient of individual IETB, to reflect the comprehensive heat transfer performance of the IETB heat exchanger is far from sufficient.

Therefore, the averageheat transfer coefficient (h_a, average values of heat transfer coefficient of the seven IETBs) and PEC are proposed to further study the effect of inlet velocity and baffle on the comprehensive heat transfer performance of the IETB heat exchanger. Fig. 8.9 shows the comparison of the average heat transfer coefficient and the PEC for different inlet velocities and baffles install conditions.

It is not difficult to see from Fig. 8.9 as follows.

(1) The average heat transfer coefficient under vibration and non-vibration both increase while increasing inlet velocity. While the value of u_{in} rises from 0.4 m/s to 1.0 m/s, and there are no baffles, h_{av} increased by 97.89%, h_a increased by 96.50%, and the PEC increased by 1.91%. After installing baffles, h_{av} increased by 92.89%, h_a increased by 86.41%, and the PEC increased by 3.45%. It can be seen that whether baffles are installed or not, increasing the inlet velocity is beneficial to significantly improve the average heat transfer coefficient of IETBs and comprehensive heat transfer performance of the IETB heat exchange.

(2) No matter what the inlet velocity is, the average heat transfer coefficient of

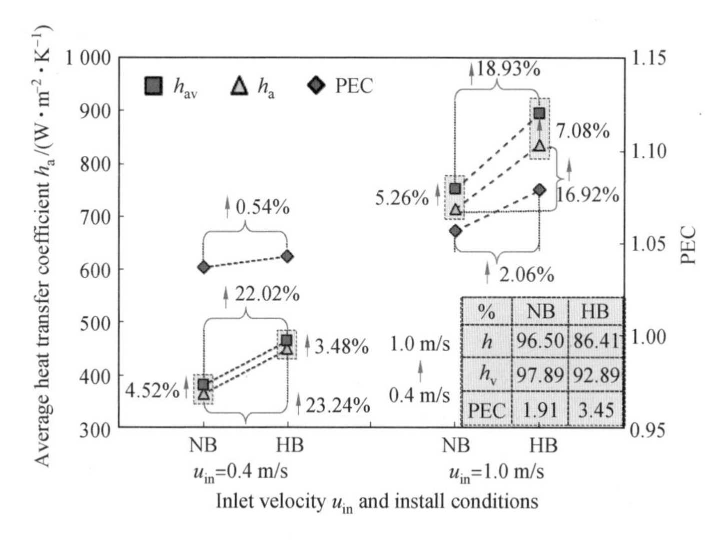

Fig. 8.9 Comparison of the average heat transfer coefficient and the PEC for diffenent inlet velocities and baffles install conditions

IETBs installing baffles is significantly higher than that of no baffles. Under the non-vibration, the relative average heat transfer coefficient is 23.24% (u_{in} = 0.4 m/s) and 16.92% (u_{in} = 1.0 m/s), respectively. Under vibration, the maximum relative average heat transfer coefficient is 22.02% (u_{in} = 0.4 m/s) and the minimum is 18.93% (u_{in} = 1.0 m/s). The maximum increase of PEC is 2.06%. It can be seen that, regardless of inlet velocity, adding baffles can significantly improve the average heat transfer coefficient of IETBs.

(3) In any cases, comparing the average heat transfer coefficient under vibration and non-vibration conditions, the former is always greater. In addition, the relative average heat transfer coefficient ($(h_{av} - h_a)/h_a$) is improved by increasing the inlet velocity. It should be noted that as u_{in} rises from 0.4 m/s to 1.0 m/s, the relative average heat transfer coefficient (IETB–NB heat exchanger) increased from 4.52% to 5.26%, and the relative heat transfer coefficient (IETB – HB heat exchanger) increased from 3.48% to 7.08%. And it can be seen that adding baffles increases the relative average heat transfer coefficient from 5.26% to 7.08% (u_{in} = 1.0 m/s). It can be seen that vibration is beneficial to improve the average heat transfer coefficient of the IETBs, that is, vibration strengthens heat transfer. The baffles installation can improve the vibration strengthening IETBs heat transfer intensity at large inlet velocity, but the effect is not conspicuous when the inlet velocity is small. So, the increase of inlet velocity and the installation of baffles have a promotion

influence on vibration strengthening heat transfer intensity, thus promoting the IETB heat exchanger comprehensive heat transfer performance.

8.4 Comparison of Heat Transfer Among Four Heat Exchangers

8.4.1 Heat Transfer Performance of ETBs

For understand the differences in the heat transfer capacity of the four heat exchangers (TETB-NB heat exchanger, TETB-HB heat exchanger, IETB-NB heat exchanger and IETB-HB heat exchanger), the shell-side temperature distribution and velocity cloud diagram at the cross-section of ETB 4 are shown in Fig. 8.10, where $u_{in} = 0.1$ m/s.

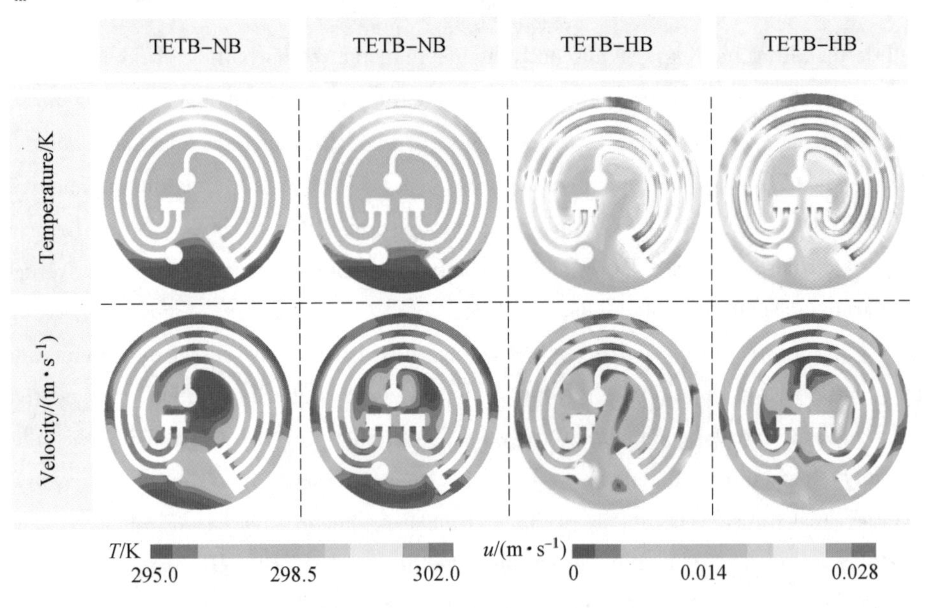

Fig. 8.10 Shell-side temperature distribution and velocity cloud diagram at the cross-section of ETB 4

In Fig. 8.10, according to the temperature distribution, low-temperature area can be observed near the stainless steel connector where the monitoring point B is located (Fig. 7.1) of the TETB-NB heat exchanger while the range of low-temperature area is reduced in comparison with the IETB-NB, and a slight increase in temperature can be observed around the IETB of the the IETB-NB heat exchanger. Therefore, it is proved that the improvement of the ETB structure is effective in enhancing the heat transfer capability of the heat exchanger. When the baffles are installed in the heat exchanger, the baffles force the fluid to impact the ETBs, which can further change the thermal boundary layer and enhance the heat transfer effect on the shell-side. As a result, the low-temperature areas in the TETB–HB heat exchanger and IETB–HB heat exchanger are eliminated, and the areas with poor heat transfer are significantly reduced. Compared with the TETB–HB heat exchanger, the temperature of the fluid around the tube in the IETB–HB heat exchanger is obviously increased.

According to the velocity contour in Fig. 8. 10, compared with the TETB–NB heat exchanger, the velocity of the fluid around the ETB in the IETB–NB heat exchanger decreases slightly, which is caused by the increase of fluid resistance caused by the improvement of the ETB structure. The same phenomenon also exists between the TETB–HB heat exchanger and the IETB–HB heat exchanger. The increase in fluid resistance leads to more sufficient heat transfer between the fluid and the ETB, which explains why the improved structure can improve the heat transfer performance. In conclusion, the improvement of the structure of the TETB and the strategy of installing the baffles in the heat exchanger are successful.

To show intuitively the influences of the improvement of the ETB structure and the addition of the baffles on the heat transfer capability of the ETB, the cloud diagram of the heat transfer coefficient of the ETB and the turbulent kinetic energy diagram of the plane in which it is located are shown in Fig. 8. 11.

It is not hard to see from Fig. 8. 11 that the wall heat transfer coefficient of the IETB–NB heat exchanger is significantly greater than that of the TETB–NB heat exchanger. The heat transfer coefficient of the ETB enhances obviously with the addition of the baffles. According to the turbulent kinetic energy diagram in Fig. 8. 11, the intensity of turbulent kinetic energy near the ETB in the TETB–NB heat exchanger, IETB–NB heat exchanger, TETB–HB heat exchanger and IETB–HB heat exchanger increases successively. This explains why the heat transfer coefficients of ETBs in the TETB–NB heat exchanger, IETB–NB heat exchanger, TETB–HB heat exchanger and IETB–HB heat exchanger also gradually increase in this order.

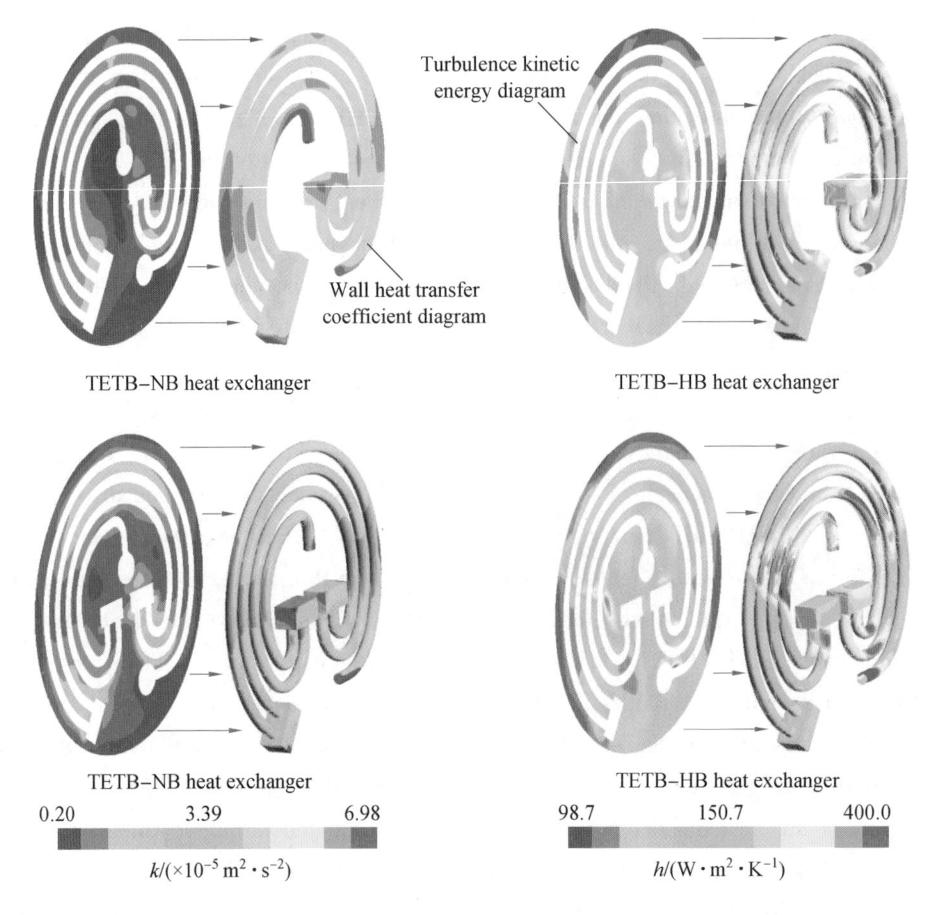

Fig. 8. 11 Cloud diagram of the heat transfer coefficjent of the ETB and the turbulent kinetic energy diagram of the plane in which it is located

8.4.2 Comprehensive Heat Transfer of Heat Exchangers

Fig. 8. 12 displays the comparison of the heat transfer capacity (average *Nu* under vibration conditions) of four heat exchangers under different inlet velocities.

From Fig. 8. 12, the general observation is that the heat transfer capability of the IETB-NB heat exchanger is better than that of the TETB-NB heat exchanger, and the heat transfer capability of the IETB-HB heat exchanger is better than that of the TETB-HB heat exchanger. In the inlet velocities range of 0. 1 m/s to 1. 0 m/s, compared with the TETB-NB heat exchanger, the heat transfer capability of the IETB-NB is improved by 8. 44%, 6. 91%, 5. 50% and 2. 41%, respectively. Compared with the TETB-HB heat exchanger, the heat transfer performance of the

IETB－HB heat exchanger is improved by 5.14%, 4.21%, 4.03% and 2.14%, respectively. Therefore, the structural improvement of the ETB in this study is successful, and the strategy of adding baffles is also effective.

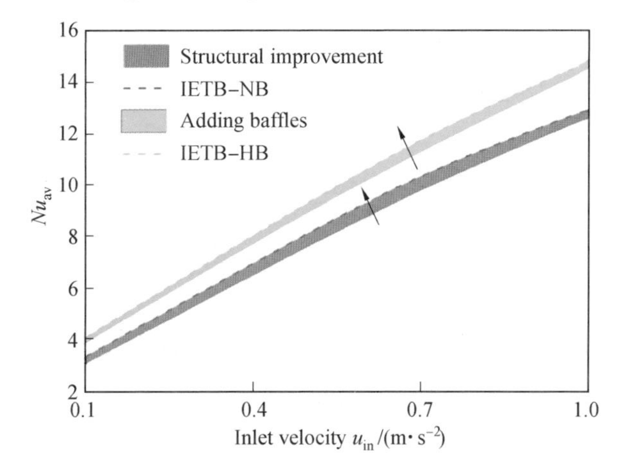

Fig. 8.12 Heat transfer performance of four heat exchangers versus inlet velocity

Improvement to the ETB and the placement of baffles in this study resulted in a larger surface area of contact between the fluid and the ETB, which resulted in a higher pressure drop. The pressure loss is an important index of heat exchanger performance. In fact, the smaller the pressure drop, the lower the operating cost.

Table 8.3 shows the comparison of the pressure drop of four heat exchangers. It is obvious from Table 8.3 that the pressure drops of the four heat exchangers all enhance with the increase of the inlet velocity. At the same inlet velocity, the pressure drop of the heat exchanger is IETB－HB heat exchanger, TETB－HB heat exchanger, IETB－NB heat exchanger and TETB－NB heat exchanger from largest to smallest. It shows that the pressure drop of the heat exchanger can be increased by improving the ETB structure and adding baffles in the heat exchanger, and the addition of baffles has a greater effect on the pressure drop than the improvement of the ETB.

Table 8.3 Comparison of the pressure drop of four heat exchangers

Heat exchangers	Pressure drop ΔP/Pa			
	u_{in}=0.1 m/s	u_{in}=0.4 m/s	u_{in}=0.7 m/s	u_{in}=1.0 m/s
TETB－NB	5.141	115.054	351.605	722.290
TETB－HB	7.571	128.051	400.356	758.507
IETB－NB	5.362	123.887	364.714	734.792
IETB－HB	7.941	135.508	410.042	810.392

Fig. 8. 13 shows the comparison of heat transfer capability and pressure drop of four heat exchangers, where u_{in} = 0. 1 m/s.

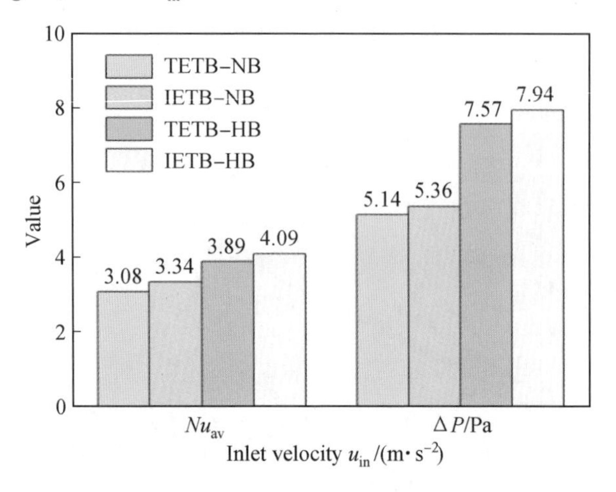

Fig. 8. 13 Comparison of heat transfer capability and pressure drop of four heat exchngers

It can be seen from Fig. 8. 13 as follows.

(1) Compared with the TETB-NB heat exchanger, the IETB-NB heat exchanger increases the Nu_a of the ETB by 8. 44% and increases the ΔP by 4. 28%. Compared with the TETB-HB heat exchanger, the IETB-HB heat exchanger increases the Nu_a of the tube bundle by 5. 14% and increases the ΔP by 4. 89%. Therefore, the improvement of the TETB to the IETB in this study has obvious advantages in energy saving.

(2) The ratios of Nu_a to ΔP of the TETB-NB heat exchanger, IETB-NB heat exchanger, TETB-HB heat exchanger and IETB-HB heat exchanger are 0. 600, 0. 623, 0. 514 and 0. 515 respectively. It is not difficult to see that compared with the TETB-NB, the Nu_a of unit pressure drop of the IETB-NB heat exchanger is improved. Moreover, compared with the TETB-HB heat exchanger, the Nu_a of unit pressure drop of the IETB-HB heat exchanger is also increased. This shows that the structural improvement of the ETB is successful. In addition, compared with the IETB-NB heat exchanger, the IETB-HB heat exchanger increases the overall Nu_a of ETBs by 22. 45% and increases the ΔP by 48. 13%.

Based on the above analysis, in industrial production, the addition of baffle can effectively enhance the heat transfer capability of the ETB, while it is difficult to avoid the disadvantages caused by pressure drop consumption, so it is necessary to fully consider the economy of both.

In order to comprehensively consider heat transfer and pressure drop, the heat transfer characteristics and flow friction characteristics of heat exchangers are expressed by factor J (Colburn factor) and factor F (Fanning friction factor) respectively.

Fig. 8. 14 demonstrates the changes of the factor J and factor F of the four heat exchangers at different inlet velocities.

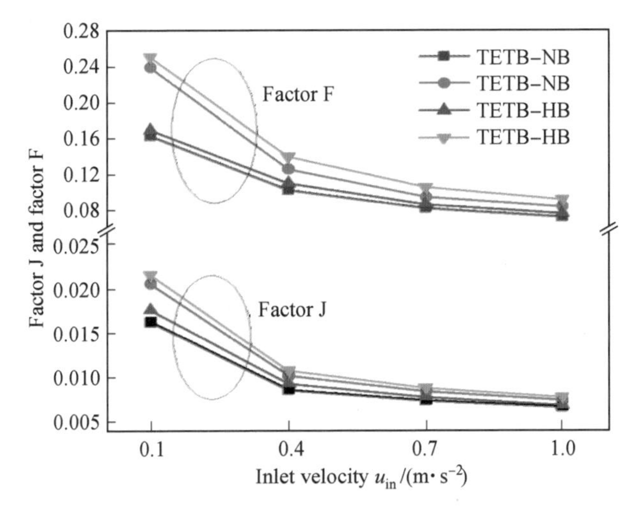

Fig. 8. 14 Factor J and factor F of the four heat exchangers at different inlet velocities

It is easy to see from Fig. 8. 14 that the factor J and factor F of the four heat exchangers at different inlet velocities are the IETB-HB heat exchanger, TETB-HB heat exchanger, IETB-NB heat exchanger and TETB-NB heat exchanger from largest to smallest. The improvement of the ETB and the addition of baffles enhance the contact surface area between the fluid and the tube bundle, which increases the heat transfer rate and improves the heat transfer capability. In addition, the improvement of the ETB and the addition of baffles increase the fluid resistance, so that the pressure drop of the heat exchanger increases.

According to the analysis in Fig. 8. 14, the higher the pressures drop in the heat exchanger, the greater the heat transfer rate. In the design of a heat exchanger, it is not easy to decrease pressure drop and enhance heat transfer simultaneously. While the purpose of designing novel heat exchangers is to enhance heat transfer and decrease pressure drop, to compare the overall heat transfer performance of two types of heat exchangers, the factor JF (dimensionless numbers related to factor J and factor F) is used, as shown in Eq. (2. 27).

Fig. 8. 15 demonstrates the changes in the factor JF of three heat exchangers (TETB – HB heat exchanger, IETB – NB heat exchanger and IETB – HB heat exchanger) at different inlet velocities (reference TETB–NB heat exchanger).

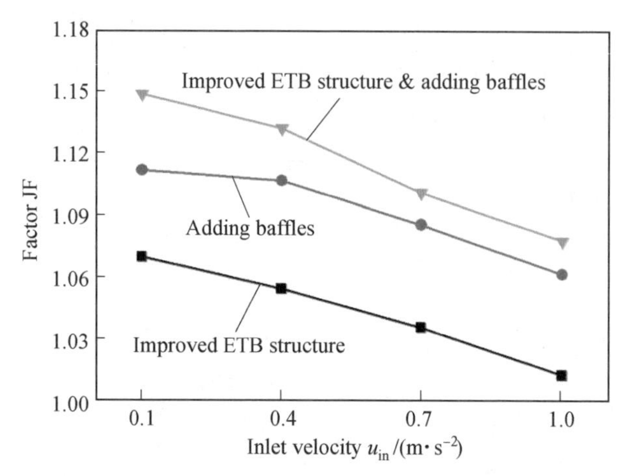

Fig. 8. 15 Changes in the factor JF of three heat exchangers at different inlet velocities

As can be seen from Fig. 8. 15, improving the ETB structure can enhance the thermohydraulic performance of the heat exchanger, and adding baffles can also enhance the thermohydraulic performance of the heat exchanger. In particular, when the ETB structure is improved and the baffles are added to the heat exchanger, it can make the heat exchanger have superior thermohydraulic performance. Therefore, the structural improvement of the ETB in this study is successful, and the strategy of adding baffles is also effective. Compared with these four heat exchangers, the IETB–HB heat exchanger has superior thermohydraulic performance.

References

[1]JI J D, ZHANG J W, LI F Y, et al. Numerical research on vibration-enhanced heat transfer of improved elastic tube bundle heat exchanger [J]. Case studies in thermal engineering, 2022, 33: 101936.

[2]DUAN D R, CHENG Y J, GE M R, et al. Experimental and numerical study on heat transfer enhancement by Flow-induced vibration in pulsating flow [J]. Applied thermal engineering, 2022, 207: 118171.

[3]DUAN D R, GE P Q, BI W B, et al. An empirical correlation for the heat transfer enhancement of planar elastic tube bundle by flow-induced vibration [J]. International journal of thermal sciences, 2020, 155: 106405.

Chapter 9 Effect of Baffle Structure on the Performance of IETB Heat Exchanger

Based on the research conclusion in Chapters 7 and 8, the addition of baffles inside the heat exchanger has a positive effect on improving its comprehensive heat transfer performance, but due to the influence of the baffle structure, the high-velocity fluid is mainly concentrated on the side close to the baffle, which cannot impact the IETB well, and the heat exchanger cannot show the optimal comprehensive performance, and it is necessary to improve the specific structural parameters of the baffles to further improve the performance of the heat exchanger.

In summary, based on the IETB–HB heat exchanger, this chapter aims to elevate the comprehensive heat transfer performance of the heat exchanger by changing the baffle structure. Using the bi-directional FSI calculation method, combined with the step-by-step calculation strategy of rough calculation and actuarial calculation, the influence of the baffle structure parameters (height and curvature) on the vibration-enhanced heat transfer performance of the IETB–HB heat exchanger was systematically studied under different inlet flow velocities.

9.1 Models and Methods

9.1.1 Calculation Domain and Method

Fig. 9.1 illustrates the IETB–HB heat exchanger structure. Multiple rows of copper IETBs are mounted on two horizontal tubes at equal spacing. The curved baffles are placed along the length of the heat exchanger at equal distances and placed between two IETBs. The shell cold fluid flows in from the bottom inlet, through the diversion of curved baffles, bends from right to left, and finally flows out from the top outlet on the left. The hot fluid flows in from the left inlet, flows into the horizontal tube through the end plate, then enters the IETBs through the horizontal tube, and finally flows out from the right outlet. The hot and cold fluids exchange heat through IETBs, which vibrate under the coupling of the shell and tube fluids, thereby realizing vibration-enhanced heat transfer. The vibration characteristics of IETBs are regulated by three identical stainless steel mass blocks (marked as A, B and C)[1]. The flow path of the shell fluid is optimized by the baffles, so that as many shell fluids as possible can impact the IETBs in the form of repeated baffling, extend the fluid flow path, and then improve the comprehensive heat transfer performance of the heat exchanger.

To investigate the vibration-enhanced heat transfer performance of the IETB-HB heat exchanger under the baffle structure improvement, different parameter settings are made from the two aspects of height H_b and curvature θ of the baffle, as shown in Fig. 9.1. To make the study informative, the height factor η (the ratio of baffle height to heat exchanger cylinder diameter, $\eta = H_b/D$) is calculated to reflect the change in baffle height, and the study parameters in this chapter are shown in Table 9.1. The remaining parameters are consistent with those shown in Table 9.1.

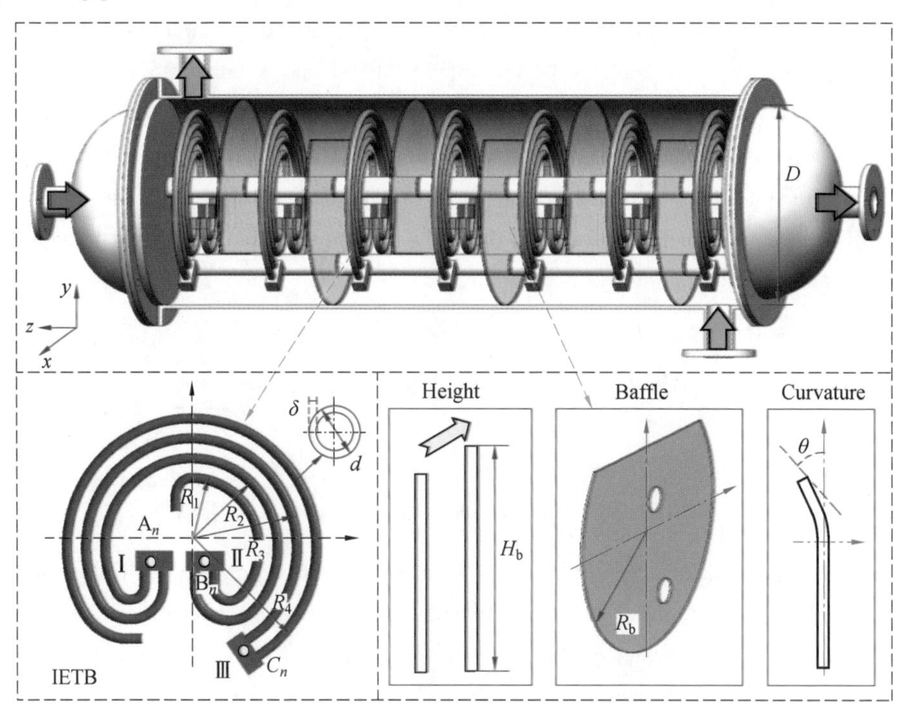

Fig. 9.1 IETB-HB heat exchanger structure

Table 9.1 Parameters and values

Project	Parameters	Values
Fluid domain	Shell-side inletdiameter D_{in}/mm	45
	Diameter D/mm	300
	Length L_1/mm	630
	Shell-side outlet diameter D_{out}/mm	45
Structural domains	Bend radius, R_1, R_2, R_3, R_4/mm	70,90,110,130
	Stainless steel connectors/mm	80×20×20
	Tube outer diameterd/mm	10
	Tube wall thickness δ/mm	1.5
Baffle	Height H_b/mm	210, 225, 240
	Curvature θ/(°)	0, 5, 10, 15, 20
	Height factor η	0.70, 0.75, 0.80

The calculation domain is shown in "8.1.2 Meshes and Boundary Conditions", and will not be repeated here. The IETBs in the heat exchanger are numbered as n, from right to left 1, 2, ..., 7. Monitoring points A_n, B_n and C_n are set on stainless steel connectors Ⅰ, Ⅱ and Ⅲ of the IETBs.

Fluid domain: The medium is set as "Water". The shell inlet is set as "Inlet". The temperature T_{in} is 293.15 K, the "Inlet velocity" $u_{in}=0.4$ m/s and 0.7 m/s. The shell outlet is set as "Outlet" with a relative pressure $P_{out}=0$ Pa. The contact surface between the fluid domain and the IETB is set to "Fluid Solid Interface" with the wall temperature $T_w=333.15$ K. The other surfaces are set to "Wall" with non-slip adiabatic wall surfaces.

Structural domain: In terms of material, the tube of IETB is set as "Copper", the stainless steel connector as "Stainless Steel". The IETB end is set to "Fixed Support". The gravity is vertical downwards. The IETB surfaces correspond to the "FSI" of the fluid domain.

Fig. 9.2 shows the specific calculation process, and the rough calculation results and computational domain mesh are displayed simultaneously.

Fig. 9.2(b) illustrates that the outlet temperatures (T_{out}) of the three heat exchangers with different baffle height factors reach a steady state after a certain calculation time, but the time to reach the steady state varies, as shown at points a, b and c in Fig. 9.2(b). At the same time, the temperature change curve of the last

50 steps is shown separately. It can be seen that the temperature fluctuations under different height factors are small. In addition, the average T_{out} of the three heat exchangers at this stage was calculated respectively, and it was concluded that as the height factor increases, the average T_{out} also increases. As the height factor increases, the time for the outlet temperature to reach the dynamic steady state becomes shorter, since as the baffle height increases, the flow gap becomes smaller and the fluid flow becomes more concentrated. During the rough calculations, the heat exchanger outlet temperatures at different height factors reach steady state values, and the fluid domain is fully developed, indicating that the numerical simulation results are more reliable when actuarial calculations are carried out on this basis.

The results of meshing can be seen in Fig. 9.2(c), with tetrahedral meshing for the fluid domain and the mass blocks of the IETB, and hexahedral meshing for the elbow part of the IETB. A six-layer boundary layer grid was set near the FSI wall, considering that the model adopted in this paper is a standard $k-\varepsilon$, and the value range of y^+ on the wall is 30–100, which meets the requirements of the turbulence model[2]. The exact number of grids is shown in Table 9.2.

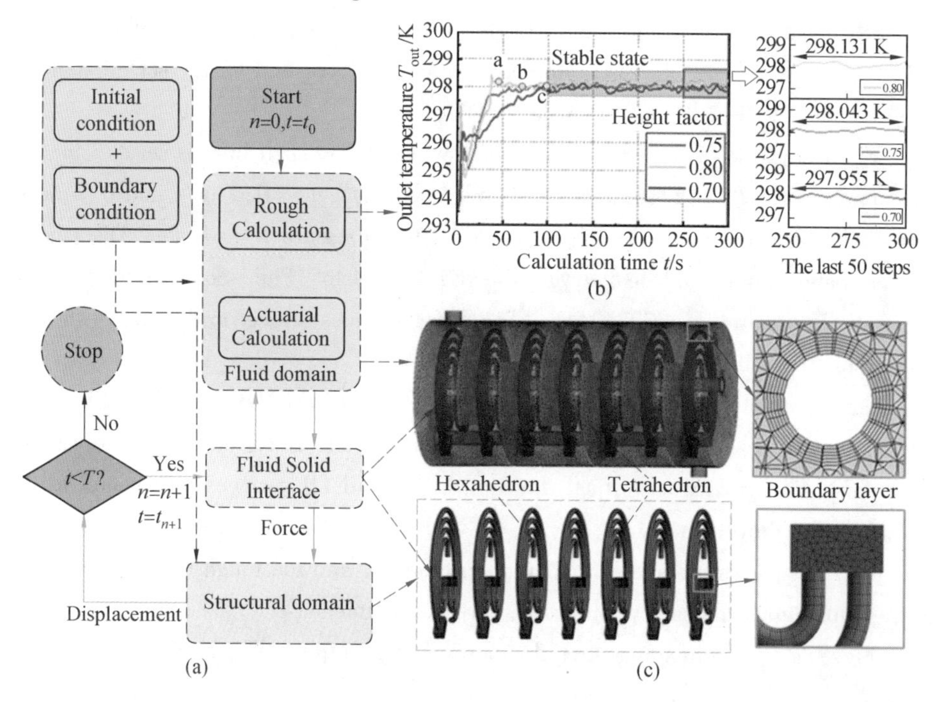

Fig. 9.2 Specific calculation process

9.1.2 Mesh independence analysis

Table 9. 2 shows a comparison of the parameters for the different meshing methods (initial conditions are baffle height factor—0. 75, baffle curvature—$\theta=0°$, inlet velocity—$u_{in}=0.70$ m/s.). The heat transfer coefficient h for IETB 1, IETB 4 and IETB 7 are calculated for the corresponding cases, and the maximum relative errors and the calculation time for each calculation example are compared, where cases 1 and 3 are obtained by reducing/increasing the size of the mesh and reducing/increasing the number of elements based on case 2.

Table 9.2 Comparison of the parameters for the different meshing methods

Case	Number of elements		$h/(W\cdot m^{-2}\cdot K^{-1})$			Maximum relative error /%	Time/h
	Structural domain	Fluid domain	IETB 1	IETB 4	IETB 7		
1	15,182	4,191,156	1,036.62	552.23	612.34	5.41	64.1
2	29,739	8,210,090	1,086.09	583.81	633.65	—	105.8
3	49,561	13,682,366	1,083.14	592.34	641.76	1.46	192.5

Note: The relative error is calculated based on the calculation results of case 2.

As can be seen from Table 9.2, based on case 2, when the number of elements is reduced to about 51.04% of the original (case 1), the maximum error of h for IETB is 5.41%, which takes 60.59% of the original; when the number of elements is increased to 1. 67 times of the original (case 3), the maximum error is only 1.46%, which takes 1.82 times of the original. Considering the calculation time and accuracy together, case 2 satisfies the independence requirement.

9.2 Effect of the Baffle Height

9.2.1 Effect of Baffle Height on Vibration

To visualize the changes in fluid flow, the y–z cross-sectional fluid velocity vector diagrams of the fluid domain at different inlet velocities (u_{in}) and different height factors (η) are shown in Fig. 9. 3.

Fig. 9. 3 illustrates that as the inlet velocity increases, the flow velocity in the fluid domain generally increases and so does the flow velocity at the outlet. At the same inlet velocity, it can be seen from the comparison of the marked areas in Fig. 9.3 that as the baffle height factor increases, the fluid around the IETB in the latter part of the fluid domain and at the gap between the baffles becomes more concentrated and the velocity increases. This also reflects the fact that the change in baffle height results in a stronger fluid flow within the heat exchanger, reinforcing the strength of the contact between the fluid and the IETBs in the second half heat exchanger.

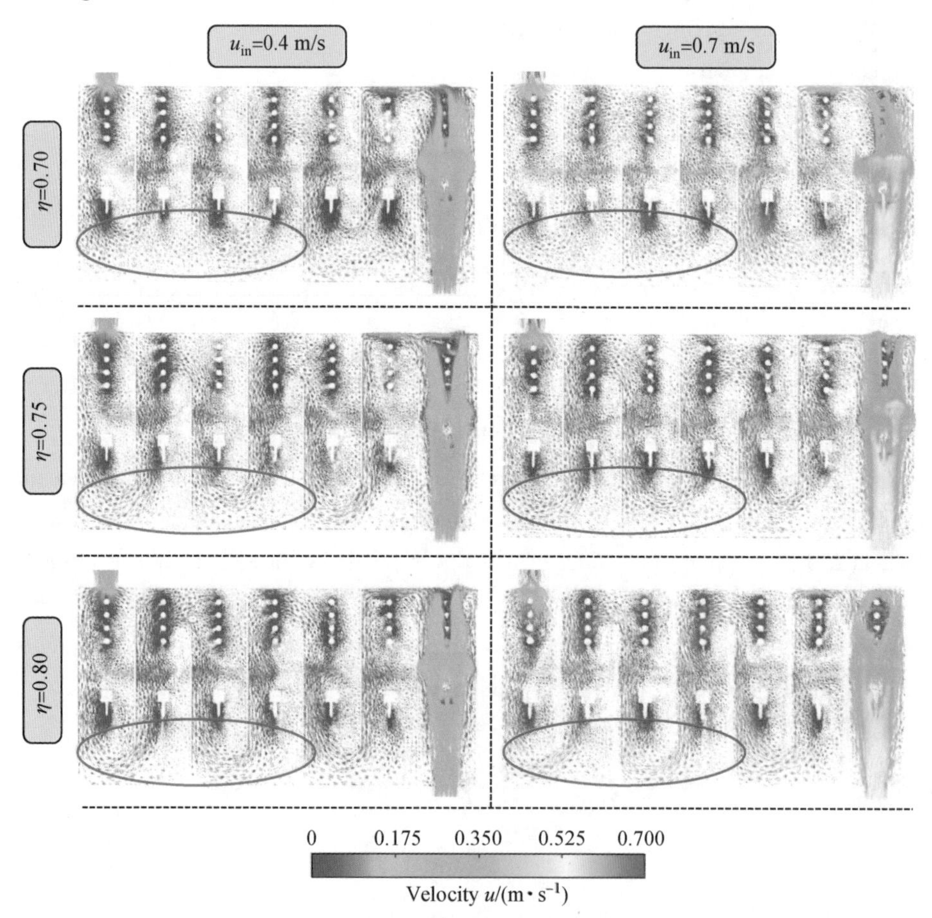

Fig. 9.3 y–z cross-sectional fluid velocity vector diagrams of the fluid domain at different inlet velocities and different height factors

The vibration characteristics of the monitoring point C_n can characterize the vibration characteristics of the IETB. The corresponding amplitudes A and frequencies f can be obtained by FFT calculation of the data of monitoring points C_n. To investigate the vibration characteristics of the IETB at different baffle heights, the amplitude variations of monitoring points C_1-C_7 at different baffle height factors are calculated and plotted, as shown in Fig. 9.4.

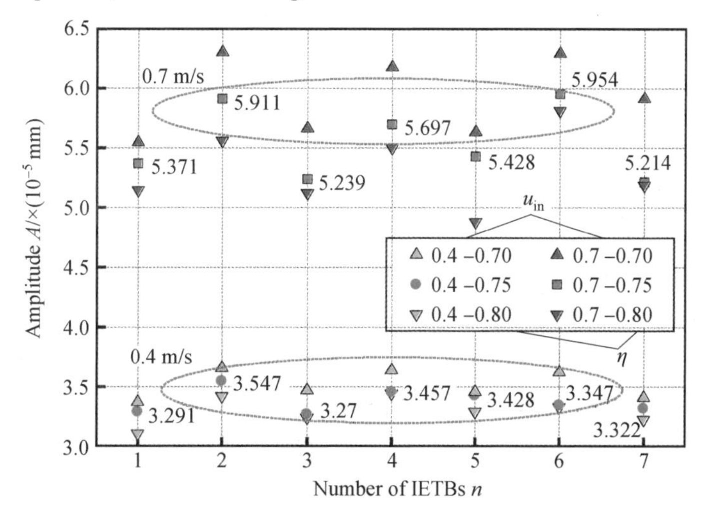

Fig. 9.4 Amplitude change at the monitoring points C_n

From Fig. 9.4, the following conclusions can be drawn.

(1) The amplitude of monitoring points C_1-C_7 varies at different inlet velocities. While $u_{in}=0.4$ m/s, the amplitude fluctuates less at C_n, and the maximum relative error in amplitude increases with the increase of the height factor, which is 8.42% ($\eta=0.70$), 8.47% ($\eta=0.75$) and 10.85% ($\eta=0.80$) respectively. While $u_{in}=0.7$ m/s, the amplitude fluctuates more and the maximum relative error increases with the increase of the height factor, which is 13.65% ($\eta=0.70$), 14.19% ($\eta=0.75$) and 18.07% ($\eta=0.80$) respectively. It can be seen that the vibration fluctuations of IETB 1–IETB 7 are small and homogeneous when inlet velocity is low.

(2) As the inlet velocity increases, the amplitudes of monitoring points C_1-C_7 at different height factors all increase. It is calculated that when the inlet velocity grows from 0.4 m/s to 0.7 m/s, the amplitude of the monitoring points C_1-C_7 increased by 62.98%–73.72% ($\eta=0.70$), 56.95%–77.89% ($\eta=0.75$) and 48.34%–74.62% ($\eta=0.80$) under three different height factors, and the amplitude of the monitoring point C_6 increased the most. For each monitoring point, as the height factor increases, the amplitude increase of the monitoring points C_n changes very little. It can be seen that the increase of U_{in} has a lifting effect on the vibration of IETB, and the increase of baffle height has no effect on the amplitude of improvement.

(3) At different inlet velocities, the amplitude of the monitoring points C_1-C_7

decreases with the increase of the height factor. It is calculated that when the height factor increased from 0.70 to 0.80, the amplitude of the monitoring points C_1-C_7 decreased by 4.83% – 8.11% (u_{in} = 0.4 m/s) and 7.23% – 13.38% (u_{in} = 0.80 m/s), respectively. It can be seen that increasing baffle height has a suppressing effect on the vibrations of the IETB.

The study based on each IETB is not sufficient to reflect the overall vibration characteristics, so the average amplitude A_a (IETB 1 – IETB 7) is calculated, thus facilitating the subsequent study of the vibration-enhanced heat transfer performance. Fig. 9.5 shows the variation of the average amplitude of IETB 1 – IETB 7 with the height factor under different inlet velocities.

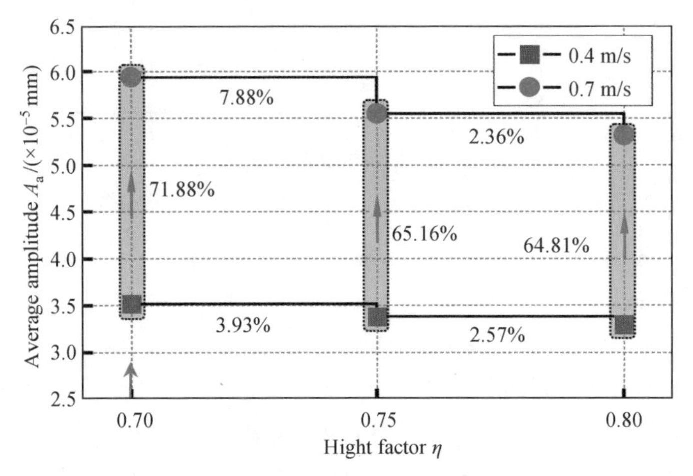

Fig. 9.5 Variation of the average amplitude of IETB 1 – IETB 7 with the height factor under different inlet velocities

From Fig. 9.5, the following conclusions can be drawn.

(1) Under different inlet velocities, the average amplitude decreases with the increase of the height factor, and the average amplitude decreases by 6.39% (u_{in} = 0.4 m/s) and 10.05% (u_{in} = 0.7 m/s) when the height factor increases from 0.70 m/s to 0.80 m/s, respectively. This is because as the baffle height increases, the gap between the IETB and the heat exchanger barrel wall decreases, so that less fluid is directly rushed into the IETB.

(2) For different height factors, when the inlet velocity increases from 0.4 m/s to 0.7 m/s, the average amplitude increases by 71.88% (η=0.70), 65.16% (η= 0.75) and 64.81% (η=0.80), respectively. It can be seen that increasing the inlet velocity can significantly increase the vibration intensity of the IETB, and increasing

the height of the baffle will reduce its lifting intensity.

In summary, increasing the baffle height has a certain vibration suppression effect on the IETBs.

9.2.2 Effect of Baffle Height on Heat Transfer

To see the effect of the baffle height on the internal temperature region of the heat exchanger, Fig. 9.6 shows the temperature contour map for the $y-z$ cross-section and the $x-y$ cross-section in the middle of the fluid domain under different height factors at $u_{in}=0.7$ m/s . In Fig. 9.6, the corresponding exit temperature values of the heat exchanger are marked.

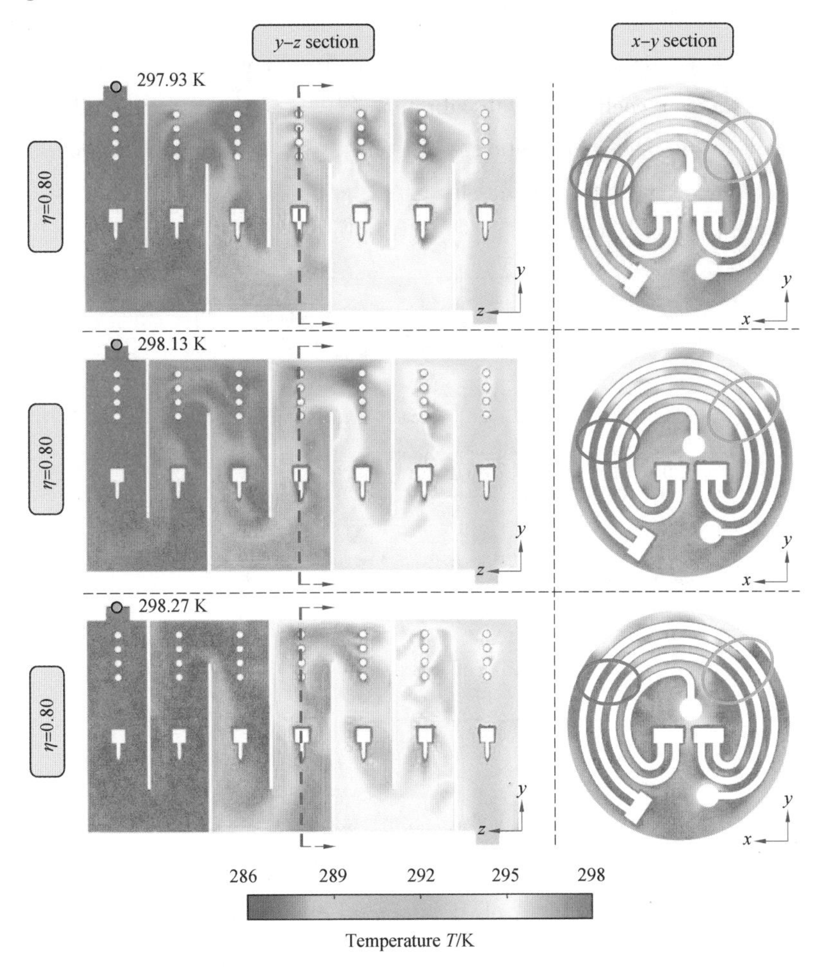

Fig. 9.6 Temperature distribution in fluid domain middle sections under different height factors

It is evident from Fig. 9. 6, the outlet temperature increases with increasing height factor. y–z cross-section shows that the high-temperature region in the middle and rear sections of the heat exchanger is more, and decreases with increasing height factor. This is due to the gap between the baffle and the heat exchanger barrel becoming smaller with increasing height factor and the fluid flowing faster, resulting in less residence time for the fluid at the front. For this type of case, the temperature distribution of the $x-y$ cross-section in the middle heat exchanger is analyzed. It is evident from as the height factor grows, the high-temperature region around the bundle also increases, due to the greater effect on the bundle as the inlet velocity increases.

To further analyze the heat transfer performance of the IETBs at different baffle heights, the heat transfer performance needs to be quantified.

Fig. 9. 7 shows the average heat transfer coefficient without vibration (h_a) and average heat transfer coefficient with vibration (h_{av}) of IETBs under different inlet velocity and height factor conditions, as well as the variation of heat transfer coefficient without vibration (h) and heat transfer coefficient with vibration (h_v) of each IETB with the number of IETBs when u_{in}=0. 7 m/s and η=0. 75.

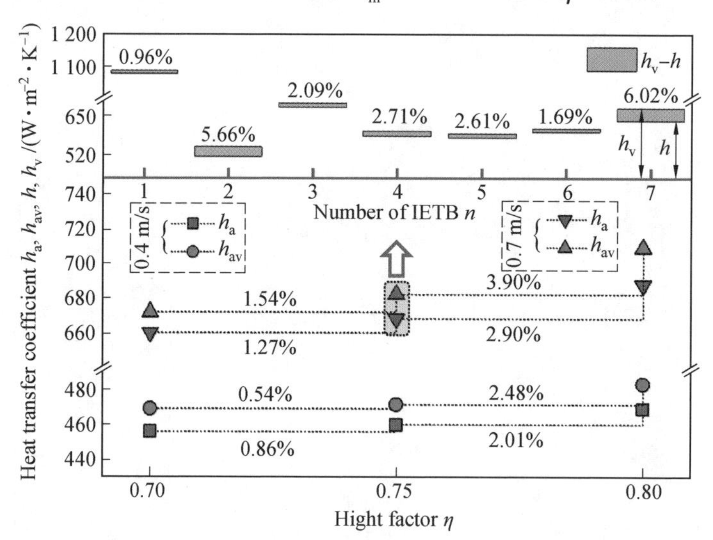

Fig. 9. 7 Heat transfer coefficient h_a, h_{av}, h, h_v under different conditions

From Fig. 9. 7, it can be seen as follows.

(1) The h and h_v values of each IETB varied, the h and h_v values of IETB 1 are the largest and the values of other IETBs are similar. The h_v of each IETB is greater than h, and the increased percentage of h_v relative to h is calculated to be 0. 96% –

6.02%. This means that vibration-enhanced heat transfer is achieved in each IETB.

(2) Both h_a and h_{av} increase with height factor growth. After calculation, while the height factor η grows from 0.70 to 0.80, the h_a increases by 2.89% (u_{in} = 0.4 m/s) and 3.04% (u_{in} = 0.7 m/s), and the h_{av} increases by 4.21% (u_{in} = 0.4 m/s) and 5.50% (u_{in} = 0.7 m/s), respectively. This suggests that increasing the baffle height has a positive effect on enhancing the h and h_v of the IETB and the enhancement of h_v is greater under higher u_{in}.

(3) The h_{av} is always greater than h_a. the h_{av} is calculated to be 2.88% (η = 0.70), 2.56% (η = 0.75) and 3.03% (η = 0.80) relative to h_a when u_{in} = 0.4 m/s. It is 1.80% (η = 0.70), 2.08% (η = 0.75) and 3.06% (η = 0.80) respectively when u_{in} = 0.7 m/s. In combination with (1), the heat transfer performance under vibration conditions is better than that under non-vibration conditions, and under higher inlet velocity, increasing the height factor improves the vibration-enhanced heat transfer performance of IETBs.

(4) Both h_a and h_{av} increase as inlet velocity increases. While the inlet velocity grows from 0.4 m/s to 0.7 m/s, h_a and h_{va} increase by 45.29% – 46.56% and 43.18% – 46.60%. So, increasing the inlet velocity has an enhancing effect on the heat transfer performance of the IETBs.

9.2.3 Comprehensive Heat Transfer Analysis

Based on the above research basis of the heat transfer performance of IETBs and combining the pressure drop under different conditions, the factor J and factor F are calculated. Using the heat exchanger model with a height factor η = 0.75 as a reference, the factor JF for heat exchangers when η = 0.70 and η = 0.80 are calculated by combining the pressure drop under different conditions, to comprehensively evaluate the overall heat transfer performance of different heat exchangers, the results of which are shown in Fig. 9.8.

It is evident from Fig. 9.8 that the difference of the factor J under different height factors is small, and it is 0.0014 (η = 0.70), 0.00142 (η = 0.75) and 0.00146 (η = 0.80) under u_{in} = 0.4 m/s, and 0.00118 (η = 0.70), 0.00116 (η = 0.75) and 0.00122 (η = 0.80) under u_{in} = 0.7 m/s respectively. This indicates that there is little difference in the heat transfer performance of the heat exchanger at different baffle heights. The value of the factor F increases with the increase of the baffle height, it is 0.00623 (η = 0.70), 0.00639 (η = 0.75) and 0.00671 (η = 0.80) under u_{in} = 0.4 m/s, and 0.0064 (η = 0.70), 0.00618 (η = 0.75) and 0.00665

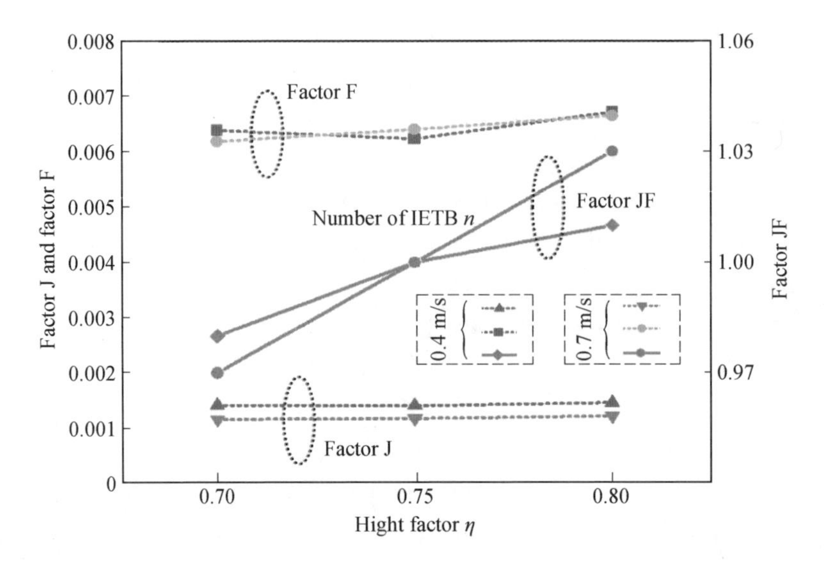

Fig. 9. 8 The variation of factors J, F and JF with height factor

(η=0. 80) under u_{in}=0. 7 m/s, respectively. This indicates that the heat exchanger has a greater resistance to flow at a height factor of η=0. 80. The factor JF calculated by the combined factor J and factor F, it shows that the IETB heat exchanger with a height factor of 0. 80 has a superior hydrothermal performance.

In conclusion, increasing the baffle height suppresses the vibration of the IETBs, but improves the comprehensive heat transfer of the heat exchanger at the same time. This reflects that vibration-enhanced heat transfer is not suitable for all types of heat exchangers and provides a better direction for subsequent research.

9.3 Effect of the Baffle Curvature

9.3.1 Effect of Baffle Curvature on Vibration

Based on the different calculation results, the fluid velocity vector diagrams of the heat exchanger y–z section under different inlet velocity are extracted for some of the baffle curvatures (θ=0°, 10°, 20°), as shown in Fig. 9. 9.

It is evident from Fig. 9. 9, increasing inlet velocity provides a significant increase in the flow velocity of the fluid domain. As the baffle curvature increases, there is a significant change in the flow direction of the fluid domain at the end of the baffle, and more vortices are generated in the vicinity of the fluid. As can be seen

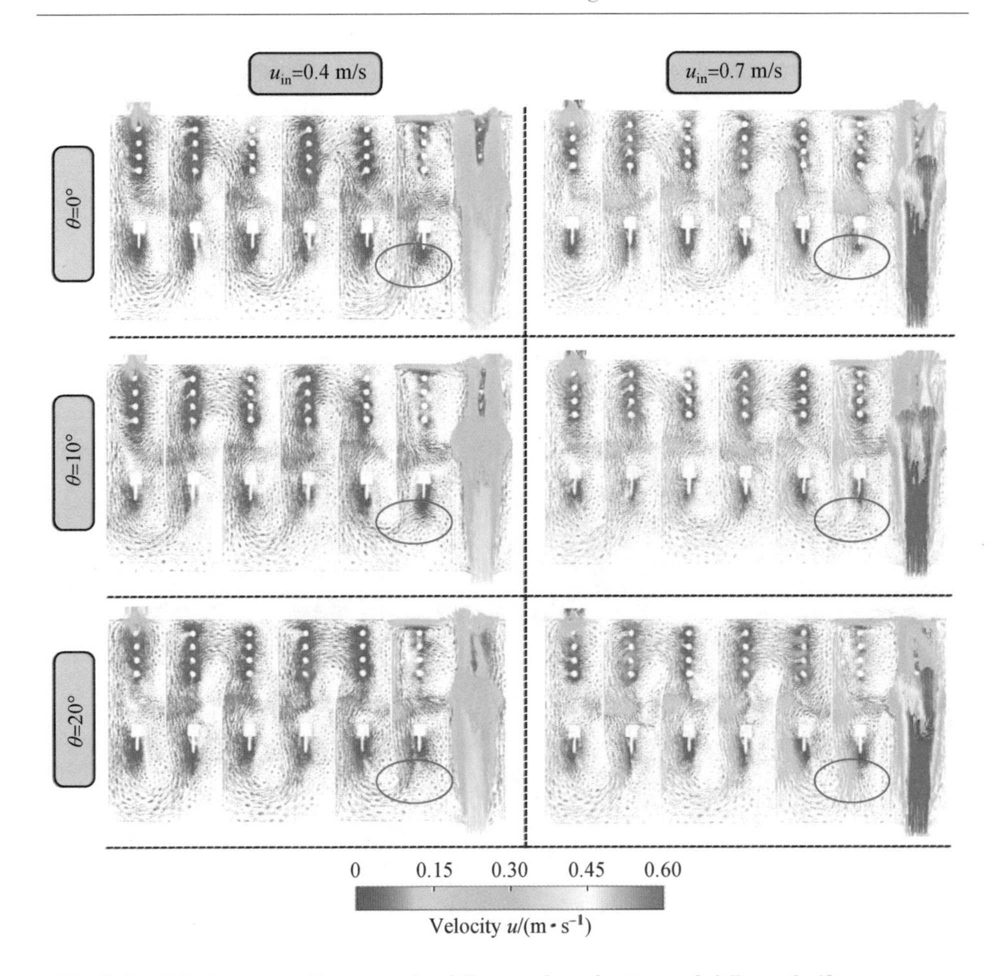

Fig. 9.9 Velocity vector diagram under different inlet velocities and different baffle curvatures

from the marked areas in the diagram, as the curvature of the baffle increases, the amount of fluid directly impacting IETB also increases. In addition, at the same inlet velocity, changing baffle curvature does not significantly influence the flow velocity of the fluid domain.

To analyze the effect of fluid impact on IETB, it is not sufficient to study the fluid domain of the heat exchanger. Fig. 9.10 shows the vibrational displacement of IETBs at different curvatures when $u_{in}=0.7$ m/s and records the $y-z$ cross-sectional flow velocity vector diagram around stainless steel connector C_n.

Fig. 9.10 illustrates that the vibrations are uniform across the IETBs and become increasingly strong from the inside to the outside of the IETB. As the baffle curvature increases, the vibration of the IETB intensifies, and the intensity of the vibration is the greatest at the baffle curvature $\theta=10°$. To further analyze the cause of this

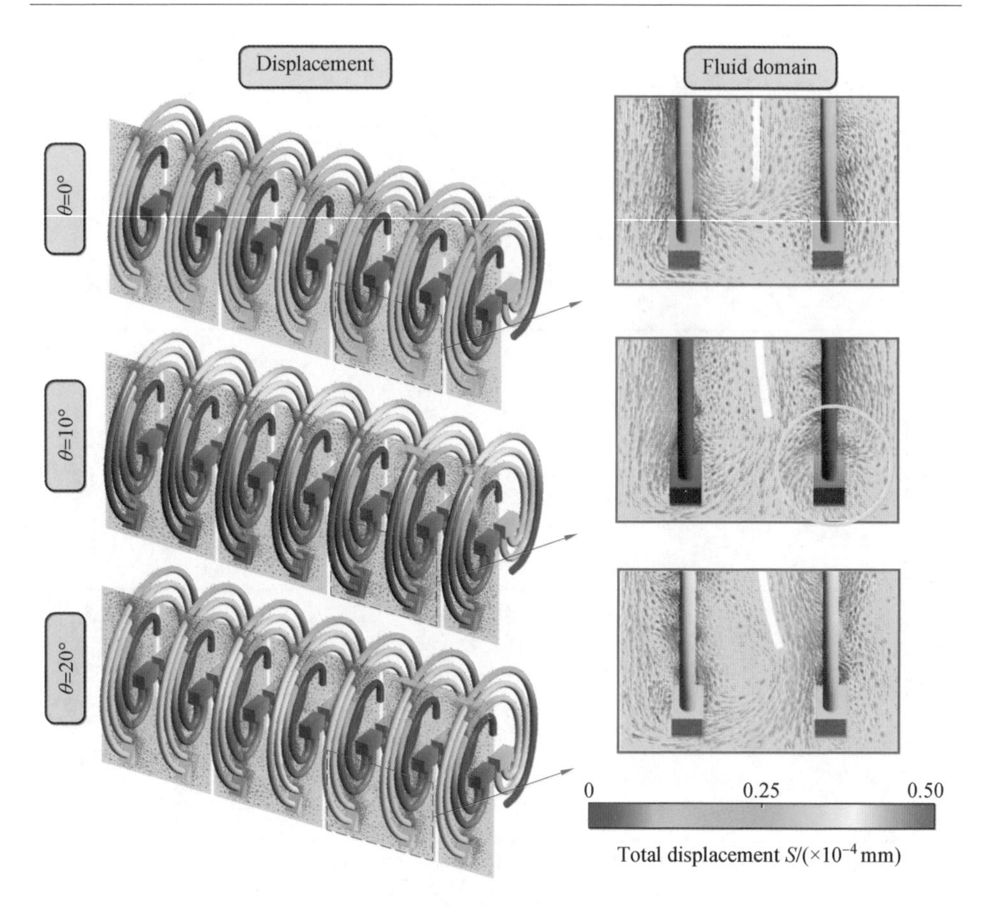

Fig. 9. 10 Vibrational displacement of IETBs and velocity vector diagram around stainless steel connector C_n

phenomenon, it can be seen that at $\theta = 10°$, more vortices are generated around the mass block C, thus contributing to its vibration. However, when the curvature is increased to $\theta = 20°$, there are fewer vortexes generated due to the greater degree of baffle bending, which causes the fluid to impact the stainless steel connector C_n on different surfaces. The change in the fluid flow path is large, the fluid flow in the back space of the baffle is less, and the fluid flow characteristics are not significantly improved.

The average amplitude of the all IETBs is calculated, Fig. 9. 11 illustrates the comparison of the average amplitude at different baffle curvatures.

From Fig. 9. 11, the following conclusions can be drawn.

(1) As inlet velocity increases, the average amplitude increases. While the inlet velocity grows from 0. 4 m/s to 0. 7 m/s, the relative average amplitude increases by

41.85%–56.81% under different baffle curvatures.

(2) The average amplitude of IETBs with a curvature of 10° is the greatest under differentinlet velocities. When $u_{in}=0.4$ m/s, compared to the other curvatures, the average amplitude of $\theta=10°$ is improved by 4.64% ($\theta=0°$), 4.00% ($\theta=5°$) and 3.78% ($\theta=15°$ and $\theta=20°$) respectively. When $u_{in}=0.7$ m/s, it is improved by 1.91% ($\theta=0°$), 10.79% ($\theta=5°$), 0.01% ($\theta=15°$) and 0.54% ($\theta=20°$) respectively. This indicates that the vibration of the IETBs is the strongest at the curvature of 10°. Combined with Fig. 9.11, it can be seen that this is because when the curvature is 10°, the fluid changes the flow direction, strengthens the impact on IETB, enhances the turbulent flow characteristics of the near-wall fluid, and enhances the vibration of IETB.

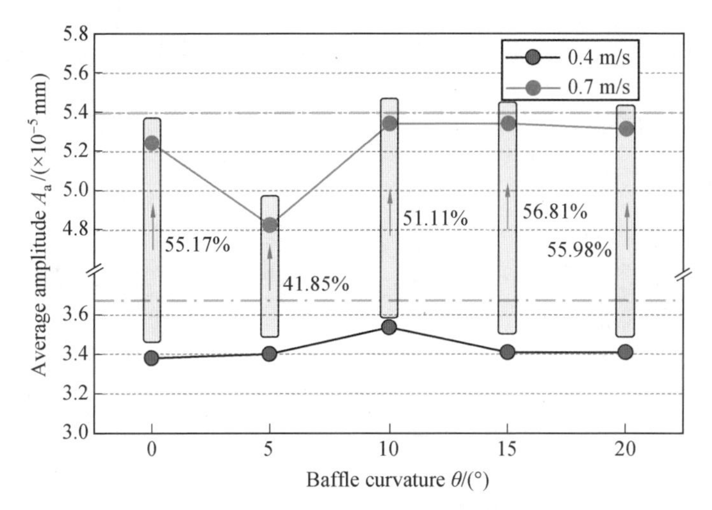

Fig. 9.11 Comparison of the average amplitude at different baffle curvatures

9.3.2 Effect of Baffle Curvature on Heat Transfer

To investigate the heat transfer performance of the IETBs under various baffle curvatures, temperature contour maps of the y–z section are recorded at various baffle curvatures ($\theta=0°$, 10° and 20°) under $u_{in}=0.7$ m/s, and a partial view of the temperature field around the IETB and the fluid flow velocity vector diagram, as shown in Fig. 9.12.

Fig. 9.12 illustrates the temperature does not vary much under different conditions, and with the increase of baffle curvature, the high-temperature region in the first half of the fluid domain increases. As can be seen from the partial view, the angle between the direction of the flow around the beam and the direction of the temperature gradient changes with the baffle curvature increases. Roughly speaking from the perspective of field synergy, when curvature $\theta = 10°$, the included angle is smaller, and the enhanced heat transfer is better.

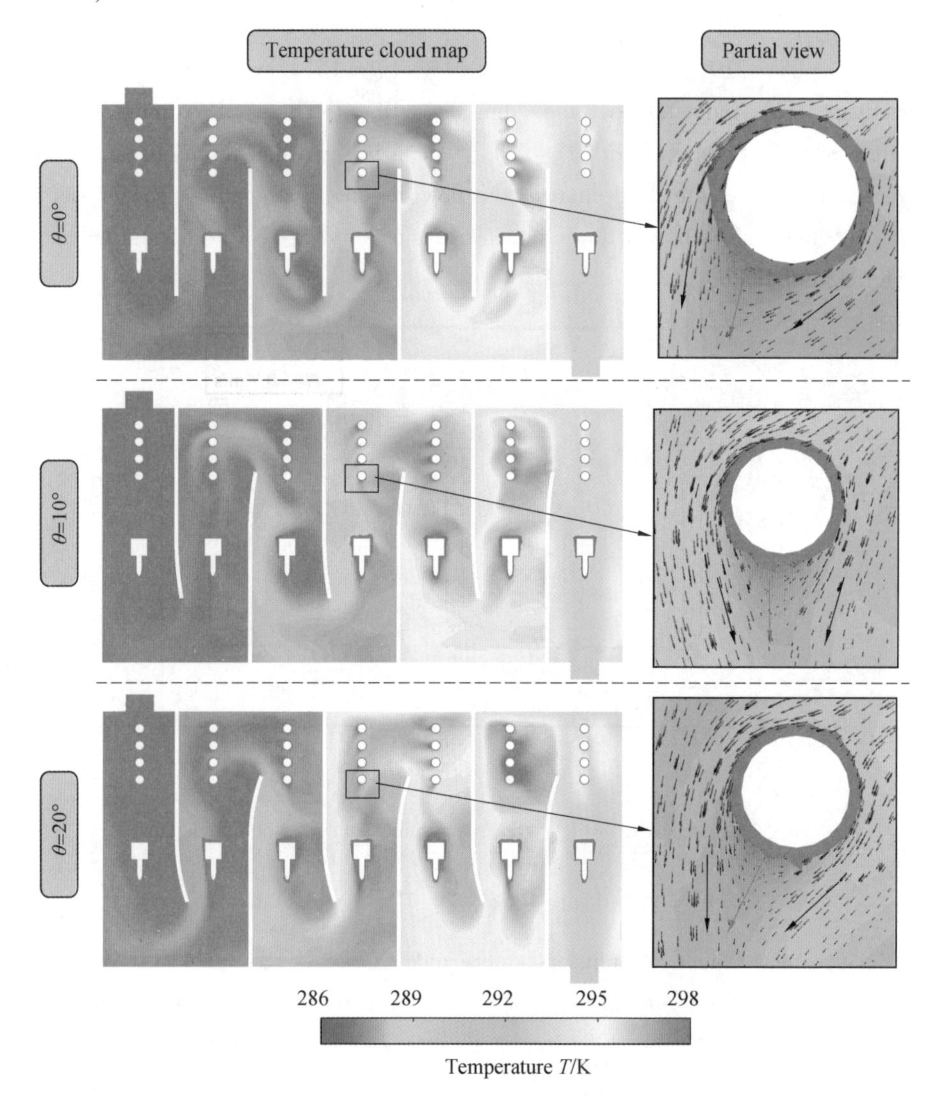

Fig. 9.12 Velocity vector diagram under different inlet velocities and different baffle curvatures

To quantitatively analyze the heat transfer performance of the IETBs, the h and h_v of all IETBs under different inlet velocities and different baffle curvatures are calculated, as shown in Fig. 9.13.

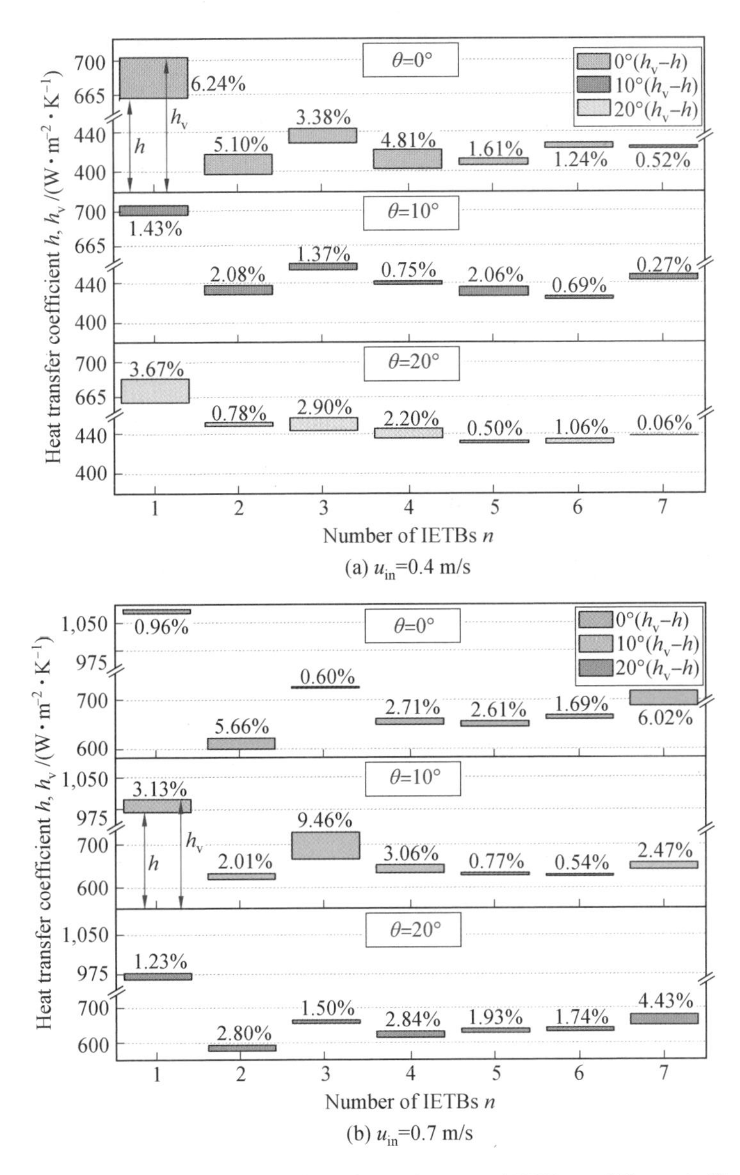

Fig. 9. 13 Heat transfer coefficients with/without vibration of IETBs at different baffle curvatures

From Fig. 9. 13, the following conclusions can be drawn.

(1) The h of all IETBs has the same change rule regardless of the inlet velocity. Under different conditions, the h and h_v of the IETB 1 are higher than that of the other IETBs, and the h and h_v of IETB 2–IETB 7 have little fluctuation, which is because the IETB 1 has the shortest distance to the inlet of a heat exchanger and is most affected by the fluid.

(2) After calulating, when $u_{in}=0.4$ m/s, the increase percentage of h_v relative to h is 0.52%–6.24% ($\theta=0°$), 0.27%–2.08% ($\theta=10°$) and 0.06%–3.67% ($\theta=20°$), respectively. When $u_{in}=0.7$ m/s, it is 0.60%–5.66% ($\theta=0°$), 0.54%–9.46% ($\theta=10°$) and 1.23%–4.43% ($\theta=20°$). This indicates that the h_v is always greater than h. That is, vibration enhances heat transfer. In addition, increasing baffle curvature can improve the h and h_v of the IETBs, but the influence on the intensity of vibration-enhanced heat transfer performance of each IETB is different.

Consistent with the above, the average heat transfer coefficients of the IETBs under vibration and non-vibration conditions h_a and h_{av} under different inlet velocities and different baffle curvatures are calculated to reflect the overall heat transfer performance of the IETBs, as shown in Fig. 9.14.

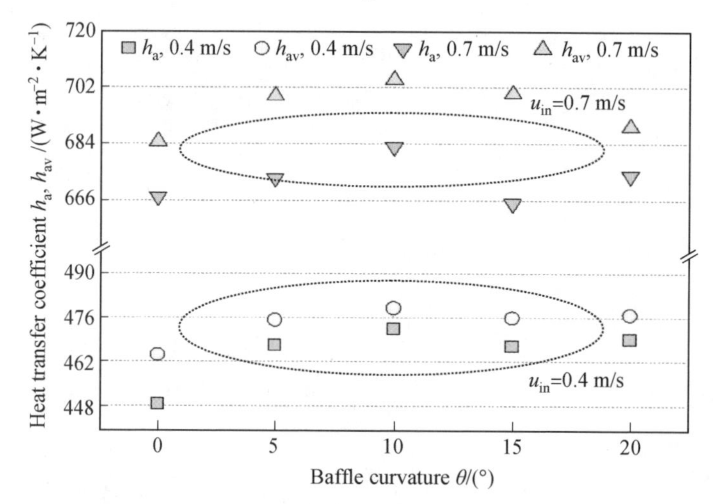

Fig. 9.14 Average heat transfer coefficients under vibration and non-vibration conditions

From Fig. 9.14, the following conclusions can be drawn.

(1) As inlet velocity increases, the h_a and h_{av} increase. While the inlet velocity grows from 0.4 m/s to 0.7 m/s, the h_a increases by 42.41%–48.67%, and the h_{av} increases by 30.73%–32.39% under different baffle curvatures. This indicates that changing baffle curvature does not affect the heat transfer performance enhancement of the IETBs by inlet velocity.

(2) The h_{av} is always greater than h_a. Through calculation, when the baffle curvature increases from 0° to 20°, the h_{av} increases by 1.38%–3.48% (u_{in} = 0.4 m/s) and 2.27%–5.86% ($u_{in}=0.7$ m/s) relative to h_a. So, changing the

baffle curvature will affect the heat transfer intensity enhanced by vibration.

(3) After calculation, when u_{in} =0.4 m/s and θ=10°, the maximum increase of h_a and h_{av} is 5.33% and 3.20% respectively compared with other curvatures. When u_{in} =0.7 m/s, it is 2.69% (h_a) and 2.90% (h_{av}), respectively. It can be seen that the heat transfer performance of the IETBs under vibration and non-vibration conditions are better than that under other conditions when the baffle curvature is 10°. This is because when the baffle curvature is 10°, the total thermal resistance drop under both IETB vibration and non-vibration conditions is greater than in other cases.

9.3.3 Comprehensive Heat Transfer Analysis

To study the influence of baffle curvature on the comprehensive heat transfer performance of the IETB heat exchanger, the $Nu/\Delta P$, $Nu_v/\Delta P$, factor JF and PEC of the IETBs under different inlet velocity are calculated to reflect the heat transfer performance of the heat exchanger, as shown in Fig. 9.15.

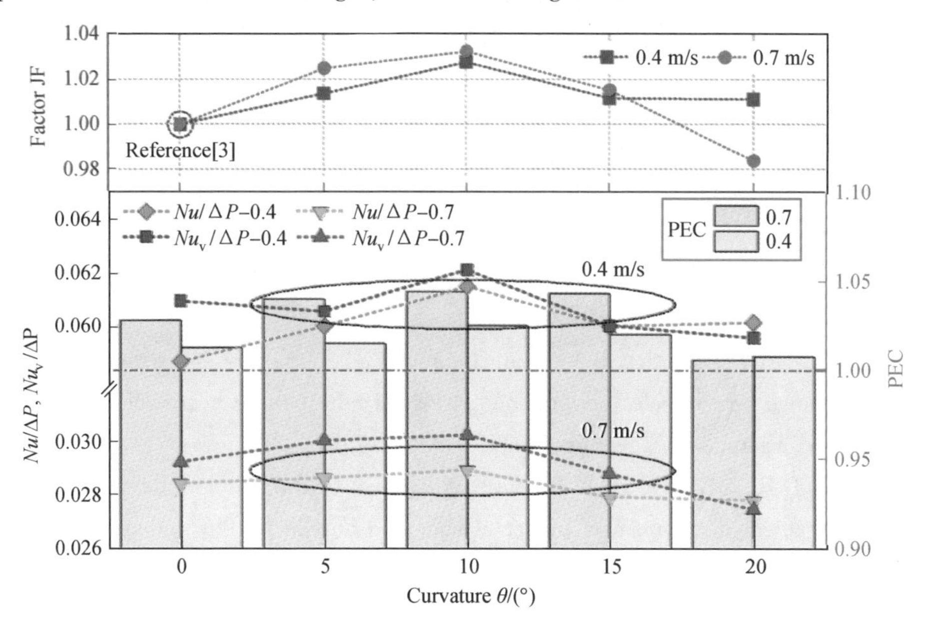

Fig. 9.15 The variation of $Nu/\Delta P$, $Nu_v/\Delta P$ factor JF and PEC with baffle curvature

From Fig. 9.15, the following conclusions can be drawn.

(1) Both $Nu/\Delta P$ and $Nu_v/\Delta P$ reach the maximum value when the baffle curvature θ=10°. Relative to other curvatures, the maximum errors of $Nu/\Delta P$ and $Nu_v/\Delta P$ are 4.76% and 4.22% (u_{in} = 0.4 m/s), 2.83% and 9.86% (u_{in} = 0.7 m/s), respectively. The values of $Nu/\Delta P$ and $Nu_v/\Delta P$ decrease as the inlet

velocity increases, this is because the pressure drop of the heat exchanger also increases. In addition, while the baffle curvature $\theta=20°$, $Nu_v/\Delta P$ is less than $Nu/\Delta P$, because the pressure drop is the largest when the baffle curvature $\theta=20°$, and the increased intensity of IETB vibration-enhanced heat transfer performance does not reach the increased intensity of pressure drop.

(2) The PEC values of vibration strengthening under different conditions are all greater than 1.00. This shows that the IETB heat exchanger has a good performance in vibration-enhanced heat transfer. No matter what the inlet velocity is, the PEC value with baffle curvature $\theta=10°$ is the largest compared with the other curvatures, but the numerical difference is not significant.

(3) As the baffle curvature increases, the value of the factor JF increases relative to the heat exchanger with the baffle curvature $\theta=0°$, except for the heat exchanger with the baffle curvature $\theta=20°$ when $u_{in}=0.7$ m/s. Combined with the conclusion (1), it can be seen that under large pressure drops, the h shows little improvement, and the comprehensive heat transfer performance of the heat exchanger is low. By comparing the factor JF under different baffle curvatures, it can be seen that the comprehensive heat transfer performance of the IETB heat exchanger is the best when the baffle curvature $\theta=10°$, indicating that the improvement of baffle curvature is successful.

References

[1]JI J D, GAO R M, SHI B J, et al. Improved tube structure and segmental baffle to enhance heat transfer performance of elastic tube bundle heat exchanger [J]. Applied Thermal Engineering, 2022(200): 117703.

[2]DUAN D R, GE P Q, BI W B, et al. Numerical investigation on the heat transfer enhancement mechanism of planar elastic tube bundle by flow-induced vibration [J]. International Journal of Thermal Sciences, 2017(112): 450-459.

[3]JI J D, ZHANG J W, LI F Y, et al. Numerical research on vibration-enhanced heat transfer of improved elastic tube bundle heat exchanger [J]. Case Studies in Thermal Engineering, 2022(33): 101936.